CITIES AND TRANSPORT

ATHENS/ GOTHENBURG/ HONG KONG/ LONDON/ LOS ANGELES
MUNICH/ NEW YORK/ OSAKA/ PARIS/ SINGAPORE

ORGANISATION FOR ECONOMIC CO-OPERATION AND DEVELOPMENT

Pursuant to article 1 of the Convention signed in Paris on 14th December, 1960, and which came into force on 30th September, 1961, the Organisation for Economic Co-operation and Development (OECD) shall promote policies designed:

- to achieve the highest sustainable economic growth and employment and a rising standard of living in Member countries, while maintaining financial stability, and thus to contribute to the development of the world economy;
- to contribute to sound economic expansion in Member as well as non-member countries in the process of economic development; and
- to contribute to the expansion of world trade on a multilateral, non-discriminatory basis in accordance with international obligations.

The original Member countries of the OECD are Austria, Belgium, Canada, Denmark, France, the Federal Republic of Germany, Greece, Iceland, Ireland, Italy, Luxembourg, the Netherlands, Norway, Portugal, Spain, Sweden, Switzerland, Turkey, the United Kingdom and the United States. The following countries became Members subsequently through accession at the dates indicated hereafter: Japan (28th April, 1964), Finland (28th January, 1969), Australia (7th June, 1971) and New Zealand (29th May, 1973).

The Socialist Federal Republic of Yugoslavia takes part in some of the work of the OECD (agreement of 28th October, 1961).

Publié en français sous le titre :

**LES VILLES
ET LEURS TRANSPORTS**

© OECD, 1988
Application for permission to reproduce or translate
all or part of this publication should be made to:
Head of Publications Service, OECD
2, rue André-Pascal, 75775 PARIS CEDEX 16, France.

Ten case studies are presented in this volume. They focus on innovative urban transport measures taken in large cities to respond to transport as well as environmental and economic concerns in Athens, Gothenburg, Hong Kong, London, Los Angeles, Munich, New York, Osaka, Paris and Singapore (Table A).

These studies are issued under the responsibility of the OECD Secretariat, but their successful completion depended largely upon efforts of many individuals who have contributed to them personally or officially. More precisely, these studies were developed by independent teams of experts, as part of the work programme of the Environment Committee and of the OECD Ad Hoc Group on Transport and the Environment. Two of them were undertaken in co-operation with the World Bank. The Group on the State of the Environment and the Environment Committee recommended the de-restriction of this report on the authority of the Secretary-General who subsequently agreed.

This publication is supplemented by a separate book *"Transport and the Environment"*, which reviews the impact of road transport on the environment, assesses innovations in urban transport management and examines technical changes.

Also Available

TRANSPORT AND THE ENVIRONMENT (March 1988)
(97 88 01 1) ISBN 92-64-13045-4 132 pages £11.20 US$21.00 F95.00 DM41.00

OECD ENVIRONMENTAL DATA/DONNÉES OCDE SUR L'ENVIRONNEMENT. COMPENDIUM 1987 (June 1987) bilingual
(97 87 05 3) ISBN 92-64-02960-5 366 pages £20.00 US$42.00 F200.00 DM86.00

DYNAMIC TRAFFIC MANAGEMENT IN URBAN AND SUBURBAN ROAD SYSTEMS. Report prepared by an OECD Scientific Expert Group (April 1987)
(77 87 02 1) ISBN 92-64-12926-X 102 pages £8.00 US$16.00 F80.00 DM36.00

THE STATE OF THE ENVIRONMENT – 1985 (June 1985)
(97 85 04 1) ISBN 92-64-12713-5 272 pages £16.50 US$33.00 F165.00 DM73.00

FIGHTING NOISE. Strengthening Noise Abatement Policies (September 1986)
(97 86 01 1) ISBN 92-64-12827-1 146 pages £7.50 US$15.00 F75.00 DM33.00

ENVIRONMENTAL EFFECTS OF AUTOMOTIVE TRANSPORT. The OECD Compass Project (October 1986)
(97 86 03 1) ISBN 92-64-12862-X 172 pages £10.00 US$20.00 F100.00 DM44.00

URBAN POLICIES IN JAPAN. A Review by the OECD Group on Urban Affairs undertaken in 1984/5 at the request of the Government of Japan (October 1986)
(97 86 05 1) ISBN 92-64-12886-7 108 pages £8.00 US$16.00 F80.00 DM35.00

MANAGING AND FINANCING URBAN SERVICES (May 1987)
(97 87 04 1) ISBN 92-64-12951-0 94 pages £6.00 US$11.00 F60.00 DM22.00

MANAGING URBAN CHANGE:
   Volume I – Policies and Finance (May 1983)
   (97 83 02 1) ISBN 92-64-12442-X 150 pages £8.50 US$17.00 F85.00 DM38.00
   Volume II – The Role of Government (November 1983)
   (97 83 04 1) ISBN 92-64-12478-0 114 pages £5.00 US$10.00 F50.00 DM25.00

Prices charged at the OECD Bookshop.

*THE OECD CATALOGUE OF PUBLICATIONS and supplements will be sent free of charge
on request addressed either to OECD Publications Service,
2, rue André-Pascal, 75775 PARIS CEDEX 16, or to the OECD Distributor in your country.*

TABLE OF CONTENTS

1.	ATHENS	9
2.	GOTHENBURG	27
3.	HONG KONG*	45
4.	LONDON	61
5.	LOS ANGELES	79
6.	MUNICH	97
7.	NEW YORK	131
8.	OSAKA	145
9.	PARIS	161
10.	SINGAPORE*	183

* Case study jointly prepared by the World Bank and OECD.

DETAILED TABLE OF CONTENTS

Chapter 1
ATHENS

I.	SUMMARY	9
II.	GENERAL CONTEXT	9
	Terrain and Climate	9
	Population and Economic Growth	9
	Physical Growth	10
III.	TRANSPORT AND TRAVEL	11
	Traffic and Car Use	11
	Public Transport Organisation and Management	13
	Public Transport Network	14
	Fares	15
	Taxis	15
IV.	STRUCTURE OF GOVERNMENT	15
V.	ENVIRONMENTAL POLLUTANTS	16
	Air Pollution	16
	Noise Survey	18
VI.	ACTIONS	18
	Urban Planning	18
	Public Transport Reorganisation	19
	Traffic Management and Control	19
	Road Construction	20
	Air Pollutant Emission Control	20
VII.	RESULTS	21
	Car Traffic	21
	Public Transport Travel	21
	Pollution and Nuisances	22
VIII.	EVALUATION	23
	Traffic Bans	23
	Inspection of In-Use Vehicles	23
	Improvements to Transit Quality	23
	Increasing Transit Capacity	23
	Fares Policy	23
	Emissions from New Vehicles	24
	Noise	24
IX.	CONCLUSIONS	24
Notes and references		25

Chapter 2
GOTHENBURG

I.	SUMMARY	27
II.	IMPLEMENTATION OF TRAFFIC CELLS IN THE CENTRAL BUSINESS DISTRICT	27
	Background	27
	The Traffic Plan	29
	Do the zones re-route or restrain traffic?	30
	Public Reactions	30
III.	DEVELOPMENTS AFTER 1975	30
	The Creation of Cells Outside the CBD	30
IV.	THE TRAFFIC AND PARKING POLICY OF 1979	32
	Traffic Measures	32
	Parking Measures	32
V.	EXTENSION OF THE CELL SYSTEM OUTSIDE THE CBD	33
	Why extend the Cell System?	33
	Design of the New Cells	34
	Implementation of New Cells	34
	Public Participation	34
VI.	RESULTS	35
	Expected Results	35
	Observed Results	35
VII.	EVALUATION	39
	Environmental Effects	39
	Effects on Safety	40
	Effects on Efficiency	40
	Effects on Energy	40
	Land Use and Economic Effects	40
	Overall Costs and Benefits	40
	Distributional Effects	40
VIII.	CONCLUSIONS	41
Annex:	Key Facts and Figures on Gothenburg and its Transportation System	42

Chapter 3
HONG KONG

I.	INTRODUCTION	45
	Population	45
	Land Use	47
II.	THE HONG KONG TRANSPORT SYSTEM	47
	Buses	47
	Minibuses (or Public Light Buses)	48
	Railways	48

5

	Trams	49
	Ferries	50
	Taxis	50
	Fares	50
	Traffic Congestion	50
III.	ELECTRONIC ROAD PRICING IN HONG KONG	51
	Road Pricing	51
	How the Electronic Road Pricing System Works	51
	The Electronic Number Plate (ENP)	52
	Data Capture and Validation	53
	Surveillance and Enforcement	53
	Accounting System	53
	The Effect of Electronic Road Pricing	53
	Cost	54
IV.	ENVIRONMENTAL PROBLEMS AND TRANSPORT	55
	Air Pollution	55
	New Towns	55
	Abandoned Vehicles	56
V.	EVALUATION	56
	Selected Features of Hong Kong Transport Policy	56
	Impacts of Policies	58
VI.	CONCLUSIONS	59
	Electronic Road Pricing	59
	Public Transport Financing	59

Chapter 4
LONDON

I.	TRANSPORT AND THE ENVIRONMENT IN LONDON	61
II.	EXPENDITURE ON TRANSPORT IN LONDON	63
III.	THE GENERAL CONTEXT FOR POLICY	64
IV.	RECENT INITIATIVES	67
	Heavy Lorry Bans	67
	The London Electric Vehicle Programme	69
	The London (GLC/LT) Travelcard	69
	Parking Enforcement - The Wheelclamps Experiment	72
	Public Transport for People with Disabilities	73
V.	EVALUATION OF RECENT INITIATIVES	74
	Heavy Lorry Bans	74
	Electric Vehicles	74
	Travelcards	75
	Wheelclamps	75
	Transport for the Disabled	75
VI.	CONCLUSIONS	76
	General	76
	Air Pollution	76
	Heavy Goods Vehicles	76
	Public Transport	76
	Wheelclamps	76
	Transport for the Disabled	76

Chapter 5
LOS ANGELES

I.	INTRODUCTION	79
	Summary	79
	General Characteristics of the Los Angeles Region	79
	Highways	79
	Public Transit	81
	Transportation System Growth Trends	81
	Highway and Transit System Funding	81
	Land Use and Employment Patterns	83
	Travel Patterns	83
	Air Quality	83
	Noise	83
	Energy Consumption	83
II.	DESCRIPTION OF INNOVATIONS	84
	Public Transport	84
	Highways	87

	Private Sector Contribution	89
	Environment	90
III.	EVALUATION AND CONCLUSIONS	93
	The Transportation System Today	93
	The Transportation System Tomorrow	93
	Recent Transportation Progress	94
	Prospects	95

Chapter 6
MUNICH

I.	THE URBAN AND REGIONAL CONTEXT	97
	Trends in Population and Housing	97
	Trends in Economic Structure and Employment	99
	Trends in Road Traffic and Environmental Conditions	100
	Transport Policy in the Munich Region	101
II.	THE DEVELOPMENT OF THE REGIONAL RAPID RAILWAY (S-BAHN/U-BAHN)	102
	The Development of the Suburban Rapid Railway (S-Bahn)	102
	The Development of the Underground Railway (U-Bahn)	104
	The Financing of the Regional Rapid Railway (S-Bahn/U-Bahn)	105
	Recent Funding Initiatives to Ensure Scheduled Development of the U-Bahn	106
III.	THE INTEGRATION AND RATIONALISATION OF PUBLIC TRANSPORT	107
	The Munich Transport and Fares Authority (MVV)	107
	The Reorganisation of Tram and Bus Networks	107
	The MVV's Common Fares System	109
	The Costs of the Unified Transport and Common Fares System	110
IV.	ROAD DEVELOPMENT POLICY	111
	The Shift in Road Development Policy	111
	The Development of Major By-passes	112
	Noise Abatement Measures along Main Roads in Munich	113
	The Financing of Road Development and Improvement	115
V.	TRAFFIC MANAGEMENT POLICY	116
	Pedestrianisation in the City Centre	116
	Gradual Traffic Restraint Following the Amendment of the 1971 Federal Highway Act	117
VI.	PARKING POLICY	119
	The Control of On-street Parking	119
	The Development of Off-street Parking	120
VII.	ENVIRONMENTAL PROTECTION POLICY	123
	Experiments Concerning the Municipal Vehicle Park	123
VIII.	EVALUATION OF ACTIONS TO IMPROVE TRANSPORT AND THE ENVIRONMENT IN THE MUNICH REGION	125
	Actions Concerning Public Transport	125
	Actions Concerning Private Transport	127
IX.	CONCLUSIONS	129
References		130

Chapter 7
NEW YORK

I.	INTRODUCTION AND CONTEXT	131
	New York City	131
	Public Transit	131
	Roads	132
	Environmental Conditions	133
II.	TRANSIT REHABILITATION PROGRAMME	133
	Origins of the Financial Crisis	133
	Changes in Policy	133
	Five Year Investment Programme	134
	Noise Abatement	135

III.	FINANCING TRANSIT REHABILITATION	135
	Lease Financing	136
	Vendor Financing	136
	Access to Capital Markets	136
	Joint Venture	137
IV.	OTHER ACTIONS	138
V.	ROAD PROGRAMMES	138
	Capacity Management	138
	Operation Clear Lanes	138
	Bond Issue	138
VI.	RESULTS	139
	Financing	139
	Obstacles	139
	Quality of Service	139
VII.	ENVIRONMENTAL IMPLICATIONS	139
	Travel Conditions	139
	Transit Improvements and Air Quality	140
VIII.	EVALUATION	141
	New York's Dependence on Transit	141
	Pros and Cons of Private Financing	141
	The Consequences of a Fall in Revenue Subsidies	141
	Beyond the Current Capital Programme	142
	Agency Co-operation	142
	Environmental Impacts	142
IX.	CONCLUSIONS	142

References 144
Annex: Key Facts and Figures on New York City and its Transportation System 143

Chapter 8
OSAKA

I.	COMPREHENSIVE TRANSPORTATION POLICY	145
	Context	145
	Aims and Objectives of the Comprehensive Transportation Policy	146
II.	TRANSPORTATION	146
	New Aspects of the Movement of People	146
	Public Transportation	148
	Motor Vehicle Traffic	151
	New International Airport	153
III.	ENVIRONMENT	154
	Pollution	154
	Measures for the Living Environment	156
IV.	EVALUATION	158
	Railways	158
	Buses	158
	Ride-and-Ride Concept	158
	Bicycles	158
	Taxis	159
	Road Networks and Motor Vehicle Traffic	159
	Pollutions	159
	Measures for the Living Environment	160
V.	CONCLUSION	160

Chapter 9
PARIS

I.	TRANSPORT IN THE ILE-DE-FRANCE REGION PAST DEVELOPMENTS AND PRESENT SITUATION	161
	Travel within the Region	161
	Transport Networks	161
	Brief Survey of Transport Policy in the Ile-de-France Region	164
	The Institutional Framework	165
	Financing Public Transport	166
II.	ACTION TO PROMOTE PUBLIC TRANSPORT, AND NOISE ABATEMENT MEASURES	168
	A Financial Measure: the "Transport Levy"	168
	A Fare Measure: the Orange Card	170
	A Promotional Measure: Improving the Image of Public Transport	173
	Constructional Measures: Traffic Noise Abatement along Motorways and Urban Expressways	175
III.	EVALUATION OF ACTION UNDERTAKEN AND CONCLUSIONS	178
	Management of the Networks	178
	Use of Economic Instruments	179
	Transport and the Quality of Life	180

Abbreviations 182
Notes and references 182

Chapter 10
SINGAPORE

I.	SUMMARY	185
II.	INTRODUCTION	186
III.	THE DEVELOPMENT OF SINGAPORE	187
	General Description	187
	The Urban Redevelopment Authority	187
	Development Plans	187
	Urban Development, 1975-1983	187
	Central Area Growth	187
	Urban Transport	188
IV.	THE DEVELOPMENT OF AREA LICENSING	189
	Origins of the Licensing Scheme	189
	The 1975 Scheme	189
	License Fees	189
	Operating Hours	190
	Parking Space and Parking Fees	190
	Park-and-Ride	191
	Administration and Enforcement	191
	Parking Availability	191
	Use of Fringe Car Parks	191
	Road Scheme and Traffic Management	191
	Vehicle Taxation	192
	Public Transport	192
V.	IMPACT OF THE AREA LICENSING SCHEME	193
	Initial Impacts – 1975	193
	Car-Ownership	193
	Changes in Travel Patterns	193
	Car Use	196
	Traffic Flow	196
	Land Values	198
	Survey of Businessmen	198
	Air Pollution	201
	Noise	201
	Traffic Safety	201
	Transport Investments	203
VI.	LESSONS OF THE AREA LICENSING SCHEME	203
	Lessons From Singapore	203
	Application of Area Licensing to Other Cities	204

Notes and references 204
LIST OF THE MEMBERS OF THE AD HOC GROUP ON TRANSPORT AND THE ENVIRONMENT 205

Table A. SELECTED STATISTICS; early 80s

	Population (Millions)	Population density (persons/km^2)	Employment[b] (Millions)	Auto population/ownership[b]
Athens	3.0[a] 0.4[d]	7 026[a] 17 391[d]	1.1	734 000 cars[a] 1 car/4.1 persons
Gothenburg	0.7[a] 0.43[c]	953[c] 1 667[d]	0.240[c] 0.025[d]	1 car/3.1 persons[c]
Hong Kong	5.3	28 000[ae]	n.a.	211 000 cars 1 car / 24 persons
London	6.7[a] 2.0[c] 0.5[d]	4 938[a]	3.65[a] 0.85[c] 1.25[d]	n.a.
Los Angeles	11.5[a]	793[a] 1 815[d]	n.a.	n.a.
Munich	2.3[a] 1.3[c]	460[a] 4 194[c]	0.771[c]	1 050 000 cars[a] 1 car/2.2 persons[a] 472 000 cars[c] 1 car/2.8 persons[c]
New York	13.2[a] 7.0[c]	6 775[a] 25 000[d]	3.7	1 514 391 cars[c] 1 car/4.7 persons[c] 161 494 cars[d] 1 car/9 persons[d]
Osaka	14.513[a] 2.626[c]	3 458[a] 12 383[c]	6.862[a] 2.473[c]	750 772 cars[c] 1 car/3.5 persons
Paris	10.0[a]	840[a]	4.58	3 610 000 cars 1 car/2.8 persons
Singapore	2.5[a] 2.25[c] 0.15[d]	4 251[a] (12 800 max)		0.270184 150 cars[a] 1 car/14 persons[a]

a) Greater urban area.
b) n.a. = Not available.
c) City.
d) Central Business District (CBD).
e) Data refer to the metropolitan zone near the harbour: average density is 5 000 persons/km^2.

Chapter 1

ATHENS*

I. SUMMARY

Athens is a large, modern city including an international port. It may be best known for the Acropolis but such antiquities and the old quarter surrounding them form only a core of an industrial metropolis of three million inhabitants covering 427 square kilometres.

For the past two decades urban growth has been accompanied by a rapid rise in road traffic. Between 1971 and 1981 the number of motor vehicles in circulation in Greater Athens increased from 255 830 to 734 175 (187 per cent). Pollutants from these and other sources have multiplied and their effects on people and property have been compounded by climate and topography. Episodes of photo-chemical smog have occurred in the city on several occasions and have recalled conditions in London in the early 1950s and Los Angeles in the 1960s.

This study of transport and environmental conditions in Greater Athens describes the contribution of road traffic to air pollution and noise and what abatement measures have been taken. It draws heavily on a recent OECD report on Greek environmental conditions[1] and indicates some possible future directions for policy.

II. GENERAL CONTEXT

Terrain and Climate

Athens lies in a basin running from South-East to North-West and has a climate that is hot and dry (Figure 1). Although the prevailing wind is from the Northeast and sea breezes from the South are not uncommon, the high pressure systems that regularly station themselves over Southeastern Europe give rise to temperature inversions. Such inversions and the location of Athens in a basin account for a tendency for air pollutants to build up over the city.

---

* Case study prepared by Messrs. T. Bendixson (United Kingdom) and A. Lombart (Belgium) with the contribution of the Greek Ministry of Environment, Housing and Physical Planning and of the Greek Ministry of Transport.

Population and Economic Growth

Between 1961 and 1981 the population of Greater Athens grew from 1.85 to 3.01 million while that of the city in its middle rose from 627 000 to 885 000 (Table 1). Recent demographic changes and decentralisation policy have led to a fall in the 3.6 per cent rate of growth experienced in the 1970s. A population of 3.6 million people in the year 2000 therefore seems likely[2].

Between 1971 and 1981 the Gross National Product of Greece increased by 42 per cent and a major part of this growth was generated in Athens. Thus the metropolis, with only 31 per cent of the country's population in 1981, accounted for over half of its GNP. Furthermore average incomes in Athens in 1981 were 60 per cent above the national average.

Figure 1. **GREATER ATHENS REGION**

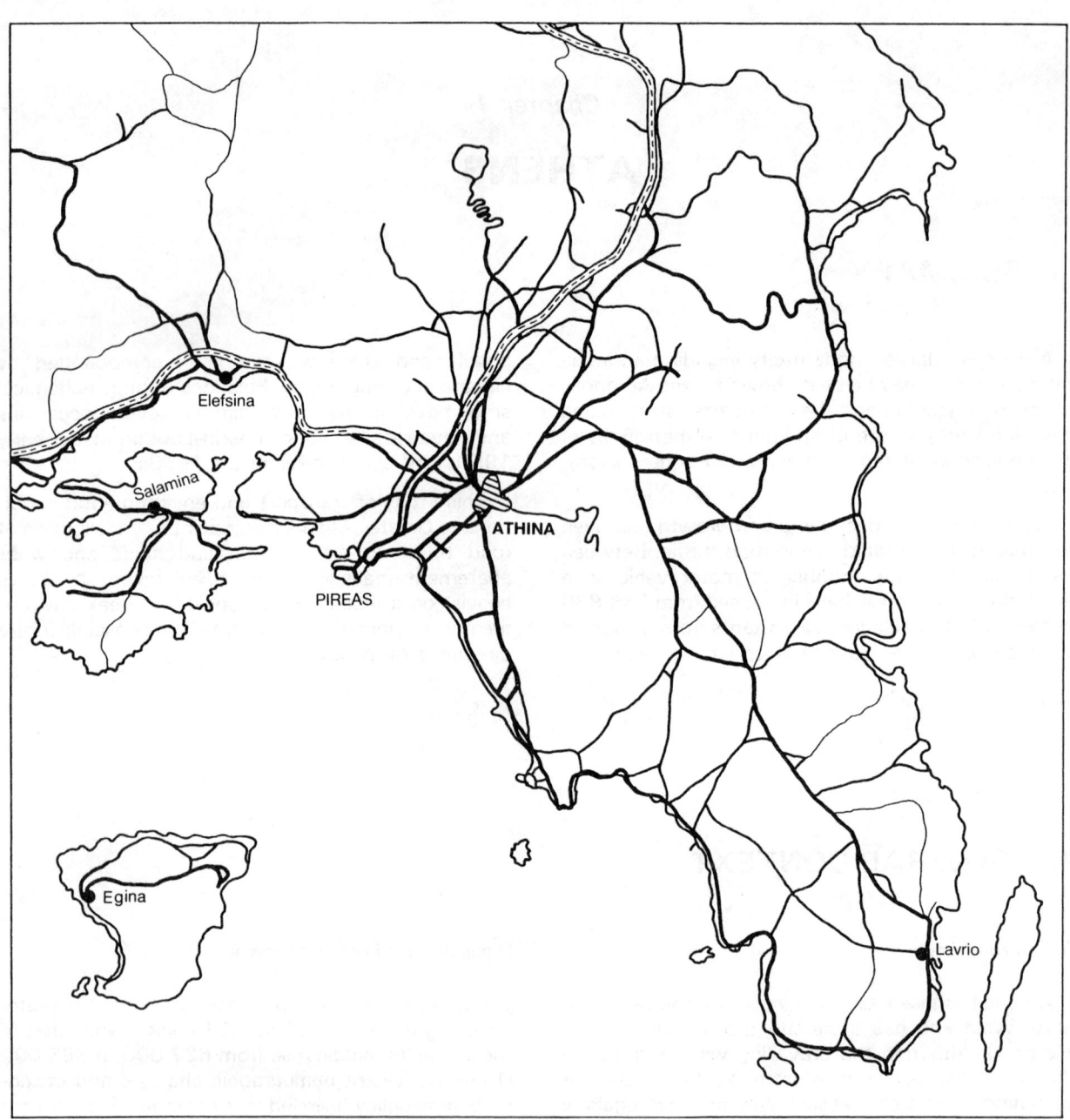

Physical Growth

Development in Athens has been rapid and largely unplanned during the past 20 years but it has been important to the country's economic growth. It has also transformed housing standards.

Within the city of Athens itself development typically fronts onto streets between six and nine meters wide including footways. In the central business district most buildings are eight floors or 25 metres high. Elsewhere in the inner city five floors is the maximum allowed and the average is four.

One result of the absence of control over development is the low proportion of ground occupied by parks and public gardens. The Western and industrial sectors of the city and the port of Piraeus are particularly deficient (Table 2).

Table 1. CHANGES IN THE POPULATION OF GREATER ATHENS, 1961-1981

	1961	1971	1981
Greater Athens	1 850 000	2 540 000	3 010 000
City of Athens	627 000	867 000	885 000
Active Population	706 000	868 000	1 075 000
	(38 %)	(34 %)	(36 %)
Actives Primary	2 %	1 %	0.4 %
Actives Secondary	38 %	42 %	39 %
Actives Tertiary	53 %	53 %	60 %

Source: Ministry of Physical Planning, Housing and Environment.

Table 2. OPEN SPACE IN ATHENS

	%
East	5.2
North-East	5.9
North-West	2.6
West	1.6
Piraeus	2.6
CBD	7.6

Source: Problem Athens. Technical Chamber of Greece, 1980.

High residential densities are another characteristic of the city. The average throughout the 31 000 hectares of legal development is 119 inhabitants per hectare. This rises to over 300 persons per hectare in the city of Athens and 450 in some areas.

A dense and expanding group of government and commercial services plus the central railway station, five subway stations, the main bus terminals and many monuments of international importance are located in the central business district, an area of 23 square kilometres. In 1981, 400 000 people lived at high density in this area.

III. TRANSPORT AND TRAVEL

Traffic and Car Use

Car Ownership

During the decade 1968 to 1978 Greece experienced the highest annual growth in car ownership in Europe. Cars registered in Attica[3] increased by 206 per cent in the decade up to 1981 and by the end of that period they accounted for 58 per cent of all cars in Greece (Table 3). Private cars constituted 72 per cent of the vehicle fleet in Greater Athens. The remaining vehicles were trucks, (15 per cent), motorcycles, (8 per cent), and buses (1 per cent).

By 1982 car ownership was at the level of one car for every 5.8 persons in Greater Athens[4] and during that year the vehicle fleet increased by six per cent even though no growth in GNP was recorded. Car ownership in Athens is nevertheless still low by European standards, although it ranges from one car to 14.4 persons in low income and one to 2.6 persons in high income districts and could therefore increase 2.5 times by the year 2000[5]. Yet the growth that has already taken place has cut transit rides per capita by more than half. Figure 2 shows that an annual increase of 10 per cent in car ownership per capita has coincided with an annual fall in transit riding of 8.7 per cent.

Traffic Congestion

Traffic speeds in the centre of Athens are estimated to average 7 to 8 km/hr at peak periods compared with 20 km/hr in London and 12 km/hr in Paris. Public transport vehicles are slower because they have to make frequent stops for passengers. Thus trolley buses, operating on very heavily travelled routes in the centres of Athens and the Piraeus, achieve 5 to 8 km/hr in peak periods and their travel times can vary by as much as 100 per cent. Over the more extensive diesel bus network speeds are 1 to 1.5 km/hr higher but they are being affected by congestion that is tending to spread out from the centre of the city. All this makes regular scheduling extremely difficult.

Table 3. VEHICLES IN ATTICA, 1971-1981

	1971	1976	1981	Growth 1971-81
Buses	5 121	6 851	10 332	50 %
Trucks	41 573	68 884	125 407	202 %
Passengers Cars	174 597	317 560	533 777	206 %
2 and 3-Wheel Vehicles	34 539	43 005	64 659 (1979)	87 %
Total	255 830	463 300	734 175	187 %

Source: National Statistical Service.

Figure 2. **GREATER ATHENS AREA: EVOLUTION OF PUBLIC TRANSPORT RIDES PER CAPITA IN RELATION TO PASSENGER CAR OWNERSHIP**

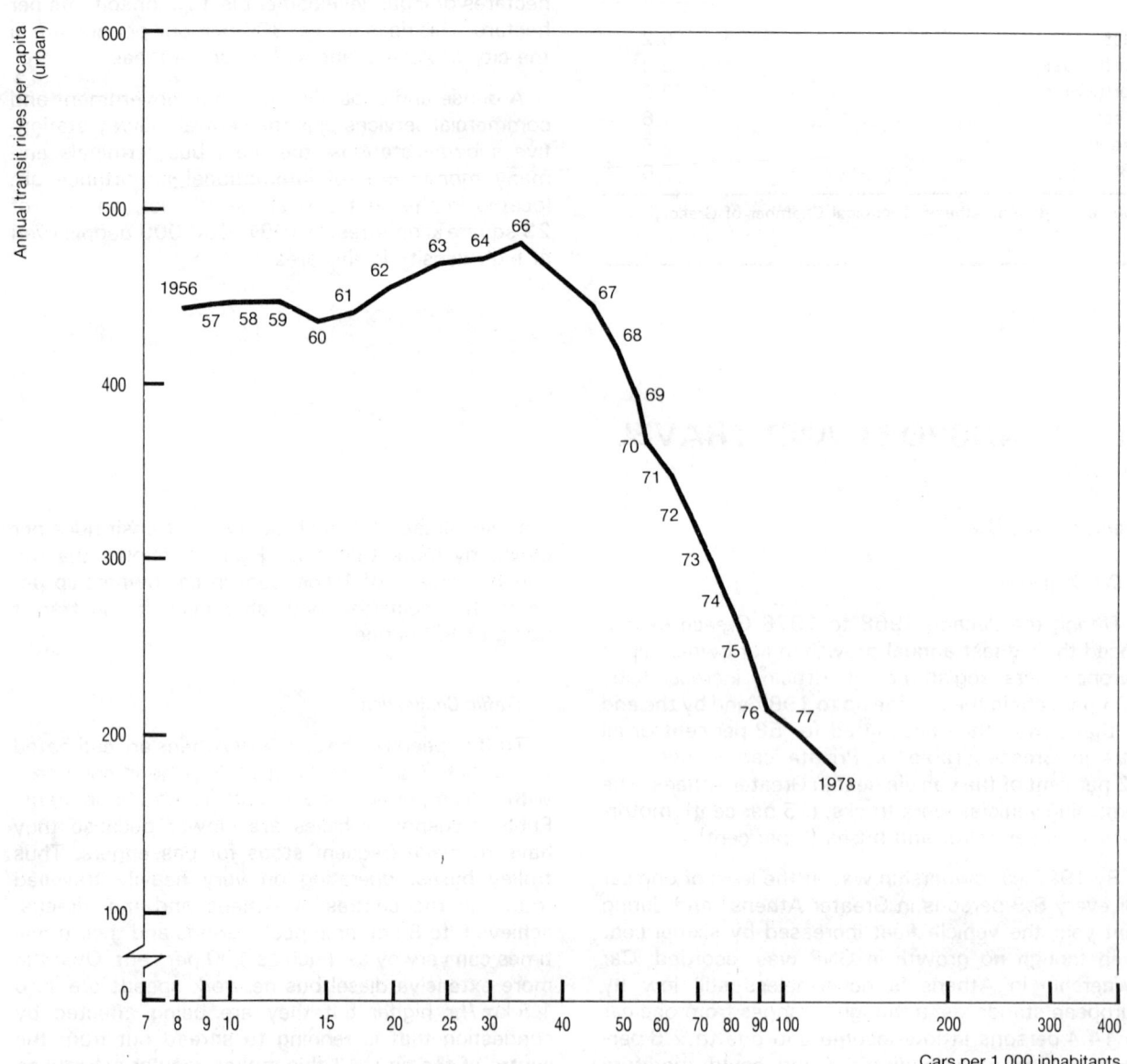

Source: "Athens Area Urban Transport Association (O.A.S.)".

Vehicle Maintenance Standards

No vehicles are manufactured entirely in Greece and as imported ones cost foreign currency, the Government imposes heavy import duties and purchase taxes on them. This slows the inflow, causes high vehicle prices, and leads to a slow rate of scrappage and to a fleet with a high average age. In 1982 about one in five cars in use in Attica was ten or more years old (Table 4). The age of the car fleet plus poor maintenance coupled with traffic congestion and low quality fuel all contribute to emissions of air pollutants and noise. Poor maintenance is a result of ill-equipped garages, lack of training, and a reluctance on the part of many car owners to pay for servicing.

Following the establishment of emission standards for carbon monoxide and soot from petrol and diesel engines in 1981, the Government set up 15 mobile exhaust measuring teams to check on emissions in Athens. Sixty per cent of vehicles were found to be sub-standard but difficulties over getting accurate readings and other problems led to the winding down of the programme.

Public Transport Organisation and Management

Public transport in Athens has undergone major reorganisation in the past decade. Today it consists of three operating companies for the underground railway, diesel buses and electric trolley buses. (The railway and trolley companies also run some diesel buses.) The Athens Area Urban Transport Organisation (OAS) co-ordinates the three operators and is responsible for research, route planning, time-tables, terminals, stops and ticketing. OAS makes recommendations to the Minister of Transport on fare levels and is responsible for allocating fares subsidies and capital funding. OAS obtains its day-to-day funding by drawing on two per cent of the ticket revenue of the three operating companies. Close relations between the operators and OAS are ensured by interlocking directorships.

Public Transport Investment

Following reorganisation in 1977, a five year passenger transport investment plan was drawn up to make good many years of neglect. In the period 1978 to 1983 US$150 million was spent on 1 402 one-hundred passenger buses, 100 articulated 160 passenger buses, 178 trolley buses and 5 subway cars. Orders were placed for a further 245 buses, 80 trolley buses and 120 subway cars.

Travel by Public Transport

The territory of the three operating companies extends to 1 440 square kilometres or considerably more than the urbanised area and the length of the network served by all modes is 905 kms.

Travel by public transport is, however, in decline, although at a rate that is decreasing. Having reached a peak of over one billion passengers a year in 1965, it is now down to 450 million and declining by five per cent a year. Furthermore this decline has taken place during a period of rapid population increase.

Transit trips per person per year fell from 470 in 1970 to 180 in 1978 and 170 in 1982/83. This decline is the outcome of at least three main influences that have reinforced the effects of rising car ownership. One was years of under investment in buses and maintenance depots. This led to frequent breakdowns, poor service and shabby vehicles at a time when living standards were rising. The second was the absence of a concerted programme of traffic management measures designed to free buses from the effects of congestion. The third was the growth of the taxi fleet.

Table 4. VEHICLE FLEET BY TYPE AND AGE: ATTICA, 1982

	0-2 years	2-4 yrs	4-6 yrs	6-8 yrs	8-10 yrs	over 10 yrs	Total
Trucks under 3.5 t	3 910	1 422	1 610	1 520	1 058	5 298	14 818
Trucks over 3.5 t	28 301	11 177	9 294	4 328	2 705	24 575	80 380
Buses	1 504	655	897	474	263	2 349	7 152
Taxis	3 085	607	1 210	2 627	1 180	5 631	14 340
Passengers Cars	168 113	71 608	42 269	34 422	22 455	86 797	425 654
Three wheeled vehicles	9 657	3 090	4 035	2 562	2 034	11 797	33 175
Total	214 570	88 559	59 315	45 933	29 695	136 447	575 519

Source: Department of Transport.

Transit, today, carries only 45 per cent of people going by vehicle to the centre of Athens and the Piraeus while between 7 and 9 a.m. every weekday morning 30 000 private cars enter the Athens central area. It is a measure of the concentration of transit demand in the metropolitan area that 60 per cent of all passengers board or alight in the centres of Athens or the Piraeus.

Public Transport Network

The Underground Railway

The underground railway consists of a single line, only the central parts of which are in tunnel, running 25.6 kms from the Piraeus waterfront in the South to the low density Northern suburb of Kifissia. The last extension, northwards, was completed in 1955.

Between 1978 and 1980 the number of underground car miles run fell slightly but the number of passengers carried rose from 85 to 89 million and then declined again by 1983 (Table 5).

Table 5. ATHENS UNDERGROUND RAILWAY, 1978-1980

	1978	1979	1980	1983
Cars	130	130	130	130
Vehicles/km (million)	94	92	89	n.a.
Passengers (million)	85	83	89	83
Revenue[1] (million drachma)	728	952	1 001	1 487
Expenditure[1] (million drachma)	914	1 021	1 246	4 074

1. Includes figures for 62 buses.
Source: OAS.

Electric Trolley Buses

The electric bus company runs the biggest fleet of trolley cars in Western Europe. This consists of 305 vehicles running on 13 routes, a service that was expanded from 204 vehicles on nine routes in 1978 (Table 6). Many of the vehicles are, however, over twenty years old and none of them incorporate energy-conserving "chopper" technology.

Most of the trolleys run on heavily travelled routes across the centre of Athens. As a result they carry 14 per cent of all metropolitan transit passengers on only 7.2 per cent of the transit network. In 1983/84 a study was made of the scope for using diesel buses as feeders to trolleys which would have alone served the central area. Interchange proved to be the major problem while the potential reduction in emissions was found to be less than expected.

Table 6. ATHENS ELECTRIC TROLLEY BUSES, 1978-1983

	1978	1979	1980	1983
Trolley buses	204	270	288	305
Routes	9	9	13	13
Length of all routes (km)	54	55	88	100
Vehicles/km (thousand)	6 776	9 336	10 902	n.a.
Passengers (thousand)	63 504	83 327	89 805	85 400
Revenue[1] (millions drachma)	365	625	764	1 003
Expenditure[1] (million drachma)	564	709	1 005	2 507

1. Includes figures for 8 buses.
Source: OAS.

Diesel Buses

The bus network consists of two sets of routes radiating from the centres of Athens and the Piraeus with few inter-suburban services and little bus-to-rail interchange. Routes range in length from 2.57 to 41.15 kms and, in the absence of a hierarchy of streets, many converge. One particular stop serves 56 of them.

Modernisation of the bus fleet and its maintenance depots have been major objectives of policy since the reorganisation of 1977. Out of 710 vehicles owned by the company in 1980, 500 were not more than four years old. The remainder were, however, pre-1965 models.

Fleet modernisation has enabled the bus company to reduce its dependence on vehicles rented from private owners although such buses still form a

Table 7. ATHENS BUSES, 1978-1983

	1978	1979	1980	1983
Buses	1 674	1 711	1 777	1 768
Bus routes	319	310	306	320
Length of routes	3 933	3 811	4 011	4 278
Bus/Kms (million)	106	109	117	n.a.
Passengers (million)	462	452	429	421
Revenue (million drachma)	2 697	3 472	3 613	4 545
Expenditure (million drachma)	4 432	5 173	6 828	15 698

Source: OAS.

majority of the 1 768 in service. As the owners of the rented buses are responsible for maintenance and as it is in their interest not to run in peak periods, a consistently scheduled service is only achieved with the utmost difficulty.

In 1977 there were no bus maintenance depots in Athens, the work being done at small yards scattered throughout the city. Four are now in operation and two more are under construction. Notwithstanding this investment in improved maintenance, smoke emitted by older vehicles is a source of dissatisfaction.

Between 1978 and 1980 the amount of bus service provided was increased and fares were kept at a very low level. The number of passengers carried nevertheless fell steeply (Table 7).

Fares

A low flat fare is charged for journeys on all forms of transit in Greater Athens, even to go 41 kms to the coast. Up until February 1983 the fare was ten drachma (US 10 cents). One effect of the Government's low fare policy is a rapid increase in the deficit on transit operations. Whereas revenue from the sale of tickets increased from 3 790 to 5 378 million drachma (from US$37 to $53 million) in the three years up to 1980, operating costs rose from 5 892 (US$58m) to 9 079 million (US$90m). And even after the fare increase of 1983, revenue accounted for 32 per cent of transit operating costs (Table 8).

Taxis

There are about 15 000 taxis in Attica. Most of them have two owners, are on the road for a large part of every day and in some central streets they account for 60 per cent of the traffic. Their contribution to congestion is compounded by their stopping habits. Improvements to traffic flow in times of taxis strikes have made this particularly noticeable.

Taxis are licensed by the regional authority and their tariffs are set by the Ministries of Transport and National Economy. The taxi trade is nonetheless, as in many countries, characterised by its independence from authority as well as by low fares. It is said of taxis both that they disappear and operate as minibuses at peak times.

Already 6 000 taxis have been converted to liquid propane gas (LPG) fuel which sells at half the equivalent price of Super grade petrol. The depreciation of LPG vehicles is also lower than that of petrol powered ones. Since emissions from LPG powered vehicles are lower than those from vehicles using other fuels, the Government is aiming to increase the supply of propane. With this objective production will be raised from 110 000 metric tons in 1982 to 250 000 metric tons in 1986. At present transport consumes 30 000 metric tons per year.

Table 8. ATHENS PUBLIC TRANSIT:
REVENUE AND EXPENDITURE, 1978 ET 1983

	1978	1979	1980	1983
Revenue (million drachmas)	3 790	5 039	5 378	7 035
Expenditure (million drachmas)	5 892	6 903	9 079	22 279

Source: OAS.

IV. STRUCTURE OF GOVERNMENT

Government in Greece is arranged in three tiers. Beneath the national government, which plays a prominent role in the management of local affairs in Athens, lies the regional authority of Attica. This is divided into four sub-regions — Eastern, Western, Athens, and Piraeus and the Islands. Activities in the different sub-regions are co-ordinated by Central Government. Throughout the Attica basin there are 61 municipalities and 89 communities. Of these 41 and 16 respectively are in the Greater Athens area.

Within the national government several Ministries have responsibilities touching transport and the environment. The Ministry of Transport is in charge of passenger transport in Athens, vehicle inspection, the control of emissions and noise, traffic accidents and, through the medium of the highway code, driver behaviour.

The Ministry of Public Works is responsible for the planning and construction of roads and other civil engineering projects and for the design of traffic management schemes.

The Ministry of Physical Planning Housing and the Environment is responsible for transportation planning and for studies of the interaction of urban development and transport with the environment. It is also in charge of research in air and noise pollution and the setting of environmental standards through the Environmental Pollution Control Project Athens (PERPA).

The Ministry of Public Security is responsible for the Traffic Police and thus for the enforcement of traffic management measures.

The Ministry of National Economy is involved in Athens in those local problems, such as the subsidy of public transport, which are of national economic importance.

A Committee of Coordination for Circulation, composed of officials of the five ministries meets weekly to decide on actions not covered by agreed plans or traffic management schemes designed by the Ministry of Public Works.

V. ENVIRONMENTAL POLLUTANTS[1]

Air Pollution

Sources of Air Pollutants

The Greater Athens area consists of the main Athens basin and, to the West, over the Egaleo hills, two areas of industry. The sources of air pollutants in the area in 1976 are shown in Table 9. At that time industry was the main source of suspended particulates and sulphur oxide (SO_2), while road vehicles were the main emitters of nitrogen oxide (NO_x), carbon monoxide (CO) and hydrocarbons (HC). Space heating made only a small contribution to total emissions.

Since then emissions of sulphur oxide from industry have been reduced by a ban on the use of high sulphur oil in 1981 and a shift from 1 to 0.7 per cent sulphur-content heavy oil in 1982.

Vehicle emissions are, however, believed to have grown in line with a 58 per cent increase in vehicles between 1976 and 1981. The main sources of NO_x, CO, and HC are thus thought to be even more strongly linked to transport today than in 1976. It will be noted that vehicle emissions are given off in close proximity to people and that they contribute to smog, especially during periods of low wind, hot sun and thermal inversion.

Levels of Pollutants

Air quality in Central Athens is not related directly to total emissions since pollutants from industrial installations near the coast are generally blown seawards by northerly winds. Sulphur oxide levels, which were once very high, have decreased to a yearly average of about 50 micrograms per cubic meter as a result of control measures. During 1983 levels of SO_2 appeared to fall slightly in the city centre while remaining relatively stable in industrial areas. Levels of suspended particulates are high throughout the Attica basin and yearly averages range from 190

Table 9. SOURCES OF POLLUTANTS IN THE GREATER ATHENS REGION, 1976

Percentage

	Particulate matter	SO^2	NO_x as NO_2	CO	HC
Industry (excluding power generation)	75.9	39.5	20.9	2.3	17.2
Power generation (Keratsini)	1.5	33.7	17.5	0.0	0.8
Transport	17.6	5.6	50.6	97.3	80.9
Space heating	5.0	21.2	11.0	0.4	1.1
Total (tons/year)	36 407	110 337	38 116	182 099	26 539

Notes: In 1982 emissions from power generation are believed to have fallen due to the closure of Keratsini power station. Emissions of SO_2 from industry and space heating are also likely to have fallen because of the use of low sulphur oil.
In 1982 (b), the distribution of SO_2 emissions in Athens was industry (54 pour cent), space heating (32 per cent), diesel-powered vehicles (11 per cent) and cars (3 per cent).
Sources: Environmental Pollution Control Project (PERPA), Technical Report, Athens, 1980 (1976 figures).
Ministry of Energy and Natural Resources (1982 figures only).

Table 10. TRENDS IN AIR POLLUTANTS IN ATHENS, 1974-1981

	1974	1975	1976	1977	1978	1979	1980	1981
Ministry								
SO$_2$	85	112	91	79	39	62	44	46
Suspended particulates	245	222	198	255	245	246	211	214
Rentis								
SO$_2$	75	74	61	50	32	54	40	36
Suspended particulates	228	206	217	215	190	210	219	219
Drapetsona								
SO$_2$	60	67	62	61	44	64	54	–
Suspended particulates	223	185	181	203	180	234	194	174

Notes: SO$_2$ and suspended particulates are given as yearly mean values (ug/m^3).
Source: Background Report to *Environmental Policies in Greece*, OECD, 1983.

to 220 micrograms per cubic meter (Table 10). Fifty per cent of this high level is due to dust from uncovered soil. Smoke levels showed a tendency to increase in 1983, particularly in the city centre (Table 11). Fully automatic measurements of nitrogen oxide began only in the second half of 1983. Measurements that year at Patission, where the highest levels are found, show that 98 per cent of average hourly values were below 200 micrograms per cubic metre — the limit suggested by the EEC. Carbon monoxide levels were not regularly measured before July 1983 and time series data over several years are not available. Measurements at Patission in the city centre in the second half of 1983 indicate, however, that the alert level of 15 micrograms per cubic meter (eight hour average) was regularly exceeded at night.

Table 11. SMOKE LEVELS AT THREE STATIONS[1]

Station	1981	1982	1983
Ministry of Health	1.44	1.78	1.98
Patission (city centre)	2.03	2.02	2.55
Drapetsona	–	0.73	0.83

1. Photometric units — average annual values.

International comparisons show that sulphur oxide levels in Athens are below those in Tokyo, Chicago and Rome and within acceptables ranges. They also meet EEC standards. Levels of suspended particulates and smoke are, however, much higher in Athens than in many OECD cities and above EEC standards. Levels of NO$_2$ are particularly high.

Ozone levels are a further indicator of air pollution in Athens. In 1977 and 1978 hourly average levels exceeded 100 parts per billion (ppb) only twice. From June to October 1982 measurements of 48-hour values gave figures over 80 ppb while six-hour values approached 100 ppb (PERPA). However, levels in the city centre are generally low. The highest levels measured in 1983 were at Liossion in July when values over 120 micrograms per cubic meter (the proposed WHO limit) were recorded seven times.

Damage from Pollutants

During air pollution episodes, Athenians observe a decrease in visibility and experience eye irritation and difficulties in breathing. Some evidence has been found to suggest that the health, particularly of children, older people and those suffering from chronic respiratory diseases is being affected and studies of hospital admissions and levels of sulphur oxide in 1976 indicated a link between air quality and morbidity. However studies made by PERPA (1983) did not show such a correlation. The contribution of air pollutants to the deterioration of building materials and national monuments and to the corrosion of structures is not disputed.

Some work is beginning in Greece to establish the economic consequences of air pollution in Athens but no results are yet available. In these circumstances some insights may be gained by transposing to Greater Athens cost estimates made in less polluted places in the US, France and the UK. The resulting figures need to be treated with caution but they point to health and property damage costs of more than 6 000 drachma (US$60) per year per inhabitant or 18 billion drachma (US$18m) per year.

Studies in Los Angeles in 1978 based on willingness to pay for clean air and on property prices suggest that the social costs of air pollution in cities, and therefore in Athens, exceed those for damage to health and property[6].

Future Air Pollution Trends

Forecasts that car-ownership in Greece will double by the end of the century point to substantial increases in future emissions from mobile sources. If such projections come to be and if stricter control measures are not taken, traffic related emissions can be expected to increase by some 60 per cent over ten years. In Athens NO_x emissions from traffic are likely to grow faster than those from other sources.

Taking industrial, space heating and automotive sources into account it is likely that in the absence of air pollution abatement actions, emissions will increase and that air quality will continue to deteriorate in Greater Athens.

Noise Survey

An extensive noise measurement programme was carried out in Athens in the period 1974-77 and led, among other things, to the production of a noise map of the central area.

Comparison of the results of the survey with findings in other cities does not reveal marked differences yet noise is a principal nuisance for Athenians. This is born out by a Noise Social Survey of 2 000 residents made in 1975/76[7]. Forty-four per cent of respondents spontaneously mentioned "noise" as the principal environmental nuisance in their lives. Thirty-one per cent mentioned air pollution.

When asked about noise, 81 per cent said they were bothered "a lot" or "a little" by it. Among those who said noise bothered them, half named "traffic" as the most annoying source and "motorcycles" the principal component of that annoyance.

VI. ACTIONS

In 1982 the Greek Government concluded that air pollution in Athens had assumed the dimensions of an "environmental crisis". It went on to define the crisis as surmountable and said it was "determined to face the problem immediately without any compromise"[1]. Although previous governments had taken abatement action, the statement of 1982 appears to mark a turning point.

Various policies aimed at stimulating economic development and raising incomes in remoter regions and the islands have been introduced during the past two years. They aim to slow migration to Athens and reduce the pace of the future growth of the capital.

Urban Planning

A new master plan for Greater Athens has been drawn up by a team within the Ministry of Physical Planning, Housing and Environment and presented to the Government. It consists of a long term urban development scenario and within it, a more specific transport plan. Implementation will be achieved via five year infrastructure programmes — the first covering 1983-1987. Monitoring and management of the transport system will be an important part of the planning process. A Law on Urban Development (1 337 of 1983) obliges the planners to work closely with "neighbourhood associations" to ensure that public views are fed into planning.

Traffic and Transport measures

In the short term effort will be concentrated on the creation of traffic cells, the modification of certain heavily used "traffic corridors", improved control of parking and the establishment of a network of bus lanes in central Athens.

Longer term proposals, on which interdepartmental consultation is continuing, include the replacement of heavily travelled bus and trolley routes by Light Rapid Transit services and the construction of off-street parking places. Nine LRT lines giving a network of 65 kms are proposed. They would be mainly on the surface, would cross the city centre between suburban centres and be fed by relocated bus and trolley services. The construction of three ring roads and a new radial route to the West is envisaged. Parking sites with a capacity of 300 to 500 places are envisaged along the ring roads and near outer LRT stops.

Air Pollution Research

A wide ranging programme of emissions studies is supervised by the Environmental Pollution Control

Project of Athens (PERPA). Scientists from the National Technical University, hospitals and other laboratories are involved. Substantial increases in understanding of the sources, distribution and effects of air pollutants in Attica are expected to flow from the work of PERPA over the next five years.

Public Transport Reorganisation

Infrastructure

i) The improvement of public transport is concentrated on redesigning the bus network to cater for the travel patterns of the growing city, modernising the vehicle fleet, improving bus maintenance and increasing the capacity of the underground line. An important aspect of this modernisation is the replacement of unreliable, rented buses by ones owned and maintained by the Athens regional bus company;

ii) The existing radial bus network is being redesigned to provide feeder services to the underground railway, thereby increasing its loadings, and to provide links between suburbs where none exist. A major study of travel demand and supply is being carried out by OAS to help define and improve transit network;

iii) Capital projects include the purchase of 100 trolley buses from the Soviet Union to be used on new and extended routes. (Unfortunately the new vehicles will not be equipped with energy saving "chopper" control.) Funds have also been allocated to the construction of two maintenance depots for buses and to increasing the capacity of the underground line from 11 600 to 20 000 passengers per hour in each direction. Platforms are being lengthened to handle longer trains. Resignalling will cut headways from 4 to 2.5 minutes.

The 1983-1987 transport infrastructure plan covers the acquisition of sites for a second subway line capable of carrying 20 per cent of the trips made in Athens but the high cost of this project makes its implementation unlikely in the next decade.

Fares

A policy for subsidising passenger transport fares in Athens was introduced in the late 1970s and by 1983 the percentage of operating costs covered by fares was about 30 per cent. In February 1983 the system-wide flat fare was increased from 10 to 20 drachma (10 to 20 US cents) but free travel was introduced between 5 and 8 a.m. in order to try and halt falling patronage. The price of an unlimited travel pass was increased to 700 drachma (US$7.0) per month. No noticeable change in patronage was noted following the 100 per cent increase in the flat fare. OAS officials consider that at the level of fare charged the choice whether to go on foot or bus is not an economic decision.

Traffic Management and Control

A Permanent Scheme

A comprehensive traffic management scheme covering 23 square kilometres of the inner city is under study. The scheme involves traffic cells, the reorganisation of movement in busy traffic "corridors", control of parking, and bus priorities.

i) *Traffic Cells*
Experience of traffic cells is being gained by means of experimental schemes the first of which was introduced, with the collaboration of residents, in the district of Neapolis in 1983 (Figure 4). If such experiments prove successful up to 24 additional cells will be implemented. It is hoped that this measure, through reducing environmental nuisances, will slow population decline in the inner city and thus slow the growth congestion and emissions due to increased commuting and traffic.

ii) *Traffic "Corridor" Reorganisation*
The creation of traffic cells will divert some vehicles back into the existing main roads. Additional capacity to handle these vehicles will be created by treating groups of parallel streets as "traffic corridors" and organising them in one-way flows.

iii) *Parking Control*
Within residential traffic cells priority is given to residents who have the right to free, on-street parking. Elsewhere the priority given to commuters and visitors and the system of pricing and time control adopted, is governed by the predominant land-use.
Generally control of parking is due to be strengthened by increasing enforcement effort and by providing additional off-street spaces close to the inner ring road.
Parking enforcement will also be assisted by a points system for driving offences introduced in 1983.

iv) *Bus Priorities*
A network of bus lanes and other priorities is under consideration. Design of the priorities is going on as part of the programme to reshape the entire bus network. Estimates suggest that the transit priorities will, in some cases, limit the capacity of the streets for other vehicles by up to one-third.

Emergency Traffic Restraint

The growing intensity of air pollution and traffic congestion in Athens first led to measures to reduce car use on weekdays in 1981. (Weekend bans had been used earlier to save energy.)

The existing restraint scheme, which is effective from 6:30 a.m. to 4:00 p.m. within the inner ring road, aims to reduce trips by private cars by 50 per cent. This objective is pursued by allowing onto the ring and the streets within it, vehicles with license numbers from 1 to 5 and from 0 to 6 on alternate days.

Restraint is exercised Monday to Friday. National holidays and days when strikes affect public transport are excluded. Christmas and Easter holidays and the three summer months from mid-June are excluded as well.

Taxis and the cars of doctors (when driven by the doctors themselves), newspaper reporters and tourists are not affected by the ban. This also applies to certain official cars.

Small signs along the ring road bounding the restraint zone remind drivers of its existence. Violations lead to about ten prosecutions per day with offenders being fined between 5 000 and 100 000 drachma (US$50 and $1000).

About 40 000 private cars are estimated to be kept within the ring road by residents and half of these are locked in every day when the ban is operative.

During an air pollution episode on the 4th-5th January 1984, an experiment was made with a more extensive and intensive fifty per cent traffic ban. Control was extended to cover an area of about twice the radius of the inner ring road and bounded by national highways and the coast. Taxis were, for the first time, included in the ban. The reasons for the wider control were preventative rather than curative and aimed at lowering emissions through reducing traffic and congestion.

Area Traffic Control

Existing area traffic control time plans and equipment are up to 25 years old, obsolete and prone to failure. New signal heads, traffic plans and control technology promise a 20 per cent increase in capacity on main roads. It is intended to take this increase in the form of higher speeds rather than added traffic flows. At the same time improvements will be made for pedestrians and to on-street parking.

Road Construction

The five year infrastructure plan for 1983 to 1987 allocates 25 billion drachma (1982) [US$250m] to road construction in Athens. This expenditure will be split so that half goes on building new roads, 30 per cent on land acquisition for future projects and 20 per cent on maintenance.

Completion of a 50-80 km/hr design speed inner ring road is a high priority. On its Western side alone it will include six grade-separated intersections. An intermediate and an outer ring and new radial roads are planned.

Air Pollutant Emission Control

Lead in Fuel

Lead in Super (high octane) petrol was reduced from 0.4 to 0.15 gr/l during 1983 in conformity with EEC regulations. Greece, as a member of the European Communities, is not, however, in a position to eliminate lead in fuel except in harmony with other Member countries. This in turn renders it impossible for the Government to require unilaterally the fitting of exhaust catalysts.

Emission Standards for New Vehicles

Under a Ministerial decision made in February 1982 type approval using the European test (ECE 15-03) was introduced for new imported petrol engined cars. EEC Directive 72/306 for new diesel motor vehicles was adopted at the same time. This opened the way to the control of NO_x as well as CO and HC. New cars have now to conform to the stricter ECE 15-04 standard. Greater reductions in emissions are required in Switzerland, Sweden, the U.S. and Japan but cannot be achieved in Greece unless adopted by the European Communities.

Emissions from Buses

During 1983 negotiations with manufacturers of catalytic exhaust gas oxidisers led to trials involving diesel buses. Equipment capable of reducing emissions of CO, HC and smoke is available and could be got into use when decisions are made.

Taxis

With an estimated 6 000 out of 15 000 taxis in Attica already converted to Liquid Propane Gas, this form of fuel appears to be economically attractive to taxi owners. Since LPG powered vehicles have low emissions and since taxis comprise up to 60 per cent of traffic in some streets in central Athens, emission "hot spots" could be reduced by the conversion of further vehicles. The wider use of diesel engines in taxis could also be beneficial if a method of reducing smoke emissions could be found.

The Government has accordingly lowered import taxes on vehicles equipped to use LPG and diesel fuel and to be used as taxis. Loans of 60 000 drachma (US$600) are available to assist intending taxi proprietors to purchase them.

Wider use of LPG is limited by a shortage of supplies of the fuel. Increased refinery capacity will become available in 1986. In the meantime experiments are to be conducted with LPG powered buses and a study is underway within PERPA of the feasibility of expanding the use of LPG.

Inspection of Vehicles In-Use

The building of national network of 62 in-use vehicle inspection stations is planned. The first four, one in Athens, came into operation in September 1984 for the inspection of buses. In Athens, 25 000 vehicles will be covered by these initial inspections.

The main emphasis of inspections will be on safety equipment such as brakes and steering, but tests will be made for CO, HC and smoke from diesels.

Buses, taxis and trucks over 3.5 tons gross weight will be obliged to undergo annual check-ups while private cars and motorcycles will be tested biannually. More frequent inspections may be introduced for older vehicles.

Smoke from diesel engines will be measured visually using the Bachara scale. The maximum acceptable degree of blackness on the test filters will be level five for buses and level six for taxis. Periodic examinations, using the Europa test, will also be made of new vehicles.

Mandatory periodic inspections may be backed up by random roadside tests for fumes and noise. Owners of vehicles found to be substandard would be required to show evidence that maintenance had been done during the preceding six months at a certified garage. Approval of such a random inspection regime, which would be operated by fifteen teams, is expected. Other ways of improving the quality of vehicle maintenance and driver behaviour are under study.

Measures are also being taken to improve the quality of maintenance work at private garages. A new law is being prepared that will require garages to have specified levels of equipment and mechanics with specified qualifications.

VII. RESULTS

Car Traffic

Traffic Restraint

No methodical measurements have yet been made of the effects on traffic volumes and speeds of banning about fifty per cent of all private cars from central Athens. However some counts were made when the measures were first introduced. These showed that the proportion of taxis in the traffic increased from 50 to 60 per cent of pre-ban volumes. The combination of this increase with a reduction in private car journeys to nearly half their previous level is believed to have led to an overall reduction in traffic of about 20 per cent. This fall has been associated with a shift of flows out of the side streets onto the main routes of the central area.

Meanwhile OAS, the transit planning company, reports that congestion is spreading out from the centre of Athens to the fringes and adding to bus travel times.

Traffic Cells

Work was proceeding during the autumn of 1983 with the creation of a first "traffic beehive" or traffic cell in Neapolis, an inner city district with 8,000 residents to the North of the CBD. Residents and the 400 people working in Neapolis were consulted on proposals for one-way streets, parking meters and a residents-only parking scheme. The displacement of a flow of 900 vehicles to streets surrounding the cell was expected. Increases in the price and a shortening in allowable parking time for non-residents was expected to deter commuters.

Public Transport Travel

A decrease in the rate of decline of travel by public transport in Greater Athens has been recorded since 1980. Amongst possible explanations are the delivery of new vehicles, increased employment in services in the central area and low fares. Lack of parking opportunities may also be a factor at play. However the low fares policy pursued by the Government is becoming increasingly costly and is beginning to be a strain on public finance.

Pollution and Nuisances

Air Pollution Trends

The control of emissions from stationary sources coupled with measures to limit traffic and improve fuel quality appear to have affected the growth of emissions during a period of rapid increases in the vehicle fleet and worsening congestion. A measure of change cited by the Greek authorities is a reduction in the number of pollution episodes calling for emergency measures. In 1982 emergency measures had to be taken on ten occasions to offset the effect of pollution episodes. No emergencies had to be declared in 1983 although one was announced on 3rd to 4th January 1984 to prevent further deterioration of bad conditions. Measurements of SO_2 levels at five stations during the first three months of 1982 and 1983 are however not encouraging. The same is true for measurements of smoke at the same locations. Measurements of NO_x emissions at Patission, however, recorded a fall (Table 12).

Lead

Biological monitoring of lead in a sample of Athenians was carried out in 1981-82. The blood level of a total of 843 residents from the city centre, selected industrial zones and the suburbs was measured. The sample included 204 children and 33 pregnant women.

The study was performed according to EEC Directive 312/77. The results showed that lead was below 20 micrograms per decilitre of blood in half of the sample, below 30 g/10 dl for 90 per cent and below 35 g/100 dl for 98 per cent.

It was concluded that blood lead of Athenians was below the critical limit set in Directive 312/77; that blood lead was highest in industrial areas and lowest in the suburbs and that these differences were more pronounced in children than adults; that automotive lead was not a negligible pollutant; and that although Athenians are not at risk from lead, biological monitoring of the metal was necessary.

Noise

No marked change has occurred in noise levels in Athens in the past eight to ten years. Congestion and its effects on local noise climates are long standing. The renewal of part of the bus fleet and recent measures to restrain private cars within the inner ring seem unlikely to have had a perceptible effect on reducing noise. On the ring road itself traffic has actually increased while within it a fifty per cent reduction in flows is unlikely to have brought more than a 3 dB(A) reduction in levels. Few residents would be likely to notice such a change.

Table 12. LEVELS OF SO_2, SMOKE AND NO_x IN 1982 AND 1983

(ug/m³)

Station	1982 Jan.	1982 Feb.	1982 Mar.	1982 Mean	1983 Jan.	1983 Feb.	1983 Mar.	1983 Mean
SO_2								
Patission	52	51	52	52	102	91	60	84
Ministry	35	45	36	39	91	98	50	80
Rentis	33	34	33	34	46	48	37	44
Pireus	40	39	40	40	52	63	44	91
Drapetsona	30	37	30	32	35	43	26	35
Smoke								
Patission	2.14	1.96	1.98	2.03	2.36	2.49	2.67	2.51
Ministry	2.49	1.95	1.65	2.03	2.53	2.40	2.09	2.34
Rentis	1.30	1.30	0.81	1.14	1.79	1.51	1.30	1.53
Drapetsona	0.87	1.14	0.89	0.97	1.41	1.07	1.03	1.17
Pireus	1.59	1.57	1.38	1.51	–	1.41	1.43	1.42
NO_x								
Patission	206	206	237	216	129	126	142	132

Source: Greek Ministry of Environment, Housing and Physical Planning.

VIII. EVALUATION

Air pollution is a major public concern in Athens and the Government has already taken numerous initiatives within the fields of traffic management and transport policy in order to abate it. Preliminary results of some of these actions are becoming available. It is however early days to measure results against objectives and this evaluation will therefore be tentative.

Traffic Bans

The ban on about half of all private cars entering the central area of Athens is a measure that led initially to a 20 per cent reduction in total traffic within the inner ring road. In the absence of repeated counts it is not possible to say whether this reduction has been maintained.

In the years ahead it is intended to supplement or replace the present ban by a combination of more conventional traffic management measures and the construction of grade separated junctions along the inner ring road. Since there is bound to be uncertainty about the effectiveness of such measures in abating emissions in a densely developed city, a case can be made for examining other forms of area traffic management. Amongst techniques meriting study are daily and weekly road-user charges of the kind levied in Singapore and for which automatic collection equipment is being developed elsewhere[8]. Such road user charges, while acting against traffic congestion, would also provide a source of revenue for subsidising transit.

Inspection of In-Use Vehicles

Regular inspection of in-use vehicles has the potential to make a major contribution to reducing exhaust and noise emissions, particularly if backed up by improved standards of maintenance at garages. Two refinements to the regime that merit study are more frequent inspections of older than newer vehicles and an obligation, following failure of a test, to return, after maintenance, for another one.

Improvements to Transit Quality

Public transport has two important environmental roles in Athens. One is to cater for movement in a city too densely built to make possible high levels of car use. The other is to contribute towards cleaner air through good vehicle maintenance and the use of vehicle types that are low polluters or free of street level emissions altogether.

These objectives have been pursued by means of investment in new buses and maintenance depots and by expansion of the trolley bus network. The value of this investment programme depends on whether the use made of public transport can be stabilised and then increased. In fact, notwithstanding the growth of population, only travel by trolley bus has grown. Use of the underground has been stable while travel by bus has fallen although the steepness of the decline has, since 1980, begun to fail off.

A reorganisation of bus routes aimed at giving better service and a network of bus priorities may arrest the fall in bus travel and open the way to road public transport making a major contribution towards environmental goals. Trials with exhaust gas catalysts designed to reduce emissions of CO, HC and smoke from diesel engines will, if successful, open the way to further reductions in transit emissions.

Increasing Transit Capacity

Some routes across the centre of Athens now covered by trolley buses are heavily enough travelled to justify public transport with higher passenger capacity. Further underground lines are one method of meeting this demand. Light rapid transit is an alternative that has recently attracted increasing attention. Such light railways cost only one-tenth of underground lines and are able to be in use three years after a decision to go ahead.

Intense line by line study of the nine LRT routes proposed by the Athens Master Plan team will be needed to establish the scale of the likely travel demand and how priority over other traffic could be most economically achieved for LRT trams. No less important will be the identification of the scope for urban development along the corridors to be served by the proposed transit lines.

Fares Policy

The cost of low fares in Athens has become an item of public expenditure of national importance. While subsidy of passenger transport has a place in urban transport policy, it can be counter-productive if it inhibits investment in capital projects designed to improve the quality of transit services. If ways could be found of raising revenue for subsidies that also helped to reduce congestion and emissions, the benefits of the subsidy policy would however be multiplied.

Emissions from New Vehicles

As car ownership levels in Athens are still low by North European standards, new vehicles tend to be additions to the existing fleet and not replacements for vehicles being scrapped. The emissions from new vehicles accordingly play an important role in raising general emission levels.

Noise

Despite the nuisance ascribed to noise by Athenians, little has so far been done to abate it. Enforcement of existing noise regulations is, at best, weak, and low importance is attached to noise abatement when designing traffic management or other road schemes. The most promising approach to abatement is a broad one covering many actions. Renewal of the bus fleet and inspection of in-use vehicles combined with increased enforcement action against motorcycles and a smoothing of traffic flows hold out most promise for reducing the nuisance of noise.

IX. CONCLUSIONS

The topography and climate of Athens combined with the effects of rapidly increasing car ownership and use are conducive to intense air pollution.

Measures more stringent than in most cities are needed if clean air is to be achieved.

No single, immediate action is capable of bringing substantial reductions in air pollutants. A wide range of short, medium and long term measures is called for. However, the high density of development in Athens gives public transport potential to contribute towards cleaner air not available in lower density, suburban cities. Higher densities tend, however, to increase the nuisance of noise.

The Greek Government's declaration of intent to reduce air pollution in Athens is a vitally important step and has already led to action.

Because it will take time for improvements to new vehicles to work through the national fleet, periodic inspections of in-use vehicles and improvements to the quality and take up of routine maintenance at garages will play major parts in achieving cleaner air up to 1990. The Government's new compulsory inspection programme, (which began January 1985), is therefore of paramount importance.

Greece, as a member of the EEC, has already tightened vehicle emission standards. Lead in fuel has already been reduced to 0.15 grams per litre in line with the Community. Still tighter standards for new vehicles are, however, needed to offset the effects of rising car ownership and bring major improvements in air quality in the 1990s. Lead-free fuel and exhaust catalysts are likely to be part of such a policy.

The Greek Government needs to play a leading role within the European Communities in pressing for tighter emission standards for new vehicles.

A shift in travel from car and taxi to transit could help to improve air quality in the medium term. This shift will only occur if there is a substantial improvement in the quality of transit service. Dependability and routes that go where people want to travel are vital. Comfortable, clean, modern vehicles staffed by helpful crews are also necessary.

The potential for improving the quality of transit service by switching funds from revenue subsidies into grants for capital improvements needs to be explored.

Road user charges of the kind already levied in Singapore and for which automatic counters are under development on the roads of Hong Kong are a potential source of revenue for transit investments.

Athens already has the largest fleet of quiet, emission-free electric trolley buses in Europe. The contribution of this system to clean air and reduced noise could be increased by raising running speeds and productivity and attracting additional passengers. The reform of transit routes following the operators' systematic survey and the application of the measures set out in the Athens Master Plan will be important in achieving this objective.

The conversion of large areas into pedestrian zones, as in Lyon and Frankfurt, has not yet been attempted in Athens but proposals for the pedestrianisation of some streets are made in the Athens Master Plan. Such environmental improvements could be introduced initially for one day a week and then expanded as experience of their effects was gained.

A dense city of over three million inhabitants needs high capacity transit. There are two main alternatives to the existing bus and trolley bus services. One is additional underground lines which would give fast, high capacity, services without interfering with road traffic on the frequently narrow streets of central Athens. The other is a main collective rapid transport system, running largely on existing streets but under or overpassing points of congestion and coupled with partial pedestrianisation in the centre of the city.

Public confidence that cleaner air can be achieved needs to be maintained. Step-by-step improvements, perhaps individually modest, but capable together of making noticeable changes are vital in keeping up public confidence.

NOTES AND REFERENCES

1. *Environmental Policies in Greece*, OECD, 1983.
2. Estimate from Greater Athens Master Plan, 1983.
3. Attica is one of 52 regions in Greece. It has an area of 3 808 square kilometres or 2.89 per cent of Greece. The Greater Athens Area (427 square kilometres) is 11.2 per cent of the Attica region.
4. OAS Survey.
5. *Long Term Perspectives for the World Automobile Industry*, OECD, 1983.
6. *Benefits of Environmental Policies as Avoided Damage*, OECD, 1983.
7. Simandonis, J.S., Noise Survey in the Metropolitan Area of Athens, University of Southampton, 1976.
8. Case studies of Singapore and Hong Kong in this programme give fuller details of the use of economic instruments in traffic management.

Chapter 2

GOTHENBURG*

I. SUMMARY

This case study presents the story of traffic cells in Gothenburg. The Central Business District (CBD) cells are still operating well, and their principles are now being applied to the city's inner suburbs. The Council's aim is to bring to areas just outside the CBD the environmental benefits already obtained within it.

These benefits are reduced noise and air pollution, increased safety, and a greater sense of well-being.

The results of the new cells are reductions of up to 8 dB(A) in noise levels for one third of affected residents, an increase in noise level up to 4 dB(A) for one third and, for the last third, no change. Through traffic was re-routed totally from the new cells, with more than 2 000 cars a day being diverted to the riverside. Traveltime by public transport was reduced mostly because of the implementation of reserved tracks. Traffic accidents were reduced by 27 per cent within the cells and by 14 per cent overall.

II. IMPLEMENTATION OF TRAFFIC CELLS IN THE CENTRAL BUSINESS DISTRICT

Background

Gothenburg is the second largest city in Sweden and has a population of 430 000. In the Gothenburg region (Figure 1) there are about 700 000 inhabitants. It was founded in the 17th Century and, as a result of the Göta river (the largest waterway in the country), has developed into Sweden's leading centre of trade and shipping. Three main factors have influenced the growth of urban transport in Gothenburg:

— The river which forms a major barrier between the extensive industrial area to the north-west and the rest of the city;
— The valleys and ridges that radiate from the city centre; and
— The layout of the old city with its wide main streets, which were once canals, and its surrounding canal, which remains.

Trips to work to and criss-crossing the city centre have long been numerous. In the 1960s rapid growth in car traffic led to the construction of two river bridges and a major overhaul of the road network. This created a substantial reserve of road capacity, at least outside the Central Urban Area (CUA). This period also saw the completion of a ring road around the CBD. However, by 1970 severe traffic and environmental problems became apparent within the CBD, particularly on Saturdays and Sundays during shopping hours and a Traffic Plan was prepared. Initially, unauthorised vehicles were banned from the commercial area at weekends. The success

---
* Case study prepared by Messrs S. Falk (Sweden), T. Bendixson (United Kingdom) and T. May (United Kingdom).

Figure 1. **GOTHENBURG REGION**

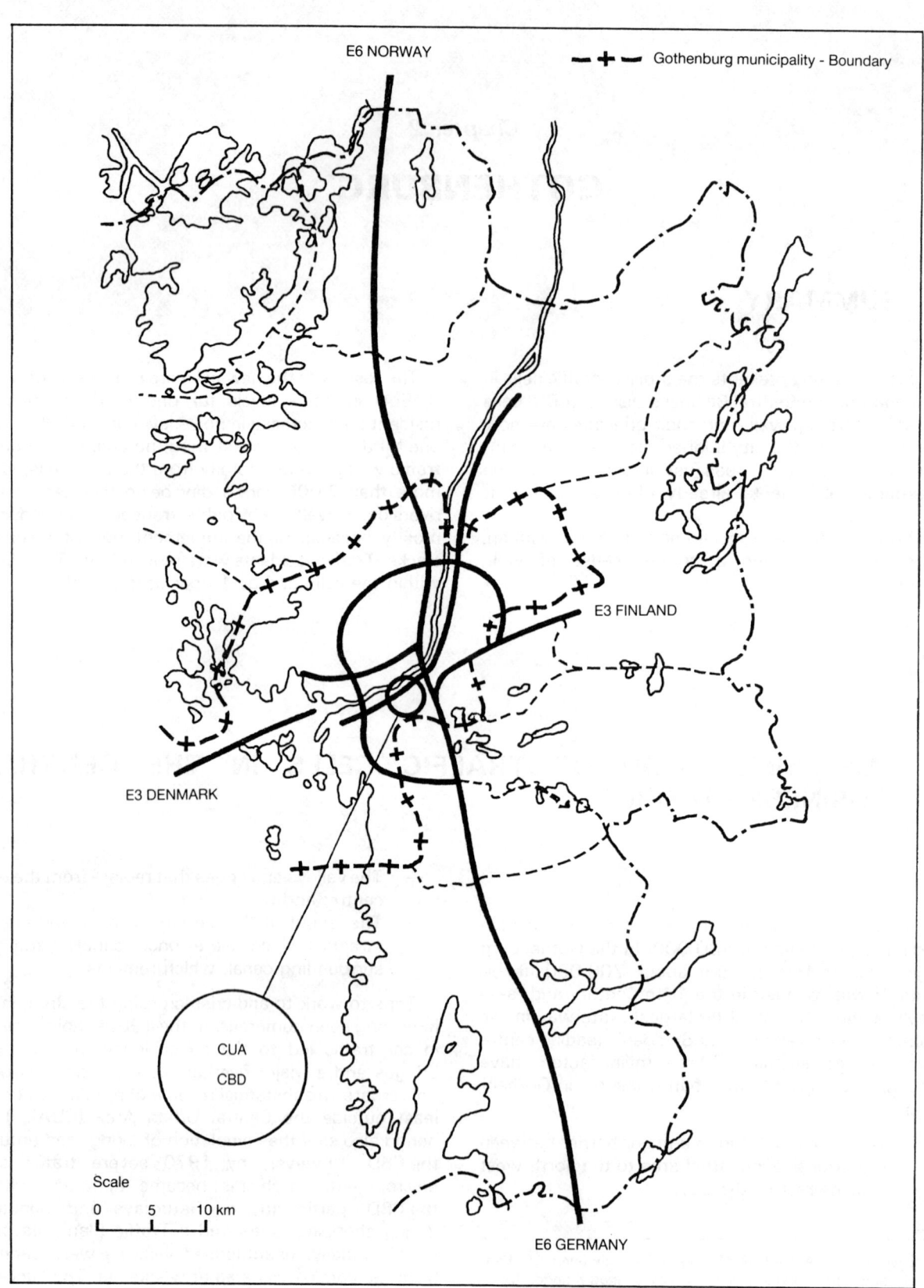

of this measure paved the way for comprehensive traffic control.

In the 1970s public transport in Gothenburg consisted of a tramway network supported by feeder buses. The tram network was radial and about 65 per cent of it was on reserved tracks. Nearly half the passenger kilometers were done on the other 35 per cent, mainly in the CUA, where delays due to congestion resulted in poor punctuality, varying journey times and low speeds.

Travel times by public transport were roughly twice those by car because of walking distances to stops, waiting time and low speeds. Except in the CBD, traffic volumes exceeded capacity at few points on the road network. Furthermore the accident rate was generally lower than for other Swedish towns of similar size and car ownership (see Annex).

The Traffic Plan

During the late 1960s the City Council was urged to adopt a traffic regulation scheme for the CBD and the following objectives were identified:

— Improved safety and a better environment for people living, working and walking in the CBD;
— Improved public transport;
— Rapid, inexpensive implementation.

The solution chosen was the division of the CBD into five cells (Figure 2). Only public transport and emergency vehicles were allowed to cross the borders between the cells. Other traffic had to use entrances and exits on the ring road with two exceptions, one because of lack of capacity on the ring roads and one because of informatory matters.

Other measures implemented between 1970 and 1980 were:

— An increase from 65 per cent to 90 per cent in the proportion of the tram network running on reserved tracks; additional priorities for public transport at signalised intersections; and a flat fare system;
— Introduction of express bus routes;
— Conversion of the CBD public, off-street parking stock to short term parking;
— A special travel service for handicapped people.

Figure 2. **TRAFFIC CELLS IN PRINCIPLE**

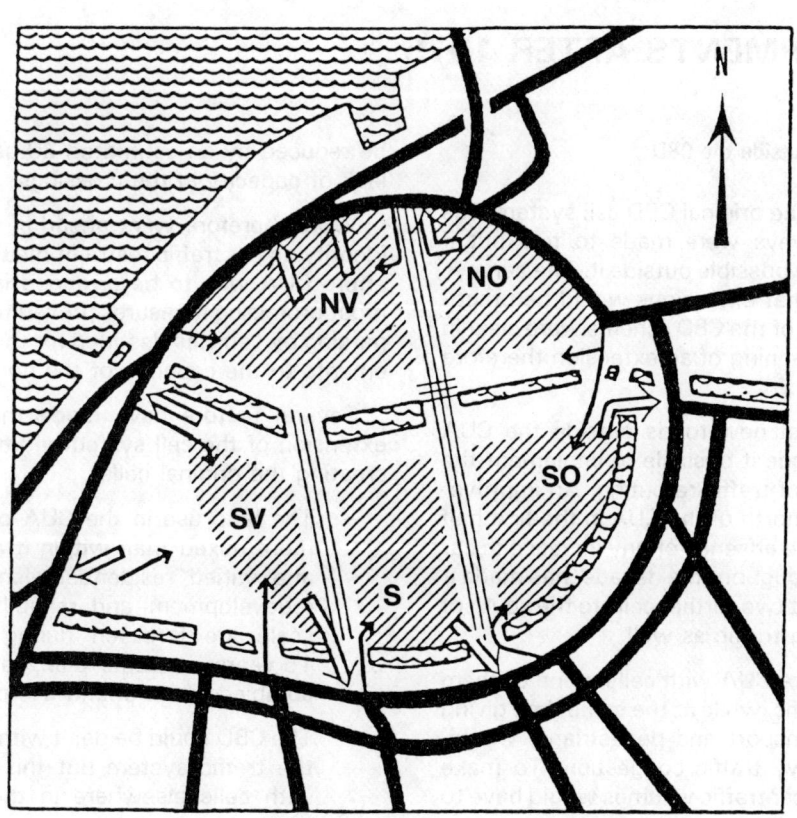

Do the zones reroute or restrain traffic?

In considering traffic management a distinction should be drawn between traffic rerouting and traffic restraint. The two concepts are not always easy to separate. What may look like traffic restraint for a small area may be traffic rerouting in a larger one. One way to clarify this issue is to look at the objectives of a scheme. If the aim is to shift traffic to other streets, then it is a rerouting scheme. The effect might be to restrain traffic too but this does not alter the description since any reduction should be seen as a by-product.

Viewed in this way the CBD zone system in Gothenburg clearly falls into the *rerouting* category. The aim was to divert through traffic from the streets in the CBD on to the ring road. The scheme had two other objectives. There were to improve the environment and to reduce accidents. It also had important by-products. These included:

— Opening the way to improved public transport;
— Improved traffic circulation;
— Opening the way to creating pedestrian streets; and
— Obtaining a limited amount of traffic restraint within the CBD.

Public Reactions

The CBD zones, which were initially opposed by traders, are nowadays accepted by everyone in the city. Politicians, downtown merchants, delivery drivers, bus and tram drivers, customers and even taxi drivers now all accept them. Taxis are treated as buses and allowed to cross most of the borders. The favourable reaction is probably a result of the continuous decrease that has taken place in traffic flows and accidents. Traffic flows across CBD border decreased by 45 per cent between 1970 and 1982 including 11 per cent during the past three years. Casualties have decreased by 45 per cent within the zones and by 40 per cent in the zones plus the boundary streets. Property damages have decreased by 55 per cent and 50 per cent respectively.

Noise and air pollution have also declined because of the reductions in traffic flow. No continuous surveys of bicycling have been made in the CBD, but generally there are today three times as many bikes as there were two years ago.

III. DEVELOPMENTS AFTER 1975

The Creation of Cells Outside the CBD

At the time when the original CBD cell system was proposed rough surveys were made to find out if further cells would be possible outside it. The surveys seemed to indicate that extensions would be practical. When the effects of the CBD scheme were clear in 1973-74 detailed planning of an extension therefore began.

The construction of new roads outside the CUA during the 1960s made it possible to implement the CBD cells by means of traffic rerouting. An improvement of the by-pass north of the CUA in the first half of the 1970s, and the absence of any increase in car traffic to the CUA throughout the decade, has made it possible to implement two further cells to the south of the CBD by rerouting traffic as well.

Equipping the entire CUA with cells, none of them with any through traffic, while at the same time giving priority to public transport and pedestrians, would, however, cause heavy traffic congestion. To make such a plan possible car traffic volumes would have to be reduced by an estimated 20 per cent because of lack of capacity in the crossings.

Work therefore went ahead in the late 1970s on identifying the traffic restraint measures needed if the entire CUA was to be arranged in traffic cells. The most promising measures proved to be restrictions on parking by vehicles destined for the CUA and improvements to the capacity of the northern by-pass.

Several factors have made the planning of the extension of the cell system more complicated than creating the original cells:

— The land use in the CUA outside the CBD is more mixed than within it and contains large and varied residential elements; in addition redevelopment and rehabilitation on a large scale are foreseen during the coming 10-15 years; a traffic plan had therefore to be combined with a land use plan.

— The CBD could be dealt with as a single item in the traffic system but this was not possible with cells elsewhere in the CUA; additional

Figure 3. **CENTRAL URBAN AREA (CUA) TRAFFIC CELLS**

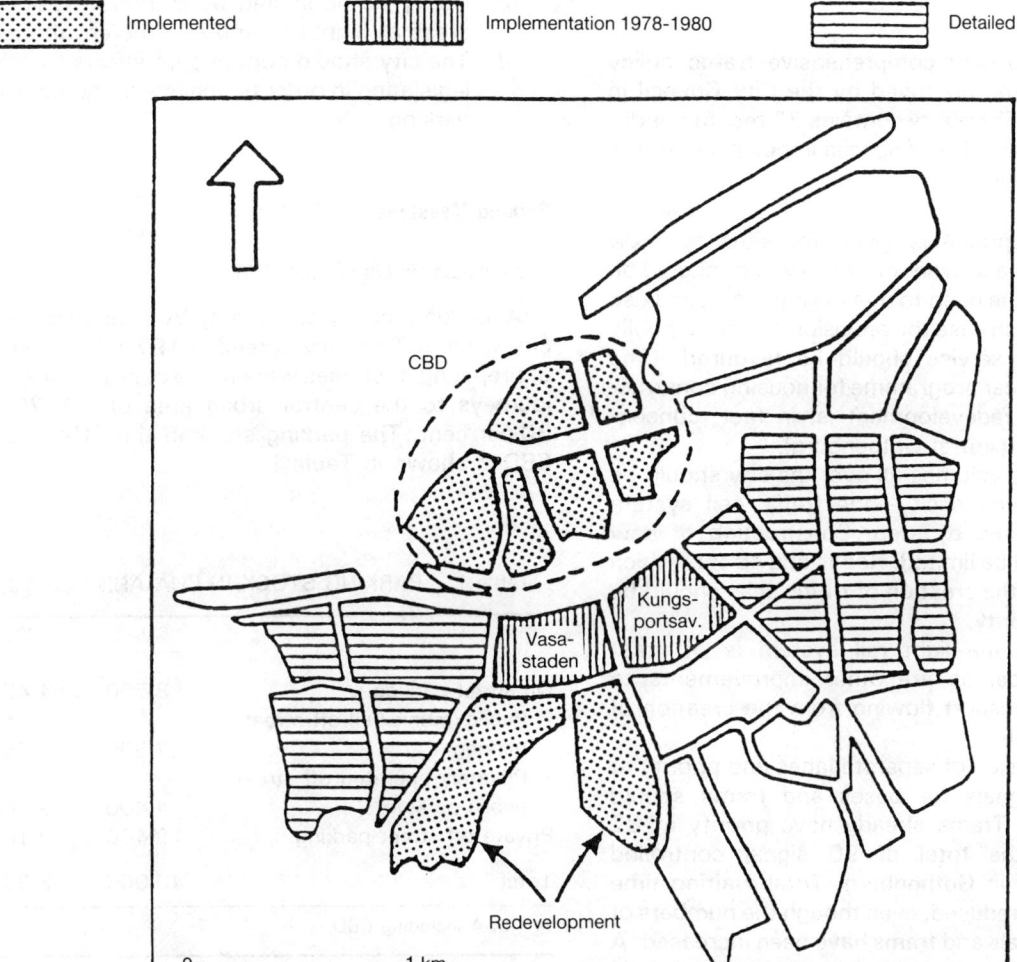

cells had therefore to be part of a traffic plan for the whole CUA.

— Whether, and when, major new roads will be constructed to the west or south of the CUA is more and more uncertain; this is due to financial stringency and to lower rates of city growth than forecasted in the 1960s; the plan had therefore to be flexible and capable of implementation step by step with each step being a functional "final solution" if necessary.

A land use and traffic plan for the CUA was approved by the City Council in 1976 as a guideline for detailed planning. It was decided at the same time that a start should be made as soon as possible on implementing cells in Vasastaden and Kungsportsavenyn (Figure 3). Studies of extensions west and east of these cells were also agreed.

IV. THE TRAFFIC AND PARKING POLICY OF 1979

Traffic Measures

In 1977-78 a new comprehensive traffic policy was drawn up and approved by the City Council in February 1979. The policy contains 32 recommendations some of which are of special importance for the Central Urban Area:

a) When approval is given for sites for new dwellings and work places, account should be taken of the need to minimise travel. Land use plans which ease the provision of a good public transport service should be favoured. (The city's 5-year programme for housing construction and redevelopment is, in fact, concentrated on central Gothenburg).

b) Car traffic volumes all over the city should be reduced and a city-wide traffic cell system created step by step. Construction of new roads will be limited. Roads and streets which facilitate the creation of traffic cells should be given priority.

c) Enlarging the light rail system is of great importance, as are other improvements to public transport flowing from the creation of traffic cells.

d) The provision of separate lanes and priority at traffic signals for buses and trams should continue. Trams already have priority at 82 out of the total of 90 signal controlled crossings in Gothenburg. Total waiting time has been reduced, even though the numbers of both signals and trams have been increased. A bus priority programme is under way and up till now, buses have priority at 50 intersections.

e) In addition to extending the traffic cell system, special measures should be taken to reduce casualties at black spots. The programme for improving ways to school and the cycle path system should be carried through. (Creation of traffic cells must not mean that accidents are moved from one spot to another).

f) The aim of parking policy in the CUA should be to reduce car traffic in order to reduce casualties and improve the environment and public transport. This should be achieved by a reduction in the number of parking spaces, the relocation of those that remain and the levying of appropriate charges. The capacity of the bus and tram system should be correspondingly increased.

g) Car-pooling and park and ride should be investigated and made the object of experiments.

h) Heavy traffic should be limited in residential areas at nights to improve the environment.

i) The city should continue its efforts to amend legislation in order to achieve better control of parking.

Parking Measures

Reducing Parking in the CUA

A parking policy for the CUA was part of the city-wide traffic policy agreed in 1979. In the course of preparing it, studies were made of reductions in car journeys to the central urban area of 10, 20 and 30 per cent. The parking stock in the CUA and the CBD is shown in Table 1.

Table 1. PARKING STOCK IN CUA AND CBD 1978

	CUA[1]	CBD
On street parking	8 700	1 450
Public (municipal) off street parking	7 500	550
Privately-operated off street public parking	4 400	3 200
Private off street parking	19 400	3 100
Total	40 000	8 300

1. CUA including CBD.

The effects of such reductions on casualty levels, the environment, travel by public transport, retail business, etc, were investigated as were different ways of minimising the decrease in accessibility to the central area.

The consequences for retail business were the subject of a special study. A market survey was undertaken to assess how turnover is divided between customers arriving by car and public transport and between people working and living in the central area. As the general aim is to increase the number of residents within the CUA, the possibility of offsetting a reduction in car customers by an increase in residents was considered. An increase in residents will not be easy to achieve since the total population of the city is expected to be constant or to decrease slightly over the coming 5 to 10 years. On the other hand, apartments in the CUA are still in demand.

Recommendations for Parking

The parking policy contains the following recommendations:

a) Predictions of the detailed effects of changes in parking policy are difficult to make; in particular the consequences of parking restrictions will be difficult to separate from changes brought about by other influences on the numbers of people living and working in the central area and on commercial activity. The implementation of parking restrictions therefore needs to be cautious and linked with studies of effects. Measures designed to achieve a reduction in car traffic of approximately 1 pour cent a year are recommended. A first goal of 10 per cent seems suitable, but a reduction of more than 20 per cent is difficult to obtain under present legislation while keeping a proper balance between publicly and privately owned parking.

b) Residential parking should be given priority though with a standard of only 0.6 parking spaces per apartment. This corresponds to approximately 300 cars per 1000 residents. The standard for the CBD is 0.4 spaces per dwelling which is 200 cars per 1000 residents. These standards are lower than elsewhere in the city.

c) Restrictions should bear harder on parking by commuters than by visitors.

d) The implementation of restrictions should be linked with the relocation to the fringes of the remaining CUA parking spaces in order to achieve the greatest reductions in traffic flows and the greatest environmental benefits in the inner parts of the central area.

e) Reductions in traffic flows should be used to create traffic cells and to give priority to public transport and pedestrians. This and improvements to the northern by-pass will also discourage through traffic.

Implementation

The following implementation programme for the years 1980-85 is in progress:

a) A reduction of commuters' parking from 21 000 to 14 000 spaces.
b) A reduction of visitors' parking from 10 500 to 9 500 spaces.
c) Relocation of 3 000 spaces towards the fringes of the CUA.
d) Experiments with special parking places for car-pools.
e) Experiments with park and ride.
f) Increased public transport service at a yearly cost of 2.5 million SKr (US$0.35 million). The investment of 20 million SKr (US$2.85 million) in new trams and buses.
g) Increases in parking charges of up to 100 per cent resulting in charges for short term parking as high as 6-8 SKr (US$1) an hour. A first increase from 3 to 5 SKr was made in 1979.

These measures are expected to reduce car traffic to the CUA by approximately 7 per cent.

V. EXTENSION OF THE CELL SYSTEM OUTSIDE THE CBD

Why Extend the Cell System?

Nearly 2000 residents live in the Vasastaden cell and 1 600 in the Kungsportsavenyn cell and there are about 3 000 and 5 200 workers respectively in the two cells.

There were four main reasons for extending the cell system outside the CBD:

First, dense, mixed traffic to the south and south-west of the CBD was associated with sixty traffic injuries a year. As cells within the CBD had been successful in reducing casualties through relatively small divergensis in traffic it was thought possible to repeat this achievement outside it.

Second, implementing a cell system means substantial internal environmental improvements but only a minor deterioration in conditions in surrounding streets if, as was the case with the new cells, these already carry heavy traffic. Traffic on one fringe residential street was, however, calculated to increase by about 40 per cent, with consequent increases in air pollutants. Implementation of the new cells was accordingly opposed by the Local Health Authority until traffic volumes had been lowered by traffic restraint. However the City Council considered that the total benefits of the scheme outweighed their disadvantages. Special care has been taken to make the traffic flow in the affected streets as smooth as possible.

Third, a traffic cell system promised to make it possible to give priority to trams and buses and create safe and comfortable conditions for pedestrians.

Fourth, there was a possibility of adding important links to the main cyclepath network in the CUA.

Design of the New Cells

The design of the new cells followed the principles of the earlier ones.

Consideration was given to design the area into a single cell but it proved impossible to make room for all the rerouted vehicles unless total traffic was reduced. A two cell system was therefore adopted. It would however be possible to introduce the one cell solution later if traffic level should fall or be restrained.

Experience of the CBD-cells has led to decisions to rebuild the foot-ways along the boundary streets so that they act as thresholds and signal to drivers that they are entering protected areas where they should be careful about pedestrians.

Implementation of the New Cells

Staged Progress

The implementation of cells outside the CBD has not been done in the same way as it was within it. That was an over-night operation using temporary concrete barriers which were gradually made permanent once the system was seen to be working.

The new cells cover smaller areas and the traffic pattern is less complicated. This, together with the experience gained in the CBD, made it possible to go at once to permanency. An urgent need to establish new bus and tram lanes and improve stopping areas supported the case for this approach. In the boundary street Engelbrektsgatan it was necessary to increase traffic capacity and improve safety through signalisation. Parking places were removed and in some blocks it was necessary to narrow the footways even though this was not popular with residents and shop-owners.

A special parking-fee was introduced for residents. For 100 SKr (US$14.3) a month they can park on specially marked lengths of street. Analyses show that 20 to 50 per cent now leave their cars at home during the day and go to work on foot or by bicycle or transit.

Implementation has, in the event, had to be staged and geared as much to the construction of new underground facilities and tram stops as to traffic management. Financial reasons too have made it necessary to spread implementation over two years. During this period measures have been taken in and outside the area that allow traffic to flow without too much inconvenience.

Public Participation

Before the comprehensive land use and traffic plan for the CUA was brought to the City Council in 1975, a week long exhibition was arranged at the City Library. It was attended by 25 000 people of whom about 1000 completed special forms. A discussion evening, attended by fewer than 50 people was held at the end of the exhibition but little came out of it.

A further, smaller exhibition was mounted before it was decided to implement the Vasastaden and Kungsportsavenyn traffic cells. This gave detailed proposals and was shown at an information centre, the post office and other frequently visited places within the proposed cells. Fewer than 100 out of the 4 000 people living or the 8 000 working in the area expressed any opinion.

Exhibitions have been found to be a poor method of encouraging public participation. Other ways of getting people interested have therefore been tried. Residents' associations have been involved from the first in traffic planning by producing a booklet of alternative plans to which they contributed a chapter.

Information about the proposed cells was also circulated by means of pamphlets. These where sent to the occupants of all apartments, offices and shops within the proposed cells. They explained the objectives of the measures and how it would be staged. Information was also broadcast through the municipal newspaper.

VI. RESULTS

Expected Results

Cell systems inevitably cause some cars to drive further than before. In the case of the Vasastaden and Kungsportsavenyn cells car kilometers in and around the area are expected to increase by 7 per cent. A corresponding increase in fuel consumption and environmental impact will be counteracted by linking traffic signals so as to improve the organisation of traffic outside the area and allow it to flow without too much inconvenience.

In the cost-benefit analysis account is taken to reduction of the total number of accidents by 20 per cent.

As a result of rerouting the number of residents living along streets where traffic flows are heavy has been reduced by 20 per cent. This and other improvements in environmental conditions are expected to encourage redevelopment or rehabilitation of houses. Shops and other commercial activities will be affected only marginally judging by the experience of the CBD-cells. Some customers are likely to be lost because of the absence of passing traffic and a decrease in number of parking spaces, but others are likely to be gained because of improved public transport and other changes designed to make the area more popular. In the long run it is expected that the changes will attract more residents. About one out of ten delivery trips to shops and offices will be extended by up to 500 metres while three times as many will have to change their routes but without going further.

Three taxi stations existed in the study area. In the new traffic system they are replaced so as to give direct access to different parts of the cells. Otherwise taxis will have to follow the same traffic rules as other cars.

The cost for implementing the first step (1978-80) of the extended cell system was 5.8 million SKr (US$0.8 million) of which 4.0 million ($0.6 million) was paid from national funds.

Observed Results

Changes in Traveltime for Public Transit

The time taken by trams and buses to pass through the area has been reduced on average by 45 seconds per trip. In most of the streets where public transport and cars are mixed, car traffic volumes have been considerably reduced. In such streets curbs have been moved out towards the tracks at stops. Cars accordingly have to wait when a tram or bus is stopped.

Changes in Traffic Flow

Traffic flows to and from the CUA have been reduced by implementing the new cells in Vasastaden and Kungsportsavenyn. The reductions on the streets going to or by-passing Vasastaden are specially noticeable. Traffic has been rerouted to the streets along the southern riverside as a result of the measures in the CUA.

Changes in traffic volume within the CUA, in the CBD and on the municipal boundary have been different. Traffic flow in the CUA decreased by 5.4 per cent between 1979 and 82 but increased by 8.0 per cent over the municipal boundary (Figure 4). Today traffic volumes in the CUA are at levels lower than ever during the 1971-82 period (Figure 5).

Estimates of the effects of the cells on traffic flows correspond exceedingly well with subsequent counts (Table 2). The traffic volume east-west was decreased by 2 400 cars per 24 hours and 2 300 of them were rerouted along the riverside north of the CBD.

The car traffic in Götaleden and Engelbrektsgatan have accordingly increased, while other streets show a decrease. The redistribution towards the southern riverside is noticeable. Götaleden's part of the thorough traffic of the CBD has increased from 40 per cent to 55 per cent. The through traffic in the other streets together has decreased with 3 125 vehicles.

Big investigations about car traffic have been made in the years of 1971 and 1983.

Changes in Traffic Accidents

Traffic signal improvements have made conditions safer and more convenient for pedestrians and cyclists both within the cells and on the surrounding streets. The safety of vehicle occupants has been improved in the cells, and to some extent in the streets around them, but not on the northern boundary road. On this street, CBD ring-road traffic volumes have increased with a marginal increase in accidents.

The overall change in traffic accidents has been favourable. Substantial reductions have occurred on Vasagatan although increases have occurred on the Nya Allén and at the crossing of Kungsportsavenyn and Kristinelundsgatan. In general terms accidents on roads surrounding the cells decreased by 9 per cent and those within the cells fell by 27 per cent. This gives a total decrease of 14 per cent which almost corresponds with calculations made before implementation. Traffic accidents involving trams and buses have also decreased.

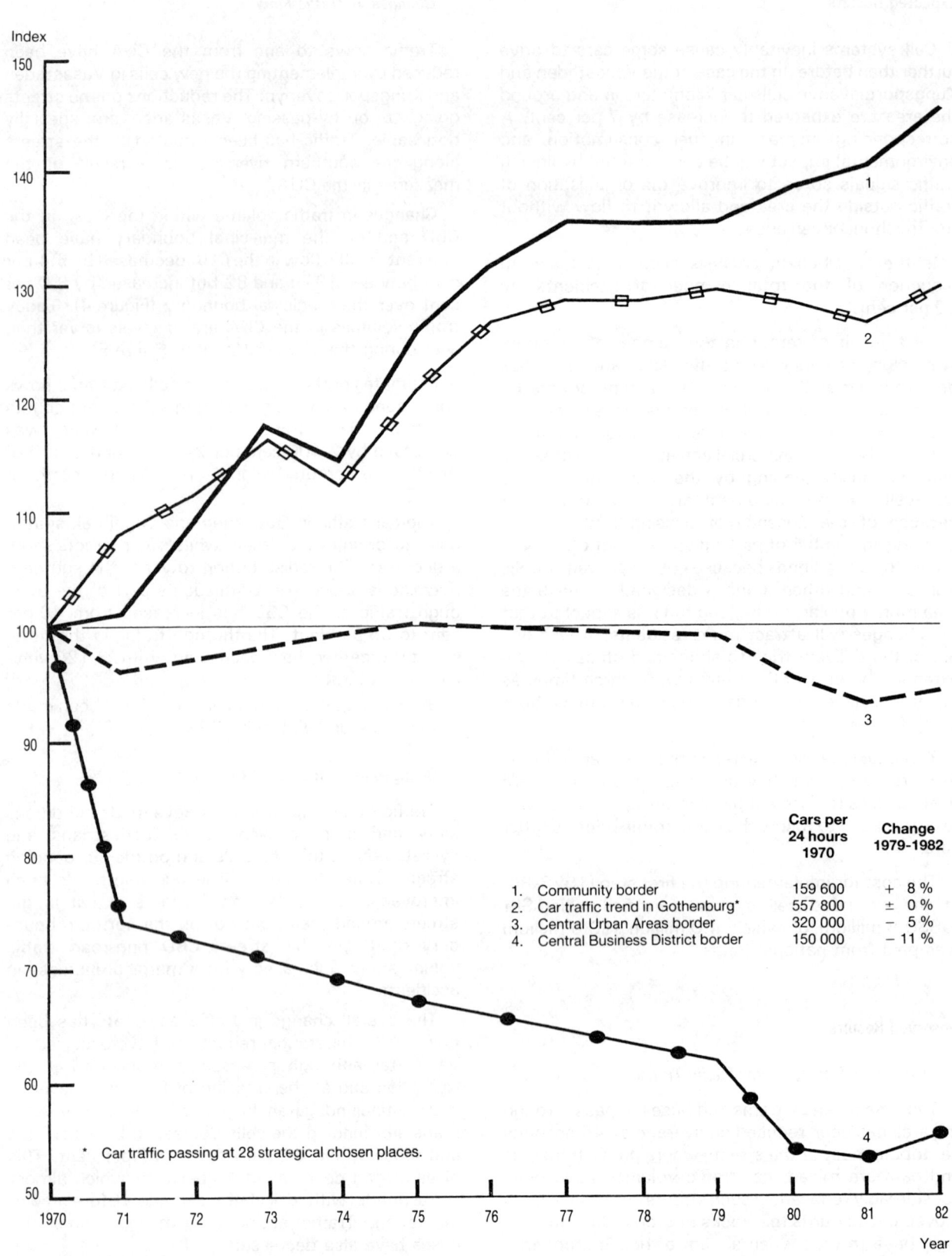
Figure 4. RELATIVE CHANGE IN TRAFFIC VOLUME IN CBD, CUA, ETC.

Figure 5. **CHANGES OF TRAFFIC FLOWS AND SPEEDS**

Changes in traffic volumes during the period 1979-1982

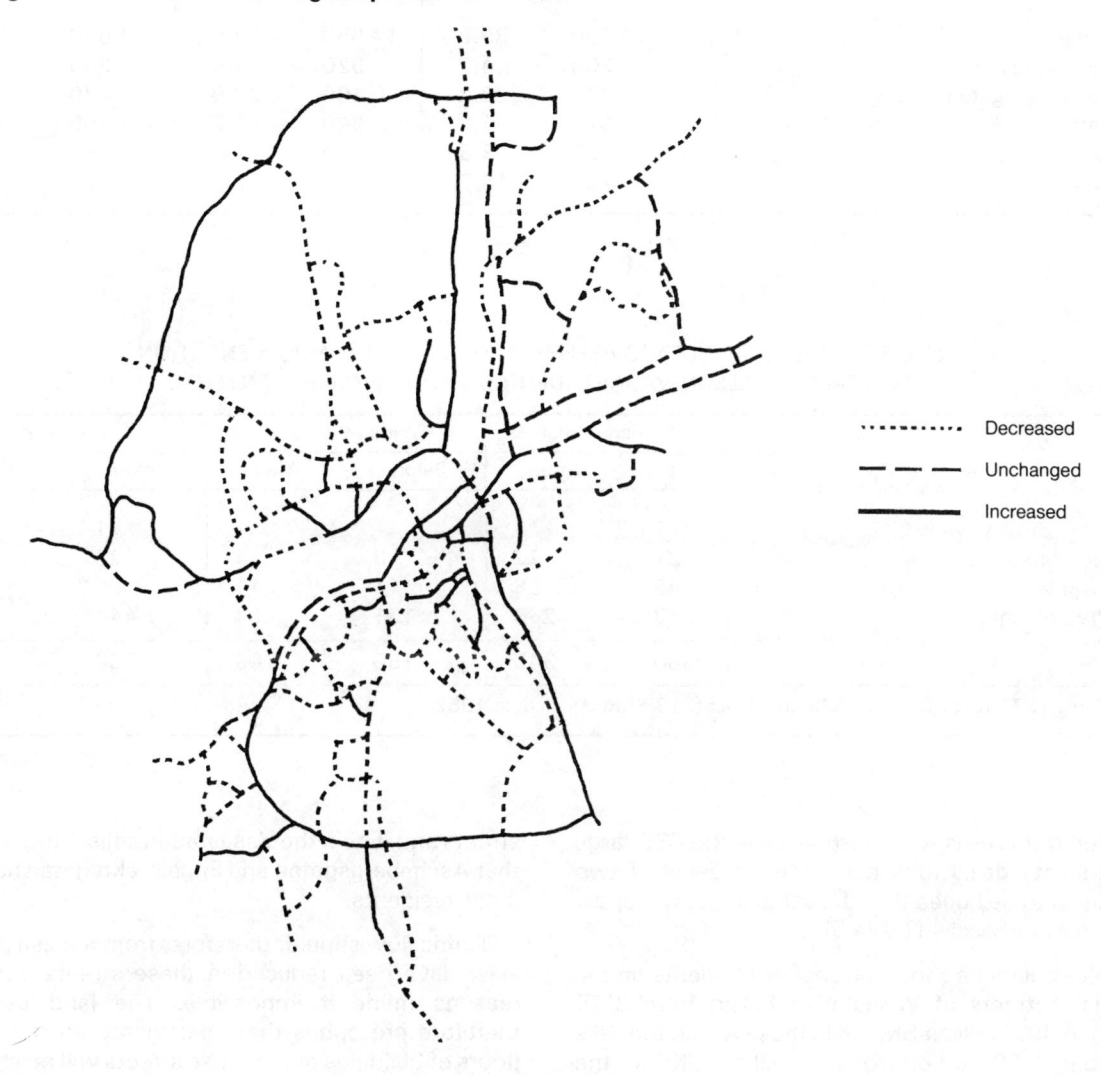

.... Decreased
– – – Unchanged
───── Increased

The proportion of streets where vehicle speeds were above 50 km/h increased from 25 % in 1976 to 48 % in 1981

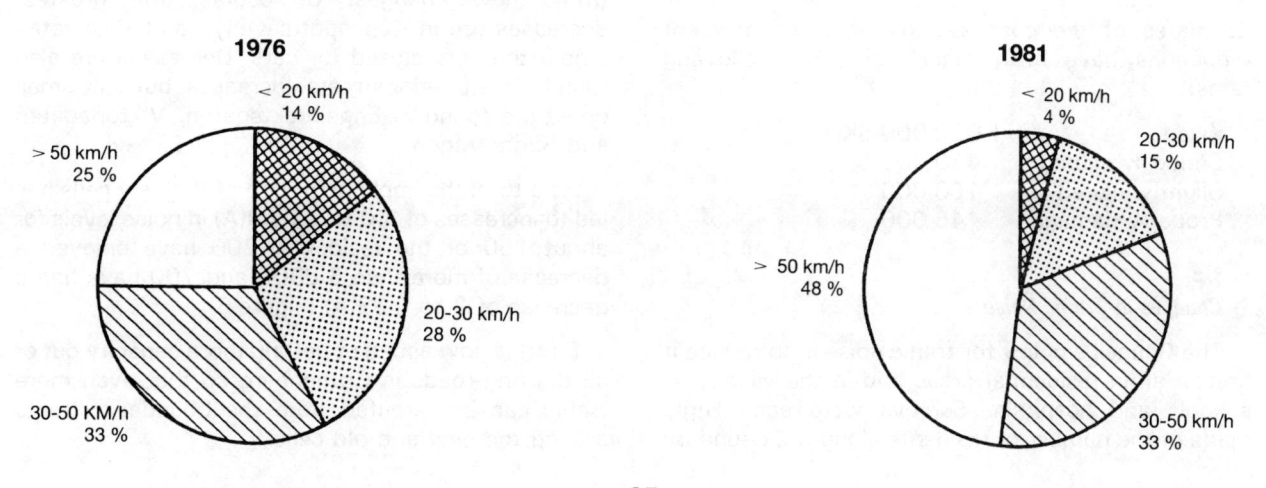

1976
- < 20 km/h: 14 %
- 20-30 km/h: 28 %
- 30-50 KM/h: 33 %
- > 50 km/h: 25 %

1981
- < 20 km/h: 4 %
- 20-30 km/h: 15 %
- 30-50 km/h: 33 %
- > 50 km/h: 48 %

Table 2. THROUGH TRAFFIC

Observation point	1971 Total	1971 %	1983 Total	1983 %	Diff. Total	Diff. %
Götaleden	10 100	39.9	14 650	54.9	+4 550	+45
Norra Hamangatan	1 200	4.7	920	3.4	−280	−23
Nya Allén/Parkgatan	7 370	29.2	6 100	22.9	1 270	−17
Vasagatan	1 415	5.6	310	1.2	−1 105	−78
Engelbrektsgatan	955	3.8	1 340	5.0	+385	+40
Läraregatan	4 225	16.7	3 370	12.6	−855	−20

Table 3. TRAFFIC ACCIDENTS BEFORE AND AFTER THE IMPLEMENTATION OF TRAFFIC CELLS IN VASASTADEN AND KUNGSPORTSAVENYEN*

	On boundary streets Before	On boundary streets After	Within the cells Before	Within the cells After	Together Before	Together After
Killed	1	—	—	—	1	—
Serious injuries	21	22	8	4	29	26
Slight injuries	45	58	11	11	56	69
Property damages	313	265	131	94	444	359
Total	380	345	150	109	530	454

* 1 Sept. 1977 to 31 August 1979 and 1 Sept. 1980 to 31 August 1982.

When traffic cells were introduced in the CBD large traffic safety gains took place after a delay of two years. It is to be hoped that the same effect will apply in the two new cells (Table 3).

The calculations show the cost of accidents on the boundary streets of Vasastaden falling from 31.5 million to 30.2 million SKr while those within the cells fell from 10.9 million to 7.6 million SKr in the two years analysed. The annual reduction in accident costs resulting from the traffic reorganisation was thus about 2.2 million SKr.

Estimates of the cost savings of these accident reductions have been made using the following rates:

Killed	1 620 000 SKr
Seriously injured	410 000 "
Slightly injured	160 000 "
Property damage	45 000 "

Changes in Traffic Noise

The Council's policy for traffic noise is to reduce it first of all in residential areas and in the vicinity of schools and hospitals. Surveys were accordingly made of the number of residents along the boundary streets and within the Vasastaden cells. They showed that Aschebergsgatan and Engelbrektsgatan have the most residents.

Traffic flow should, therefore, from a noise point of view have been reduced in these streets but other reasons made it impossible. The land use plan therefore prescribes that apartments on the lowest floors of buildings along those streets will be changed into offices.

Changes in traffic noise have been calculated from traffic flow changes. Of course, the greatest decreases are in Kungsportsavenyn and Vasagatan where they are closed for cars. Decreases are also found in the interior streets. Increases, but only small ones, are found in Engelbrektsgatan, Viktoriagatan and Södra Vägen.

As a total the implementation of the new cells has led to increases of 3.0 to 5.0 dB(A) in noise levels for about 150 of the residents. 200 have enjoyed a decrease of more than 6 dB(A) and 700 have had a decrease of 3 to 5 dB(A) (Figure 6).

Effort is now aimed at getting more capacity out of all the ring-roads in Gothenburg so that even more traffic can be rerouted from the boundary streets around the new and old cells.

Figure 6. **CHANGES IN TRAFFIC NOISE**

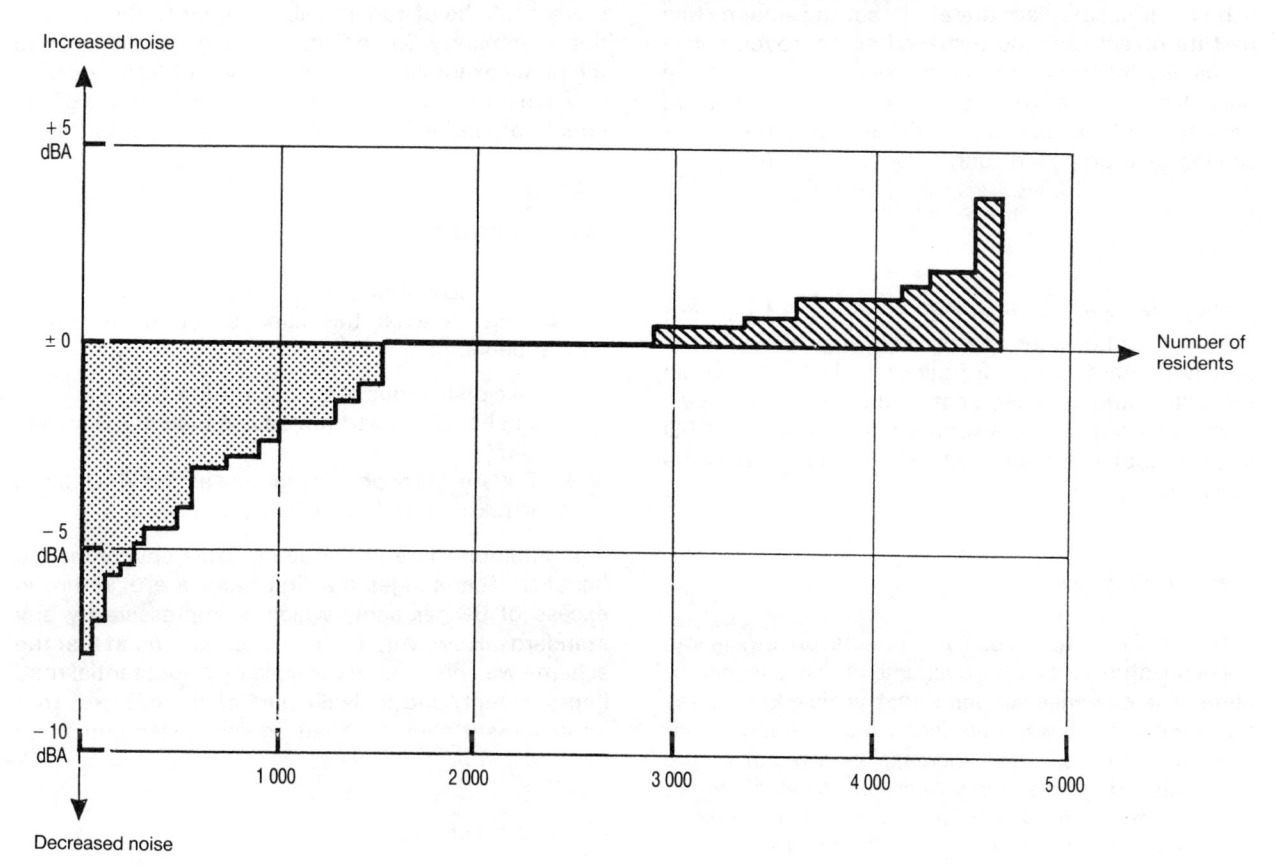

	Decreased noise	Increased noise	
≥ 6 dBA	200	0	residents
3 - 5 dBA	700	150	residents
1 - 2 dBA	600	1 600	residents
Total	1 500	1 750	residents

VII. EVALUATION

Traffic cells were the main measure carried out in Gothenburg. However it is important to note that their implementation was made possible by a substantial road improvement on the south side of the river. Moreover, changes in parking control, providing preferential terms for residents who kept their cars at home during the day, and increased priority for trams and buses were also implemented, and will have affected travel patterns. It is difficult to isolate the effects of these different measures, although it is clear that the main cause of the changes described in this study have been the cells and associated road improvements.

Environmental Effects

The main objectives of the scheme were to reduce noise and air pollution, increase safety and induce a greater sense of well-being. Estimates of noise from traffic flow suggest that there has been a net reduction of noise, and that one third of residents have experienced reductions of up to 8 dB(A). It seems likely that reductions in air pollution will also have occurred. It is to be hoped that these improvements, the reductions in accidents and the improvements to public transport (see below) will have enhanced the sense of well-being of the residents, although there is no direct evidence of reactions to the scheme. It seems clear therefore that the scheme has met its objectives, and achieved an improvement in the environment. A greater improvement could have been achieved, with no residential roads experiencing worsened conditions, if traffic flows had been restrained or a proposed tunnel had been built.

Effects on Safety

The effects on safety are impressive. A 14 per cent overall reduction in accidents has been achieved in the first two years, with a 27 per cent reduction within the cells, and a 9 per cent reduction on the surrounding roads. There is some evidence from the CBD scheme that a further long term reduction may be expected.

Effects on Efficiency

It is difficult to assess the effects on efficiency, because information on travel time changes is incomplete. The evidence suggests that vehicle kilometres for private traffic will have increased by about 7 per cent; however, a similar increase as a result of the CBD cells did not result in an increase in vehicle hours in the system. This would imply only a minor loss in efficiency for private transport. On the other hand public transport movements through the area have experienced a 45 sec. reduction in travel time. It seems likely therefore that the total costs of travel through the area have fallen slightly.

Effects on Energy

Effects on energy consumption seem likely to have been insignificant. There will have been a slight increase as a result of the increase in vehicle kilometres, but improvements in junction operation will have reduced this.

Land Use and Economic Effects

While traders generally have expressed themselves satisfied with the CBD cells, there has still been no detailed study of the effects on trade. This is unfortunate since one study, in Groningen, has suggested an adverse effect on trade, and traders' opposition still acts as one of the major barriers to innovation elsewhere. It is not too late to conduct a retrospective assessment of the effects of the new cells on trade and other land use effects, and such a study could be of considerable benefit to the internatinal community, since Gothenburg is usually cited as the prime example of the application of traffic cells. It is worth noting, too, the value of a similar study of the effects of Gothenburg's proposed parking policy.

Overall Costs and Benefits

The new cells and associated measures cost 5.8 million SKr. Against this can be set the following annual benefits:

- Accident reductions 2.2 million SKr;
- Public transport time savings (say) 2.0 million SKr;
- Private transport time losses (say) 0 extra veh.km (say) 0.5 million SKr.

In addition there are unquantifiable environmental benefits. This suggests a first year rate of return in excess of 60 per cent, which is impressive by any standards. However, it must be borne in mind that the scheme was only made possible by a substantial road improvement, and at least part of the costs of that improvement should be set against these benefits.

Distributional Effects

As already noted, traffic cells are primarily traffic re-routing rather than restraining techniques and therefore involve the transfer of traffic induced problems to other locations. If these locations are better able to accommodate the traffic, or less sensitive to it, then there is likely to be a substantial net benefit. In the main this has happened in Gothenburg, with the construction of a purpose-built road in the dockland area. However, there are some areas, particuraly in Engelbrektsgatan, where residential roads have suffered from the transfer of traffic and from parking restrictions to accommodate that traffic. It is notable that the city intends to encourage a change of land use in those places to make them less sensitive, and hence in the longer term reduce the distributional effects of the scheme. Rerouting of private vehicles also implies a redistribution from private to public transport. It appears that this has been achieved without any significant disbenefit to the majority of private users. However, some local access movements have inevitably suffered. Overall, the distributional disbenefits are small, and probably smaller than they would have been with a policy of either greater traffic restraint or more road building.

VIII. CONCLUSIONS

Gothenburg has demonstrated, in both its central area and its inner suburbs that traffic cells are a highly effective way of reducing environmental intrusion, accidents and disruption to public transport. However, it is clear that a vital pre-requisite is spare capacity on a less sensitive part of the road system, since cells serve to re-route rather than restrain traffic. Moreover, the alternative route must not involve a long detour, or increased vehicle kilometres may offset the benefits of the cells. The spare capacity may be provided by building new roads, as in Gothenburg, or by restraining traffic. Either will introduce its own costs and additional benefits.

The major benefits are likely to be in terms of reduced environmental intrusion and accidents and improved conditions for public transport. Disbenefits will accrue to those on any routes which experience increased traffic and to certain access movements. However, careful design should be able to reduce these disbenefits to a minimum. The effects on trade are as yet uncertain, and require further study.

Implementation of the CUA cells took far longer than for those in the CBD partly because of the need to link them to other infrastructure changes, partly because the network and land use patterns were more complicated, and partly because of an increased emphasis on public participation. With hindsight it appears that the need for consultation was exaggerated and that it would have been preferable to implement the scheme more rapidly, modifying it later as necessary in the light of experience. The last few years have seen the development of several traffic management analysis tools to help the design errors, and it is interesting to note that the results of the cell scheme accorded well with predictions.

There is no reason why the Gothenburg experience should not be tranferable to other cities which meet the requirement of spare road capacity, although provision of this is likely to be a significant barrier in some cities. An approach based on selective traffic restraint provides the only low cost way of releasing such spare capacity. As yet there has been no experience of a combination of restraint and cells to distribute the benefits of that restraint. This is an important area for further practical experimentation, and it will be extremely interesting to monitor the effects of Gothenburg's proposed parking policy.

Annex

KEY FACTS AND FIGURES ON GOTHENBURG AND ITS TRANSPORTATION SYSTEM

FACTOR/AREA	FACT/FIGURE	COMMENTS/UNITS
Population		
Gothenburg metropolitan area	695 000	
Gothenburg	430 000	451 km^2
Central Urban Area (CUA)	41 000	7 km^2
Central Business District	2 000	1.2 km^2
Employment		
Göthenburg	240 000	
CUA	65 000	
CBD	25 000	
Travel pattern		
Total No. of trips:		
by car	1.6	Per person per weekday
by public transport	0.6	
Total traffic mileage:		
by car	7 000 000	Persons × km/day
by public transport	1 800 000	
Modal split. All trips		Public transport/cars "based": origin and/or destination in the area
CUA – based	35/65	
CBD – based	45/55	
Modal split. Work trips		10 per cent of all works trips to CBD are on foot, by cycle or moped
CUA – based	55/45	
CBD – based	65-35	
Travel distance. Work trips		Little difference between public t. and cars in the actual pattern today fir inhabitants living within the municipality
25 % less than	4 km	
50 % less than	4 km	
90 % less than	14 km	
Travel time-Work trips		
by car:		
25 % less than	15 min	Time for parking, walking and waiting included
50 % less than	20 min	
90 % less than	30 min	
by public transport:		
25 % less than	30 min	
50 % less than	40 min	
90 % less than	60 min	
Public transport system		
Tramways:		
No. of routes	8	
Total length of routes	131 km	
Travel speed peak periods		
CUA	10-15 km/h	
Buses:		
No. of routes	40	25 routes morning/afternoon for industrial workers excluded
Total length of routes	625 km	
Car ownership transport system		
Per 1 000 inhabitants:		
Gothenburg	320	Business registered private cars included
Suburban area	360	

Annex (cont'd)
ECONOMICS, 1980

in million SKr

	Public transport	Private road transport
Expenditures on urban transport:		
infrastructure development	5	80
functioning costs	600	4 200
TOTAL	605	4 280
Financing of urban transport:		
Central government	20	100
Regional or local government	550	100
Users	4 100	230
% of operating costs covered by fares	40 %	–
Operating deficit per passenger	2.75 SKr	–

ENVIRONMENT, 1980

	Petrol	Diesel Fuel	Electricity
Energy consumption by urban transport (TOE)			
Private motor vehicles (cars and lorries)	140 000	70 000	–
Buses	–	10 000	–
Other public transport	–	–	8 000

	Mobile Sources	Stationary Sources	Total
Air pollutant emissions in urban area: (thousand metric tonnes)			
SO_2	0.5	20	20.5
Suspended particulates	0.5	0.5	1
NO_x	7	3	10
HC	8	1	9
CO	80	3	83
Lead	0.05	0.05	0.1

	Outdoor Noise Levels (dBA)				Total
Population exposed	55	60	65	70	
to daytime noise levels (06.00–22.00) (number of persons)	30 000	25 000	25 000	8 000	90 000

Chapter 3

HONG KONG*

I. INTRODUCTION

Hong Kong is a small British dependent territory on the south coast of China, adjacent to Guangdong Province. Most of the territory is on lease from China until 1997. In 1984 the UK and Chinese Governments reached agreement that sovereignty over the whole of Hong Kong would revert to China in 1997. At that time Hong Kong will become a Special Administrative Region of the People's Republic of China. It will enjoy a high degree of autonomy and, for instance, will retain its status as a free port, an international financial centre and a separate customs territory. The joint agreement between the UK and Chinese Governments is aimed at preserving the stability and prosperity of Hong Kong.

Currently the territory is administered by the Hong Kong Government, with a Governor advised by an Executive Council. Legislation is considered and enacted by a Legislative Council.

In making decisions on transport matters the Government is advised by a number of committees. The principal committee is the Transport Advisory Committee, chaired by a non-Government member which advises on all major transport matters. Local issues are considered by 19 District Boards, recently established and first directly elected in April 1982. The Government would normally consult District Boards on any major transport issues affecting them, and the District Boards can raise matters of concern, for example with local public transport operation.

Population

Hong Kong has a population of 5.5 million, of whom 98 per cent are ethnic Chinese. The remainder of the population comes from many countries, but the largest groups are from the UK, North America, the Philippines and India. Immigration from China has been a significant element in population growth, and less than 60 per cent of the territory's population were born in Hong Kong.

With a large inflow of immigrants from China and boat refugees from Vietnam, the average annual growth rate in population was almost 4 per cent between 1978 and 1980. However, the annual growth rate between 1981 and 1985 came down to 1.3 per cent because of a stricter immigration policy. The rate of natural increase has also been falling as a result of the decline of the birth rate from 18 per thousand in 1975 to 14 per thousand in 1985, and the death rate remaining stable at about 5 per thousand.

With a land area of only 1 068 square kilometres, Hong Kong is one of the most densely populated places in the world. At the end of 1985 the overall density was slightly more than 5 000 people per square kilometre. However, this figure conceals the fact that most of the population is concentrated in the metropolitan area around the harbour, with an average density of over 28 000 people per square kilometre which is comparable to the density of the city of Paris. The most densely populated district, Sham Shui Po, has over 165 000 people per square kilometre. For comparison the UK has a population density of only 231 people per square kilometre.

There have been changes in the distribution of population because of a New Town development programme in the New Territories. In 1987, eight new towns in the New Territories are in various stages of development. They are planned to accommodate a total of 2.3 million people in the mid-1990s. This was designed to alleviate the high density in the urban areas, and to help provide the increasing population with better housing and an improved living environment. The plan envisaged balanced development,

---

* Case study carried out jointly by the World Bank and OECD, and prepared by Messrs. M. Clancy (Hong Kong), R. Darbera (France), A. May (United Kingdom) and A. Amstrong – Wright (Wold Bank).

Figure 1. **MAP OF HONG KONG TERRITORY**

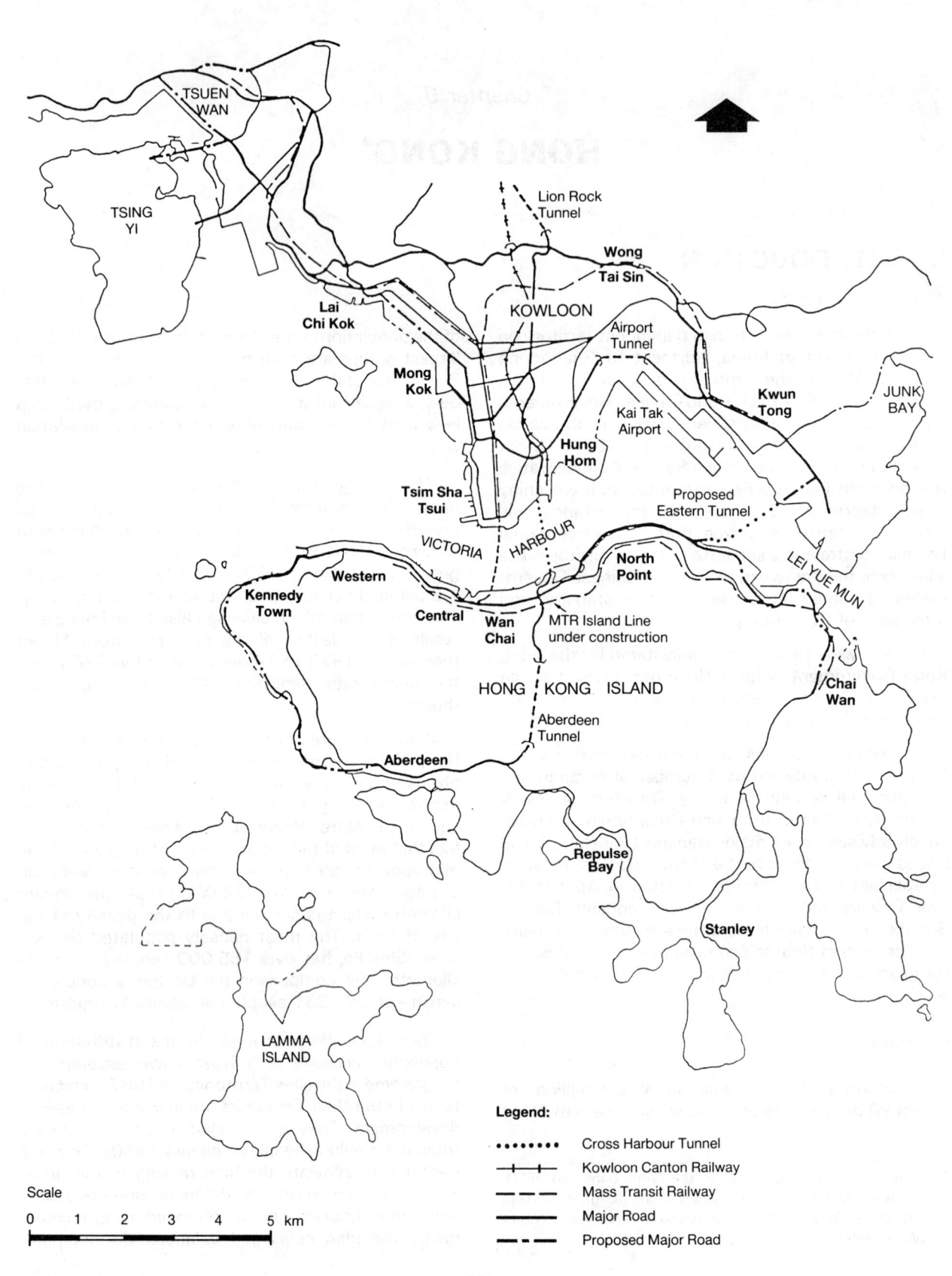

with factory developments complementing the residential development. However, most jobs have remained within the central urban area, and this has implications for transport.

Land Use

Of the total land area almost 75 per cent is marginal land. Only 9 per cent is used for farming and built up areas comprise the remaining 16 per cent.

Much of the land outside the Kowloon Peninsula and the northern shore of Hong Kong Island is hilly with steep slopes. Weathered granite is common and makes for difficult development (Figure 1).

Although in New Towns there has been urban planning which separates industrial development from residential areas, a feature of the older urban areas is the mixture of high rises commercial, industrial and residential uses not just in close juxtaposition, but sometimes within the same building.

II. THE HONG KONG TRANSPORT SYSTEM

Hong Kong is characterised by a diverse public transport system. Modern transport is well represented by an underground mass transit railway, but the tram service which dates back to the turn of the century continues to run. Ferries take passengers and vehicles across the harbour, and the largest buses in the world have been specially produced to meet the high peak hour loadings in Hong Kong. Transport is run on commercial principles. Indeed, many of the transport modes such as buses, trams and ferries are operated by private companies at a profit. There is a great reliance on public transport. About 90 per cent of the 10 million passenger trips per day are made on public transport.

The trends in passenger transport and the market shares are indicated in Table 1 which lists weekday boardings by mode of travel in 1974 and 1985.

The main public transport modes are: buses (more than 14 seats, the average capacity is above 100 passengers), minibuses (limited to a maximum of 14 passengers), railways, tramways, ferries and taxis.

Buses

Buses play a vital role in moving people around Hong Kong. In 1985 the larger buses which are operated mainly by the franchised bus companies handled 45 per cent of all public transport passenger journeys (Table 1). If minibuses are also taken into account, then over 60 per cent of all public transport passenger journeys are made by buses.

Most of the bus services are provided by two companies under franchises. The Kowloon Motor Bus Company (1933) Limited (KMB) operates 187 daily bus routes in Kowloon and the New Territories, and 20 cross harbour routes jointly with the China Motor

Table 1. HONG KONG PUBLIC TRANSPORT PASSENGER JOURNEYS BY MODE 1974-85
Daily average in thousands passengers and %

	1974		1982		1985	
Franchised Bus	2 046	41 %	3 439	44 %	3 908	45 %
Public Light Bus	1 332	27 %	1 460	19 %	1 480	17 %
Mass Transit Rail		0 %	962	12 %	1 268	14 %
Kowloon Canton Rlwy	38	1 %	61	1 %	283	3 %
Tramway	410	8 %	398	5 %	337	4 %
Ferry	573	11 %	435	6 %	348	4 %
Taxi	608	12 %	1 018	13 %	1 155	13 %
Total	5 007	100 %	7 773	100 %	8 779	100 %

Source: Commissioner for Transport, Department Report 1984.

Bus Company. At the end of 1985 its total fleet was 2 511 buses comprising 2 368 double deckers, 105 single deck buses and 38 coaches. In 1985 it carried 1 078 million passengers and operated 188 million kilometres — increases of 1 per cent and 9 per cent respectively over the previous year.

On Hong Kong Island the China Motor Bus Company (CMB) operates 80 daily bus routes and shares the 20 cross harbour routes with KMB. In 1985 its fleet of 1 054 double decker buses carried 344 million passengers and operated 55 million kilometres.

A third much smaller franchised bus company operates 58 buses over 10 routes on the relatively sparsely populated and undeveloped outlying island of Lantau.

All the franchised bus companies operate without Government subsidy. In return for their franchises the Government imposes standards for safety, maintenance, depot provision, etc... and specifies the fares such that the companies will be permitted a profit equivalent to a 15 to 16 per cent return on the companies' net fixed assets (except for the company that operates in Lantau). Excess profit in any one year is carried forward and added to general reserve as an equalisation asset in following years. If the profit falls below the permitted level in any one year this can be recovered from the reserve fund in subsequent years.

Since 1985 the companies have been required by law to undertake annually a forward planning study to determine the anticipated demand for bus travel in the short/medium term and to estimate requirements to meet this demand such as fleet, depot and terminus provisions. The results of this study are discussed in detail with government in order to arrive at an agreed Route Development Programme. The implications of this Programme in terms of costs and revenues are calculated and used as a basis for setting the fares.

Minibuses (or Public Light Buses)

These are 14 seat buses which in the past have helped to deal with the travel demands which the more traditional forms of public transport could not cope with in the short term or at the heights of the peak period. Since 1976 the Government has limited the minibus fleet to 4 350 buses and no new licenses are issued. However, existing licenses are transferable. The minibuses remain popular with passengers prepared to pay higher fares for a quicker, more direct and more comfortable (often air-conditioned) service than the franchised buses. There is another advantage that, with the exception of clearway restrictions, passengers can board or alight anywhere along the minibus route.

There are two types of minibuses. The green minibuses provide scheduled services at fixed fares. The red minibuses, on the other hand, provide non-scheduled services. The Government controls neither fares nor the routes of the majority of minibuses. Pricing is very flexible with premium fares for morning and evening peaks, and special trips to, say, the horse race course.

However there has been concern over the disrupting effect on other traffic and the inadequate coverage of such services. Since Public Light Buses were introduced in 1969 there has been a policy of containment which has been exercised through i) restricting the maximum number of vehicles to 4 350, ii) restricting the areas of activity within established patterns by preventing extension to new areas (e.g. new towns or express operation on limited access roads) and by imposition of local restrictions, clearways and Public Light Bus no-stopping zones, to reduce congestion, to prevent the practice of "poaching" passengers at bus stops and to protect scheduled services; and iii) encouraging light bus operators to provide franchised services to areas outside the main corridors, for example to residential areas where the demand would not justify nor the road environment permit the operation of a franchised bus. This conversion of minibuses is known as the green minibus scheme, and in return for a route 'franchise', a minibus operator agrees to provide a certain frequency of service and accepts Government control of a fare ceiling. By the end of 1985, 137 green minibus routes utilising 1 050 minibuses were in operation throughout the territory and carried about 467 000 passengers daily.

Railways

Hong Kong has two heavy rail systems:

Mass Transit Railway

The Mass Transit Railway (MTR) is managed by the wholly Government owned MTR Corporation. Construction of the present system began in 1975, and the railway, as well as providing a cross harbour route in a 'bored' tunnel, runs through some of the most densely populated parts of the urban area (the lines are shown in Figure 1). It is an underground system for much of its length and is essential in relieving the demands on public transport on the roads. In 1985 it carried 1.3 million passengers each weekday, making it the heaviest carrier per track kilometre in the world. It now operates on 38.6 kilometres of route with 37 stations. Excluding finance costs, about HK$10 000 million* was spent on the construction of

---

* 1HK$ is approximately equal to 0.125 US$

the first two lines: the Kwun Tong to Yau Ma Tei and the Tsuen Wan to Central lines. There are interchange facilities at Prince Edward and Mong Kok stations.

The Corporation's third line, the Island Line, runs along the Island's northern shore linking Chai Wan and Sheung Wan with 14 stations. The greater part of the line was opened in May 1985.

A feature of this railway development is that the project is financed by a mixture of export credits covering construction and equipment contracts placed with overseas companies, profits from property development and commercial borrowings. Revenue from property development accounted for about 10 per cent of the construction cost of the system. According to MTRC's latest financial forecast, the railway will start operating at a profit in 1993 and will pay off its debts in 2001.

The trains are electrically powered with completely automatic signalling and control systems. The cars, stations and tunnels are air-conditioned to cope with Hong Kong's sub-tropical climate. All stations have a fully automatic fare collection system.

The next extension of the MTR system to be built will be a second cross-harbour link between Kwun Tong in Kowloon and Quarry Bay on Hong Kong Island. This link is expected to be completed by early 1990. Unlike previous MTR lines, construction of this link will not be undertaken by the Corporation itself. Instead, the MTR link will be financed and constructed by the New Hong Kong Tunnel Company Ltd. as part of the Eastern Harbour Crossing which is a rail cum road configuration. The rail portion of the project will be leased to the MTRC upon completion.

The Kowloon-Cantoon Railway

From its inception in 1910 until 1982 the Kowloon-Canton Railway, for its British section up to the border with China, was run as a Government department. However, in 1983 the Kowloon-Canton Railway Corporation (KCRC) assumed responsibility for operating the railway. KCRC is wholly owned by the Government and is required to operate on normal commercial principles. The 34 kilometre route is now completely electrified, and the railway provides an increasingly popular commuter service between the developing New Territories and the centre of the urban area. For instance, as electrification was completed during 1983, daily patronage grew from 80 000 at the beginning of the year to almost 200 000 by the end of the year. In 1986, KCR's patronage is about 300 000 per day.

The Tuen Mun Light Rail Transit System

A 34-kilometre light rail transit (LRT) network has been planned to form the backbone of a comprehensive and integrated public transport system in the western New Territories in the 1990s, serving close to a million people in the new towns in the region.

The system will be built by the Kowloon Canton Railway Corporation (KCRC) as its owner/operator. The LRT system will be an at-grade electric light railway operating frequent services with single light rail vehicles or two-car trains. Phase I of the system is scheduled to commence operation in August 1988, and the full network is expected to be in operation by the early 1990s.

The full LRT system will be worked by about 150 cars on 15 services. About 90 per cent of the system will be laid on roadside reservations, with the rest being located in the street of highway medians. Suitable priority will be afforded to light rail vehicles at highway intersections and on street-running sections. The KCRC will also assume control of the local feeder bus services in the region to be served by the LRT.

Trams

Hong Kong Tramway Limited is a private corporation that runs tram services over 30 kilometres of track along the densely populated north shore of Hong Kong Island. During 1985 the fleet of 161 double deck tram cars carried a daily average of 331 000 passengers. There are bus and minibus services operating along the same corridor, but the trams remain popular because of their very low fares and for short journeys.

Closure of Hong Kong Tramway Limited has been mooted and investigated several times since 1925, however this unique system managed to survive quite well. The infrastructure was first laid down by the turn of the century and the present generation of vehicles appeared as early as 1949, but the system has been adapted to fit the demand and to reduce costs (convertion to a one-class operation in 1973 and one-man operation in 1978). Stops are closely spaced and meet the demand for short rides along the busiest streets of the island corridor. Ridership fell by 23 per cent in the early 80's when the Tramway Company introduced several hefty fare increases. The fare increases of 1981 and 1983 had actually doubled the previous 30 cents fare. The opening of the Chai Wan to Admiralty section of the MTR Island Line in May 1985, however, only caused a marginal decrease in ridership. The daily patronage has only decreased by 0.3 per cent from 335 400 in March 1985 to 334 000 in December 1985. The HK tramway still has the heaviest boardings per route-km of any system in the world and the tramcar still is the most intensely used public transport vehicle in the Territory.

Ferries

In a city built around a harbour ferry services have long been important. Indeed until 1972 they provided the only means of crossing the harbour for both passengers and vehicles. In 1972 the Cross Harbour Tunnel was opened, and the MTR line under the harbour opened in 1980. Thus during the 1970s and 1980s the ferries' patronage has declined. Nevertheless in 1983 the Star Ferry Co. with its fleet of 10 vessels carried 38 million passengers across the harbour between the Kowloon Peninsula and Hong Kong Island. The Hong Kong and Yaumati Ferry Company with its fleet of more than 80 vessels carried 83.7 million passengers and 4.5 million vehicles in 1985. As well as operating cross harbour services, HYF serves some of the main outlying islands and also runs excursion services.

Taxis

Hong Kong has 3 types of taxis. One type is allowed to operate anywhere in the territory but primarily serves the urban area. The other two are limited to operating in the New Territories and Lantau Island respectively. At the end of 1985 there were 13 809 urban taxis and 2 388 New Territories taxis. By the end of 1985, taxis carried a total of 1.14 million passengers per day.

Fares

The fare structures differ for each mode, reflecting their relative attractiveness, most are graduated by distance and some (eg. on MTR) have peak surcharges. A new fare has to be paid for each stage of a multi-modal journey. Table 2 illustrates the range of fares for two typical journeys in the congested area.

Traffic congestion

The Government currently spends about HK$1.6 billion per year on the construction and maintenance of the roads. However, in the tightly packed urban area the government has to resort to unusual ways of increasing the raod network. Thus there are roads which have been double decked, flyovers across overloaded junctions and a large dual carriageway which for a significant part of its length runs on stilts in the harbour, and elevated roads weaving between high rise buildings. However, road building in the urban area is becoming increasingly costly, difficult and open to objections on environmental grounds.

Table 2. TYPICAL FARES
In HK$

Mode	Causeway Bay to Central (3 km)	Thim Sha Tsui to Central (3 km by rail, ferry, 5 km via tunnel)
Tram	$0.60	n.a.
Star Ferry:		
2nd class	n.a.	50 c
1st class	n.a.	70 c
Franchised bus	$1.00	$2.50
PLB	$2.50 ($1.50–$2.00 off peak)	n.a.
MTR	$2.00	$3.50
Taxi	$7–$8	$40[1]

1. Includes tunnel toll.

Although almost all the through roads are Government owned and maintained, an interesting exception is the road in the submerged tube tunnel which links Hong Kong Island with the Kowloon Peninsula. Although the Government made a contribution to the capital invested in this project and has a 25 per cent shareholding, the dual two lane tunnel is run as a commercial enterprise. The privately owned Cross Harbour Tunnel Company, under the terms of its franchise from the Government, is responsible for operating and maintaining the tunnel and collecting tolls. The second cross harbour tunnel, which will be built by the private sector, is now under construction.

Despite the large investment in improving the road network and also in measures such as reserved bus lanes, congestion in Hong Kong has become a major problem. An increase in congestion is forecast even with the large road building programme.

The congestion problem gradually intensified throughout the latter part of the 1970s. For instance, from 1976 to 1981 the numbers of private cars grew at over 13 per cent per annum, but even though at constant prices expenditure on road building increased at over 10 per cent per annum, the length of traffickable roads increased by no more than 2 per cent per year. Thus by 1981 Hong Kong had the unenviable place in the Guinness Book of Records as the territory with the highest traffic density in the world. At that time there were just some 4 metres of road length for each vehicle. A number of transport studies carried out for the Government forecast further congestion as a result of the widening gap between the growth in travel demands and the expansion of the road network which could reasonably be provided.

The Government had acknowledged the need for traffic restraint, particularly of private cars, in its 1979 White Paper and had considered various methods of traffic restraint (e.g. parking fees, fuel taxes, area license scheme). By 1982 the Government had decided that action was necessary if the problem of congestion was to be tackled before it got out of hand. In the short term it seemed that the most suitable method of traffic restraint was to increase the costs of owning a car. In May 1982 the Government doubled the tax payable on the first registration of a private car or a motorcycle, and trebled the annual license fees for such vehicles. These fiscal measures coincided with the effects of the world recession on Hong Kong, and have had a dramatic effect on car ownership. Today, the first registration tax is comparable to the c.i.f. price (cost, insurance, freight) of a new car and the annual license fee roughly amounts to 1/10 of this price. Since the increases were implemented, the number of licensed private cars has fallen from over 190 000 to about 140,000. This has had its effect on car usage and congestion has been relieved in several areas. Despite the effect of these measures in controlling car ownership and relieving congestion, the Government has also been exploring the feasibility of systems to tackle road usage directly. The Singapore Area Licensing system was closely examined, but it was considered unsuitable for Hong Kong where industries, offices and residential uses are intermingled and where congestion arises over a wide area. The use of parking control was considered, but rejected because of the problems of controlling private sites and the difficulty of relating parking charges and controls to the amount of use made of congested roads. A feasibility study in November 1982 recommended that technology had so advanced that it was worth giving consideration to an electronic road pricing system, with the potential of controlling road usage throughout the main urban area. Against this background the Government engaged consultants to carry out a two year project which would not only test the electronic technology, but also design a road pricing system for the urban area and assess the cost and benefits of proceeding to a full system.

III. ELECTRONIC ROAD PRICING IN HONG KONG

Road Pricing

Road pricing has a solid basis in economic theory. It works by attaching a price to the use of a commodity in limited supply such as the roads in a densely populated urban area. Road pricing can tackle congestion directly and fairly because different charges can be set according to the congestion on different routes and varied according to the time of day. Road pricing can also be used to support transport policies favouring public transport. In Hong Kong there is no intention to charge buses for their road use. Goods vehicles, too, would be exempted from charges. On the other hand, private cars which, for instance, in the business district, Central comprise up to 45 per cent of the peak hour traffic flows, would need to be charged. Both taxis and private cars are inefficient road users compared to the double deck buses. In Hong Kong Central buses comprise only 4 per cent of the traffic flow, but a typical occupancy would be about 70 passengers compared with the private car and taxi occupancies of about 1.8 persons.

Congestion in Hong Kong arises because many of the junctions in the urban area are operating at or near capacity. For instance, considering a typical junction: if the flow is 900 vehicles per hour, the average delay is 1.2 minutes. If the demand is reduced by 100 vehicles, the time savings are dramatic. The average delay for each of the 800 vehicles continuing to use the junction falls to 36 seconds. The objective of road pricing in Hong Kong is to reduce private vehicle usage by about 10 per cent at peak periods because this would give significant improvements in traffic flows.

Road pricing systems depend on vehicles being charged accurately according to their road usage. There are advantages in using a meter in the vehicle which would show congestion charges being accumulated or a stored value being used up. However, there are difficulties in making such meters secure. Therefore the road pricing system tested in Hong Kong uses a secure unit to identify each vehicle entering a congested area, and the charges are "metered" at a central office. The Hong Kong system is covered in more detail below.

How the Electronic Road Pricing System Works

The electronic road pricing pilot system in Hong Kong comprises 4 main components:

Figure 2. **SCHEMATIC ILLUSTRATION OF THE ROAD PRICING SYSTEM**

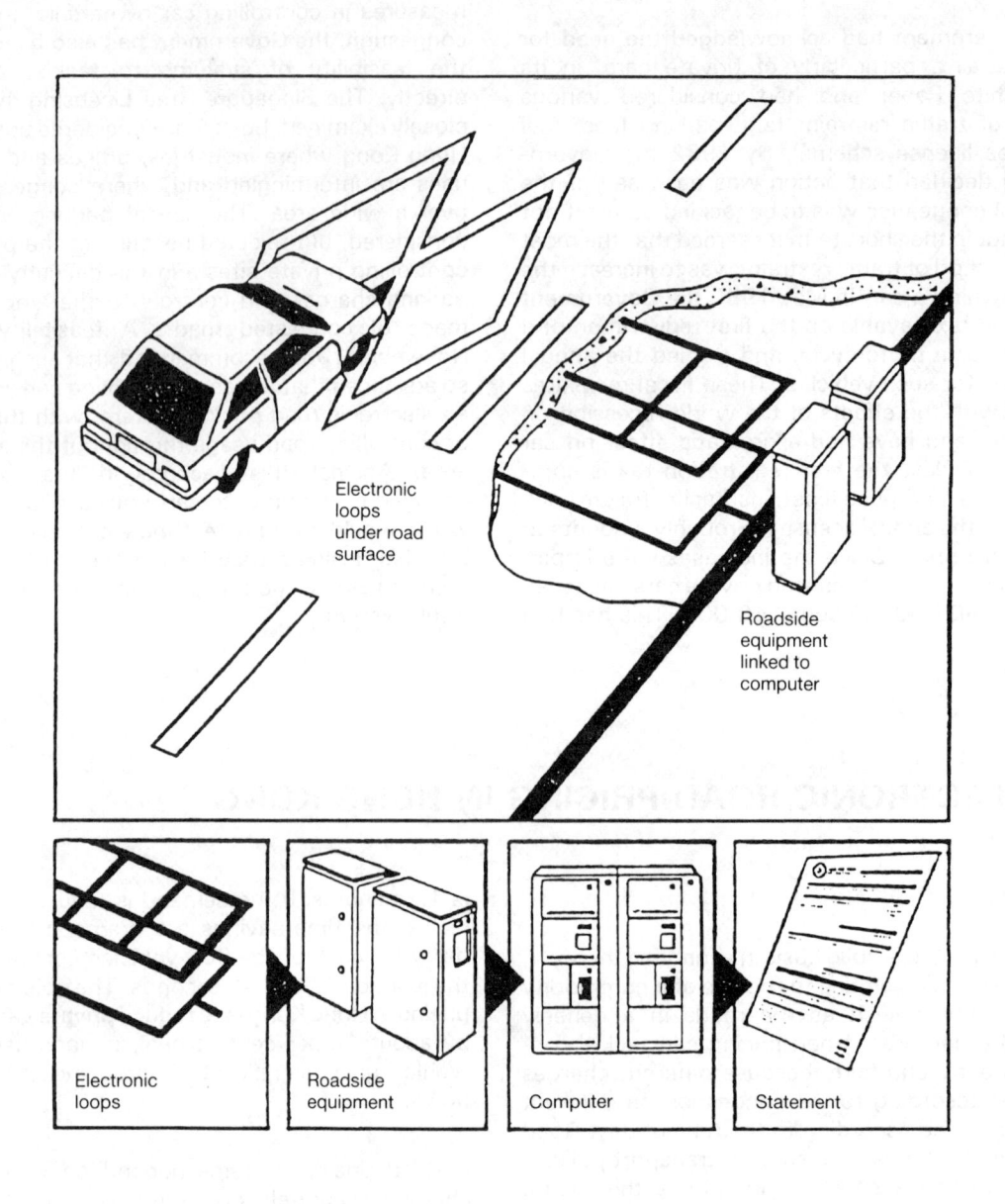

Electronic loops under road surface

Roadside equipment linked to computer

Electronic loops → Roadside equipment → Computer → Statement

a) A unit to identify unambiguously individual vehicles — the electronic number plate;
b) Equipment to register the passage of vehicles together with the time and location — data capture and validation;
c) Equipment to detect and identify vehicles trying to avoid road charges — surveillance and enforcement equipment; and
d) An accounting system to bill accurately motorists.

A schematic illustration of this system is shown in Figure 2.

The characteristics and functions of the main components are as follows:

The Electronic Number Plate (ENP)

This is a robust box about the size and shape of a video cassette, which is welded beneath each vehicle. The ENP, which should last the life of a vehicle, is given a unique and confidential electronic code in its integrated circuit at the time of manufacture. By subsequently matching this code on computer records with a vehicle registration mark, it is possible to identify unambiguously individual vehicles. The

ENP has no battery nor is it connected to the vehicle's battery. However, it contains antennae which receive radio waves broadcast from loops buried in the road. The radio waves are converted into power for the integrated circuit, and on being momentarily powered the ENP transmits its code to be received by another set of loops in the road.

Data Capture and Validation

Figure 2 also shows how the power and receiver loops are arranged in the road. One power loop can cover a carriageway comprising 3 to 4 lanes, but each lane contains its own receiver loop. This enables vehicles straddling lanes or crossing the site diagonally to be identified. The electronic loops are controlled from a roadside cabinet. The vehicle codes together with the times and locations they were received are passed to a computer known as a communications controller. This computer controls a set of outstations in defined geographic areas. The code numbers are then checked by another computer, the data validator. If the data is properly validated, it is passed to the accounts processor. At this stage the system knows the owner of the vehicle (because the confidential code number is linked to information on the registration and licensing of vehicles) and the zone and time the vehicle was used. In the accounting computer a tariff is applied to this information so that the owner is accurately billed according to his use of busy roads at peak periods.

As the above description indicates, Hong Kong's electronic road pricing system is made as automatic and reliable as possible. A series of relatively small computers is used rather than one large main frame computer. As well as the cheaper cost of running such a system compared to having a main frame computer with a back-up spare, the system can be readily expanded by adding more computers to the local area network.

Surveillance and Enforcement

Inevitably with any system involving charges, there will be some people who will try to cheat the system. This has been recognised from the outset in the design of the electronic road pricing system, and detection of such offenders has been made as automatic as possible. As well as identifying code numbers, the system also detects passing vehicles with damaged or shielded ENPs. At selected strategic sites the passage of such a vehicle would trigger a camera to take a photograph is then digitised, sent over a telephone line and re-constructed in the control room. At this stage the registration mark can be read and the first recourse would be to call up the vehicle for further examination. If an owner did not comply, then the transfer and relicensing of his vehicle could be blocked and other appropriate enforcement action taken. Photographs are taken from the rear, and there is not and the Government would not wish to have sufficient clarity to try to identify vehicle occupants.

Accounting System

This would be similar to the systems used by utility companies. A motorist would be given a regular, probably monthly, bill setting out the charges for road use which he had incurred. It would be technically possible to list all the sites which the motorist had possessed, broken down by date and time. However, it is envisaged that post motorists would prefer a less detailed statement giving a daily breakdown of the charges incurred. There would be inquiry centres where a motorist with a doubt about his bill (after producing sufficient identification) could receive a full record of his vehicle's movements for the period in question.

The Effect of Electronic Road Pricing

When implemented at full scale, the major effect of road pricing would be to reduce congestion and give substantial savings in journey times to all road users. The variability of journey times should also be significantly reduced, making it easier for people to keep appointments without allowing themselves large margins for possible delays. The buses which are essential for transporting people on Hong Kong's roads would be better able to keep to schedule. Thus public transport passengers would receive major benefits.

Those motorists who are prepared to pay for the use of the road at busy times would also benefit from less congestion and reduced journey times. The benefits for motorists do not stop there.

Electronic road pricing is a fair way of controlling road use since it charges people according to their use and gives them a choice. It is estimated that over 60 per cent of motorists' mileage would not be subject to any road pricing charges since any full road pricing system would be limited to the main urban areas, and apply only on weekday between 7 a.m. and 7 p.m. Since the cost of owning and operating a car was increased in May 1982, car ownership in Hong Kong has fallen, but if ERP were introduced more people would be able to own cars. The Secretary for Transport has said that with road pricing charges the Government would not need to retain the harsh deterrent of high taxes on ownership. For instance, it is estimated that if ERP were introduced today at full scale, annual license fees which range from HK$2 300 to HK$6 700 could be halved. Motorists in the New Territories whose cars are used mainly outside the urban area and those motorists who have a car largely for evening and weekend use would be more fairly treated under an electronic road pricing system.

There is very little formal flexitime working in Hong Kong, and it is not expected that there would be a major change in hours of work to avoid ERP charges. Rather the main response is expected to be a switch to public transport modes which are generally readily available. Park and ride arrangements might be made attractive, especially for the relatively longer trips from the New Territories to the urban area.

ERP would need to be applied throughout the main urban area, so there is not a danger of development or congestion being squeezed from one charged area to an adjacent uncharged area. Some businesses might be tempted to move to the New Territories where there would be no road use charges. This would not be an unwelcome trend since the infrastructure exists for the New Towns to absorb more industries, offices and jobs.

Commuting habits would certainly be affected. 50 per cent of car use is for commuting only.

Cost

Three different zoning schemes for ERP have been evaluated:

— Scheme A with 5 large zones requiring 130 toll sites;
— Scheme B where tidal charging is introduced with 5 large zones requiring 115 toll sites;
— Scheme C where more trips in the busy area, even short ones, are charged and influenced with 13 smaller zones requiring 185 toll sites.

In Table 3 the benefits and costs are evaluated for different Electronic Road Pricing schemes and an equivalent set of car ownership measures.

The restraint of car ownership through high license fees and taxes not only costs the Government nothing but also boosts its revenue. The cost falls on the community. By contrast the Government would have to invest HK$240 million in capital cost for an ERP system. Running cost would not exceed HK$20 million a year. Over a system life of 10 years, this means a total annualised cost of about HK$50 million.

The electronic road pricing pilot scheme began in July 1983, and was completed in mid-1985 after some six months' running of the electronic equipment. Its operation shows that motor vehicles are recorded accurately and consistently as they cross boundaries between urban area zones. Such recordings have been used to compile accurate mock bills for volunteers, without actually changing them.

The consultant has concluded that the technical, administrative, accounting and legal aspects of ERP are satisfactory. Despite its advantages, the scheme attracted widespread criticism from the District Boards. Among the criticisms raised were the suggestion that conditions did not merit restraint, that public transport could not cope, that congestion would simply be transferred, that it would be unfair on car users, that it was an unacceptable restriction on freedom to travel, that it would invade privacy, and that it was yet another means by which the Government could tax the population and sell British technology. It was extremely difficult to judge the relative strength of three criticisms. The Government in the light of public comments on the proposed scheme, the decline in car ownership and improved traffic conditions, has decided that a decision on the implementation of the scheme in place of the high car ownership cost policy should be postponed and re-considered only after a period of sustained growth in vehicle ownership and/or increased congestion.

Table 3. BENEFITS AND COSTS OF DIFFERENT ELECTRONIC ROAD PRICING SCHEMES AND CAR OWNERSHIP MEASURES

	Electronic road pricing schemes			Restraints one car ownership
	A	B	C	
Cost to the government (HK$m/year)	50	50	50	0
Gross revenue ($m/year)	395	465	540	1 200
% change in congestion	−16	−14	−17	−11
Daily total travel time savings (000'hours)	98	113	124	83
Community benefits (HK$m/year)	730	870	920	300

Source: Electronic road pricing pilot scheme — Results brief — Consultation document.

IV. ENVIRONMENTAL PROBLEMS AND TRANSPORT

In 1981 the Government set up a free standing Environmental Protection Agency (EPA) which has a central coordinating role in formulating and carrying out the Government's environmental protection policy. The environmental protection programme comprises 5 main elements:

1. Planning and environmental impact assessments aimed at pre-empting future problems;
2. Legislation to provide a statutory framework for planning as well as routine control of emissions;
3. Construction and operation programmes for public sector environmental control and waste disposal facilities, such as sewage treatment works and incinerators;
4. Monitoring of environmental quality to check the effectiveness of existing measures and the need for new ones; and
5. Consultations both with the community and with the representatives of industry and commerce likely to be affected by protection measures.

The Government also tackles environmental problems through its transport policies:

Air Pollution

Emissions from vehicles make a significant contribution to air pollution in Hong Kong (Table 4).

At the strategic level the Government's aim is to reduce air pollution from vehicle emmissions by reducing congestion. Much pollution arises when a vehicle is idling or in stop and start motoring. The fiscal measures introduced to control the growth of car ownership have had some success in controlling congestion and hence air pollution.

While the road pricing scheme was not proposed as a solution to environmental problems, it is anticipated that the resulting reduction in congestion would significantly reduce CO emissions (14-17 per cent) particularly by reducing idling time. While these effects could be slightly offset by a growth in car ownership and use away from the congested areas, road pricing could always be extended if these problems became serious. The developments of railway systems have also played their part in reducing congestion on the roads.

Pollution from vehicle emissions is also tackled by the Police carrying out checks on smoky vehicles. An offender has to pay a fixed penalty of HK$200. Proper maintenance is encouraged by a vehicle examination programme in which all goods vehicles manufactured before 1976 are examined annually. Second-hand imported vehicles are also subject to vehicle examination. As from 1986 private cars over 6 years old have to pass the examination before license can be renewed. The emission system is within the scope of this examination. New vehicles must meet specified European or equivalent regulations on emission controls.

New Towns

Many of the environmental problems arise from having such heavy densities of population on the North shore of Hong Kong Island and in the Kowloon

Table 4. ESTIMATED EMISSIONS OF AIR POLLUTANTS BY ROAD TRAFFIC IN HONG KONG

Tons/year

	Sulfur oxides	Carbon monoxide	Nitrogen oxides	Hydrocarbons	Particulates
Car	47	126 809	5 166	12 681	775
Taxi	424	606	550	160	314
Van	6	16 089	656	1 609	98
Public light bus	303	433	394	114	224
Truck	2 692	18 428	13 459	3 002	1 025
Bus	407	2 066	2 098	391	155
Coach	236	1 200	1 218	227	90
Motor cycle	3	3 270	19	1 114	32
Total	4 118	168 901	23 560	19 298	2 713

Source: E.P.A. Hong Kong.

Peninsula. A better distribution of population has been encouraged by the development of New Towns in the New Territories. Improvements in transport have played an essential role in linking these New Towns with the main urban area.

Abandoned Vehicles

In the past vehicles abandoned on quiet streets, in residential blocks or on Crown land have been a problem. However, the problem is now under control. An owner may now "dump" his vehicle at a vehicle surrender centre. The Government also has a contract with a private contractor for the disposal of abandoned vehicles. The Government limits its role to identifying the abandoned vehicles and giving the owner an opportunity to reclaim it. If the vehicle is not claimed, removals are effected by a scrap metal dealer who pays the Government for each vehicle collected and then crushes it for export.

V. EVALUATION

Selected Features of Hong Kong Transport Policy

Although no global transport policy had ever been stated prior to the publication of the "White paper on Internal Transport policy" in 1979, the policy decisions, taken in Hong Kong since the late 60's exhibit several common features. Most of these features are quite unusual in the transport policies implemented in the cities of comparable size in the OECD countries, but are similar to the policies pursued by the World Bank in assisting Third World countries.

Reliance on Market Mechanisms

Market mechanisms are used to provide information about the demand for and the profitability of different transport services as exemplified by the way new licenses for taxi operations are issued. Tenders reflect the demand for taxi transport and selling the licenses at this price is a way for the Government to recapture the capitalised excess profit that arises from the limitation of the total fleet size.

Although unrestrained free competition between different modes is not permitted, organised competition between different modes is an incentive for operators to keep their fares low. No single mode has a complete monopoly over a segment of the demand. Franchised Buses feel competition from both Public Light Buses and the Mass Transit Railway that offer a more rapid and more comfortable service along many of their routes for a price that is barely 2 to 3 times higher. Public Light Buses compete between themselves but also with taxis and Mass Transit Railway, Maxi-cabs very often do not apply the ceiling fare they are allowed to because they feel the competition from the taxis and the private cars, etc.

Willingness to Make Transport Users Bear the Transport Costs

Perhaps the most striking feature of Hong Kong transport policy decisions is the will to have the users bear the economic costs of their trips. These include not only user costs and public expenditure on infrastructure construction and maintenance, but also the cost imposed by users on other users through congestion, as exemplified by the Electronic Road Pricing project.

Of course, and this is specially true for urban transport, no policy instrument allows all transport users to bear at every moment the full economic cost of their trip, and in any case, it is practically impossible to assess or even to define this cost precisely. On the other hand, equity considerations and political realism may impose some exception to the implementation of this principle. However, many transport policy decisions in Hong Kong are based upon this principle.

Some tunnels have been and will be built and are operated by private corporation under franchise from the government and use the tolls collected topay back the investment and ensure maintenance and operation.

The Mass Transit Railway Corporation is publicly owned, but is expected to run a commercial operation. it is entrusted with the task of covering costs, meeting interest charges on loans and amortising capital cost out of revenue. It is also expected to produce a "reasonable return", for its shareholder on the capital invested which equals approximately one third of the loans contracted, although the timetable for making the return is longer than would be tolerated for most projects in the private sector. The Mass Transit Railway operates at a profit and is expected to have all its debts repaid by 2003, according to recent estimates based on a set of assumption on interest rates, ridership and costs. As early as its second year of operation in 1982, its revenues covered its operating cost. These achievements are quite remarkable in comparison with the financing of heavy rail projects elsewhere in the world. For most of them, not only construction but also operation must be subsidised. The Hong Kong Mass Transit Railway Corporation gets its revenues from the fares charged to its passengers and also, for a substantial portion, from property developments. The fares vary according to the distance travelled, and they are higher when part of the trip goes under the harbour to the island. The property developments conducted by the Mass Transit Railway are realised on the sites associated with its depots and stations. With them, the Mass Transit Railway is able to recapture at least a portion of the betterment induced by increased accessibility through mass transit.

The Electronic Road Pricing Project has been designed along the same lines. When car ownership rose from 26 to 42 cars per thousand people between 1977 and 1981, congestion expanded dramatically over the busiest areas of Hong Kong and Kowloon, because of its extremely dense land use pattern. It was obvious that the share of private car passengers in the modal split (10 per cent) was far exceeded by their responsibility in the total congestion, i.e. the minority of private car users were not bearing the cost they imposed on the majority of bus riders. The fastest way to deal with the problem was to stop the rise of car ownership. This was done in 1982 through increased taxes on car ownership. However, the tax affected indiscriminately traffic in congested areas at peak hours as well as journeys on uncongested roads when the number of registered cars dropped by 25 per cent. The Electronic Road Pricing project is aimed at tackling more directly the congestion problem by charging the road users the congestion cost they impose on others.

Concern for Providing Maximum Flexibility

In a rapidly changing environment of economic growth and urban development, flexibility and adaptability are necessary attributes of transport policies.

As we have seen, in Hong Kong, this flexibility has been provided by involving many different private firms operating transport services in largely overlapping markets, although the government does exercise a degree of co-ordination to restrict competition between modes.

An other instance of this concern for flexibility is the way by which fare increases are granted to the franchised bus companies. The time interval between two increases has not been fixed and no indexation formula is used to calculate the new fare. Instead, the firm asking for an increase must submit a develoment plan, estimating future costs and ridership, and including proposals for new routes to be open. If the plans and cost estimates appear reasonable to the Government, the fare is set such that as it covers the cost and permits about 15 per cent return on capital invested.

This system was slow to be put into practice and until recently, fare increases have been rare and drastic but increases are becoming more reasonable as their frequency is improving as can be seen in Table 5.

Fare increases are granted in a way to allow a fair return on capital invested. Under the legislation, the franchised bus companies are required to submit forward planning programmes which must be agreed with Government so as to ensure that levels of service and of investment are commensurate with a proper and efficient standard of service. Other incentives for the bus companies to operate at maximum efficiency are fistly competition by other modes, secondly the imposition of financial penalties if they do not comply with franchise obligations, and thirdly Governments option not to renew the franchise if it is felt that they are not run efficiently. This franchise is a ten years franchise and is usually rolled over every 2 years. Roll over and fare increases have been delayed in the recent years and this seems to reflect the will of the Government to put some pressure on the companies so that they will increase their efficiency.

Pragmatism

Transport policy decisions in Hong Kong do not reflect any dogma, and sometimes, they may appear to be based on somehow contradictory principles. As

Table 5. EVOLUTION OF THE FARES OF FRANCHISED BUSES SINCE 1970

	1970	1976	1980	1981	1983	1984	1985
Consumer price index B	100	162	217	247	309		
KMB fares	100	100	100	143	167	167	199
CMB fares	100	150	240	312	312	378	378

experience is gained, prior measures are corrected and policies are adjusted to the new realities. This can be illustrated by the evolution of the Government's attitude towards the Public Light Buses or by the proposed first stage for Electronic Road Pricing implementation.

Public Light Buses are granted total freedom to fix fares and to choose their routes. However, when the Government realised that they were flocking in the busiest streets creating congestion and slowing down the flow of the franchised buses, it decided to forbid their stopping in certain streets and to issue no more licenses so that their number would remain constant. Limiting the total number of supplyers of a commodity in growing demand gives room for excess profit. This is exactly what happened as is shown by the fact that the premium for buying a license rose to HK$150 000, almost twice the price of a new vehicle. Unlike with taxis, the government did not try to recapture this capitalised future rent, probably because there was no easy way to do so, technically and politically.

Another instance is given by the Electronic Road Pricing project where it has been decided that, at least in its initial stage, only private cars would pay the congestion charge. Economic theory would recommend that all the road users should pay proportionately to their use of the road. Actually, this possibility has been left aside for various reasons. For goods vehicles and for Public Light Buses, the main reason is certainly political acceptability. Charging the regular franchised buses would have probably only resulted in a very small increase in fares. For practical purposes and to make the policy objectives easier to understand, they have been exempted. The reason for exempting the taxis from the congestion charge is to protect the livelihood and investment of those holding urban taxi licenses since it is felt that there is already an over supply of taxis, and that increases in fares would result in a substantial drop in the demand. However it has been proposed to equip the taxis with an electronic number plate linked to their metre so that their travelling pattern could be better known and a special charging system might be designed for a second phase of the Electronic Road Pricing implementation.

Impacts of Policies

A Better Supply of Public Transport Allowed for a Substantial Increase in Mobility

The supply of public transport in Hong Kong is characterised by its wide variety in terms of speed, comfort and price. For almost any given destination, riders have the choice between three or more different modes among franchised buses, Public Light Buses, taxis, trams, ferries, suburban train. Over the last ten years, this choice has been widened even further with the introduction of the maxi-cabs and the opening of 3 Mass Transit Railway lines.

With the exceptions of the tramways and of the light buses, the number of which has been limited to 4 350 since 1976, the supply of public transport measured in terms of seats-Km has greatly increased over the same period, at a rate much faster than population growth.

On the average, the prices of the journeys have also increased in real terms (Table 5); but for the riders, this increase has been offset by the growth of their disposable income.

This betterment of the supply of transport, in terms of quality, quantity and price has permitted an increase of the average mobility of Hong Kong inhabitants. Between 1973 and 1981, the increase in trips per person has been estimated at about 25 per cent. Today mobility in Hong Kong has almost reached an average of two vehicle trips per person per day. In most of the big cities of the developing countries with comparable income per capita, mobility hangs around one trip per person per day.

Transport is a Net Source of Revenue to Hong Kong Government

Except the two railways, i.e. the MTR and the KCR, the Hong Kong Government generally refrains from direct involvement in the financing of public transport operations The Government is the only shareholder of the Mass Transit Railway corporation capital. This participation represents about one third of the total assets of the system.

As a matter of fact Hong Kong Government collects revenues from public transport operations through annual license fees on buses, Public Light Buses and taxis. However, the royalties it used to collect from the two franchised bus companies (15 per cent of KMB turnover and 46 per cent of CMB net profits) were abolished altogether in 1971 when the companies were put under fierce competition by the Public Light Buses.

Franchised buses in Hong Kong are run at a profit and the only financial contribution they receive from the Government is a compensation for lower student fares and not a subsidy. It should be said, however, that the franchised buses are indirectly subsidised. The reason is that the damage they cause to the roads and associated road maintenance costs exceed their contribution to the Government budget through license fees and fuel taxes. In addition, bus stations are provided by the Government free of charge to the operators. The objective of this policy is not so much to provide cheap transport to the poorest as to promote the mode that is the most efficient user of road space.

Although the revenue that the Government collects for road usage through fuel taxes and for vehicle ownership through annual licence fees are not hypothecated for road building and maintenance, it can be shown that the government expenditures on the latter are largely offset by its revenues from the former resources. The different schemes of Electronic Road Pricing proposed will not alter this equilibrium if, as proposed by the consultant, they replace the present heavy registration charges imposed on private cars (Table 3).

VI. CONCLUSIONS

Electronic Road Pricing

The pilot project has clearly demonstrated that the technology is now available to permit electronic road pricing to be implemented and problems of administration and enforcement to be overcome. This result should be wholly transferable.

The project has also demonstrated the substantial benefits to be gained from ERP in Hong Kong. In general terms this result will also be transferable, although the scale of benefit will depend particularly on the level of existing congestion, the current modal split, the opportunities for long term land use changes, and the size of the fleet to be equipped for either regular or casual use in the controlled area.

One particularly interesting result is the simplicity of the controls recommended for Hong Kong. Even the most complex scheme involved only 13 zones and 185 toll sites. Such a system, with fixed charges independent of route for most journeys, would make the controls much easier to understand, and remove one of the anticipated disadvantages of ERP over simpler cordon pricing schemes.

Perhaps surprisingly, Hong Kong exhibited very similar public opposition to that expressed in other cities which have proposed restraint schemes. Most of the arguments adduced have been levelled against other restraint proposals in Europe and the US. The only arguments which appear specific to Hong Kong are those associated with the impending transfer to Chinese control. There was also some concern expressed over the potential of ERP for invading privacy, but little over impacts on economic activity. The comments on inequity from District Boards in the New Territories were particularly surprising, since New Territories residents in particular were likely to benefit from the suggested reduction in annual license fee.

It is difficult to judge the strength of the criticisms raised. It does seem, however, that it is important to demonstrate that congestion is sufficiently serious to merit restraint before embarking upon it. In practice the reduction of congestion following the increase in annual license fee, the doubling of cross-harbour tunnel dues, and the opening of the Island Eastern Corridor, suggested that the problem was surmountable in other ways. Equally, it is apparent that many of the criticisms will always be matters of political dimension and that this needs to be taken into account and resolved in addition to the transport matters involved, when drawing up proposals for a scheme.

Public Transport Financing

There can be no doubt that Hong Kong has developed an extremely effective and financially viable mix of public transport modes which, between them, cater hierarchically for the full range of users. The main reasons for this success are the very high levels of corridor flow to be accommodated and the combination of competition and regulation which is operated under the franchise system. This operating framework is certainly worthy of consideration in other cities, because of the stability which it provides for service planning, even if corridor flows there make profit-making impractical.

Despite the general success of the public transport operation, some problems have arisen. The first concerns the government's inability to improve all the controls that it would wish. A taxi drivers' strike thwarted attempts to increase the costs of taxi operation, transfer from Public Light Bus to Maxi-cab has been slower than hoped. Further, some difficulty was experienced in limiting competition between express buses on the recently opened Island Eastern Corridor and the recently opened MTR Island Line. This was one of the factors which led to the promulgation of a newly announced policy designed to co-ordinate public transport within rail corridors so as to optimise the heavy investment in the railways and avoid unnecessary or wasteful bus competition.

The second problem has arisen from one of the potential successes of the funding of rapid transit development in Hong Kong: the input from generated property development. The funding for the MTR Island Line was somewhat strained by the sudden slump in the property market in 1983/4. However, it should be pointed out that only 10 per cent of the cost of the MTR is expected to come from property development and that MTRC has always enjoyed high credit ratings in the international financial community and has never experienced any difficulty in raising loans to finance the construction of the Island Line at favourable terms.

Among the transport modes, some are clearly less efficient in the use of congested road space than others. In particular taxis, and to a lesser extent Public Light Buses, add substantially to congestion while seeking custom; on some streets in Hong Kong Central, taxis are 60 per cent of the total flow. Experience on the ERP project demonstrates how difficult it is to control such modes once they have been encouraged by market forces. There is therefore a clear justification for resisting the introduction of new low capacity modes into congested cities.

Chapter 4

LONDON*

I. TRANSPORT AND THE ENVIRONMENT IN LONDON

About 17 per cent of London's surface area is given over to transport uses, of which 12 per cent is for highways, footways and parking, but the impact on the environment is disproportionately greater than this. Traffic noise is the greatest source of environmental nuisance in the UK and is a particular problem in urban areas where the great bulk of the population lives. The present soiling of the fabric of London, like many other cities, is attributable in large part to the motor vehicle and ground and airborne traffic induced vibration contribute to the deterioration of some historic buildings. From multi-storey car parks and elevated highways to street-side traffic signs and devises, the trappings of today's transport are very present aspects of the townscape but the most pervasive environmental degradation is that of atmospheric pollution.

Transport has been a major conditioner of London's environment (Tables 1 to 4). Transport operations occupy land, degrade the air and create noise and vibration. The importance of these effects has been increasing over recent years for three main reasons. First, because of the growth in road transport. Second, because the relative importance of the environmental effects of transport has grown as other sources of pollutants have been reduced. (Reductions have been achieved in domestic and industrial noise and in atmospheric emissions as well as by a tightening of pollution standards). Third, because the general public has come to expect higher environmental standards as a result of greater affluence and the impact of modern technology on consumer durables.

The contribution of vehicle emissions to atmospheric pollutants is given in Table 4. Motor vehicles annually add about 1.05m tonnes out of a total of about 1.3m tonnes of pollutants and are the major source of carbon monoxide (CO) and nitrogen oxides (NO_x). Over the years emissions of sulphur (SO_2) and smoke from fixed sources have fallen with reduced use of coal while emissions from moving sources have grown. Motor vehicles are now also the main source of smoke and lead.

Smoke from the growing number of diesel engines in use is a potential hazard to health, a powerful soiling agent and a contributor to reduced visibility. Stricter controls over it have had accordingly been urged on the UK Government by the Royal Commission on Environmental Pollution.

Evidence is also accumulating of the contribution of Nitrogen Oxides and Hydrocarbons to the formation

Table 1. ENERGY CONSUMPTION BY TRANSPORT

	Petrol (10^6 litres)	Diesel Fuel (10^6 litres)	Electricity (10^6 Kwh)
Private motor vehicles (cars and lorries) (1981)	2 546	423	–
Buses (1982)	–	106	–
Other public transport: LT underground (1982)	–	–	866

---

* Case study prepared by Messrs. D. Bayliss (United Kingdom), T. Bendixson (United Kingdom) and J.R. Fradin (France).

Table 2. SULPHUR-DIOXIDE EMISSIONS TRENDS IN LONDON
1965-1980

1 000 tonnes/an

	Source				
	Domestic	Industrial	Commercial	Power Stations	Total
1965	47	122	56	170	395
1970	30	111	58	159	358
1975	10	45	57	58	167
1980	7	25	32	17	81

NB: Road transport contributes about 1% of SO_2 emissions.
Source: GLC Report No DG/SSB/ESD/R129.

Table 3. SULPHUR DIOXIDE AND SMOKE CONCENTRATION TRENDS IN LONDON

1974/75 — 1981/82

ug/m^{3-1}

	74/75	79/80	80/81	81/82
SO_2	117	81	69	77
Smoke	35	28	22	25

1. Average daily concentrations.
Source: Digest of Environmental Protection and Water Statistics.

of ozone and to concentrations well above the GLC guidelines on hot days in summer. In 1983 the highest ozone concentration recorded was 140 ppb (1 hr mean) and the GLC's guideline for ozone of 80 ppb was exceeded on sixteen days in July at Teddington and reached 110 ppb on two days at Chigwell on the other side of London.

The maximum levels of noise and some exhaust emissions are proscribed by national legislation (the Road Traffic Acts of 1972 and 1974) and a number of European Community regulations issued since 1972. Because these apply to new vehicles rather than those existing prior to the regulations' introduction; because they have to avoid being unfair to any

Table 4. ESTIMATED POLLUTANT EMISSIONS FOLLOWING COMBUSTION OF FOSSIL FUELS IN LONDON (1978)

1 000 tonnes/year

	Carbon	Sulphur	Nitrogen	Smoke	Hydro-carbons	Lead
VEHICLES						
Petrol	900	2.2	37	3.4	44	0.8
Diesel	20	3.1	16	3.1	3.5	—
Sub-total	920	5.3	53	6.5	47	0.8
DOMESTIC						
Solid Fuel	24	8.6	0.8	6.0	1.6	—
Oil	0.2	1.0	0.7	0.1	—	—
Gas	0.9	—	3.6	—	0.4	—
Sub-total	25.1	9.6	5.1	6.1	2.0	—
COMMERCIAL AND INDUSTRIAL						
Solid Fuel	0.4	10.8	3.0	0.4	0.2	—
Oil	1.6	80	13.9	1.4	0.3	—
Gas	0.6	—	5.7	—	0.2	—
Sub-total	2.6	90.8	22.6	1.8	0.7	—
Grand total	950	105	81	14	50	0.8

Note: Power station emissions are not included in this table.
Rounding errors may be present.

particular manufacturer and because they reflect average situations rather than the more acute problems of large cities against a background of a growing road vehicle population this legislations' contribution to improving air quality in London has been limited. It should also be remembered that motor vehicle emissions tend to be concentrated at ground level and are often most prevalent in places with high population densities. Their effects on human health tend therefore to be relatively more important than emissions from industrial and power station flues.

On London's main roads maximum daily eight hour average concentrations of CO range between 10 mg/m^3. Whilst these levels rarely lead to immediate problems of toxicity the long term health effects especially on vulnerable people in 'exposed' situations can only be adverse. The EEC limit value of atmospheric lead concentration (2 ug/m^3) is frequently exceeded in the environs of London's busiest roads and there is little doubt that the most cost effective solution lies in the improvements to vehicle engine technology. Whilst the average concentration of these and other atmospheric pollutants are generally below the health hazard levels localised incidents and high pollution episodes can give rise to real problems. Whilst the deadly levels of the 1952 episodes have never been equalled since, as recently as 1975 smoke concentrations in excess of 500 ug/m^3 and SO_2 concentrations in excess of 1000 ug/m^3 have been recorded.

Transport's relative importance as a source of both atmospheric pollution and noise is expected to continue. Transport policies are geared to minimising this contribution by facilitating access without motor vehicles and concentrating journeys on rail and bus services with their limited environmental effects of transport are lorry bans, electric vehicles and the Travelcard.

II. EXPENDITURE ON TRANSPORT IN LONDON

In Great Britain most expenditure on transport is by households and firms rather than government (Table 5). London is no exception although the percentage of public expenditure on transport is closer to fifteen percent compared with the national average of about ten per cent. At present public expenditure on surface transport in London is about £900 million annually (Table 6). Complexities of accounting and, in particular, difficulties in isolating the London element of British Rails' accounts hinder the achievement of precision in arriving at public financial support for public transport in London. A best estimate of trends in recent years is given in Table 7. The trend has been upwards although different policy regimes have produced variations. Compared with other industrialised cities the level of expenditure on both public transport and highways is rather low and the need to manage historic assets has been that much greater.

Table 5. EXPENDITURE ON INLAND TRANSPORT IN GREAT BRITAIN 1977-1982

Million £

	1977	1978	1979	1980	1981	1982
User						
Bus & coaches	1 174	1 294	1 438	1 655	1 790	2 005
Motoring	10 789	12 884	16 673	19 515	21 158	22 981
Rail	747	890	1 017	1 221	1 294	1 219
Road freight	10 393	10 448	14 607	16 708	17 366	18 500
Rail freight	470	517	578	606	632	581
Waterways	30	40	40	70	60	70
Total user	22 603	24 902	33 078	38 338	40 754	43 946
Public	2 610	2 890	3 310	3 870	4 410	4 820
Total	25 213	27 792	36 388	42 208	45 164	48 766
Percentage Public	10.4	10.4	9.1	9.2	9.8	9.9
User taxes	4 209	4 645	5 612	7 017	8 426	9 798
Taxes/Public	1.61	1.61	1.69	1.81	1.91	2.03

Source: Department of Transport Report—Transport Statistics in Great Britain 1972-1982.

Table 6. EXPENDITURE ON URBAN TRANSPORT IN GREATER LONDON (£ million)

£ million

	Infrastructure development 1978	Infrastructure development 1980	Functioning costs[1] 1978	Functioning costs[1] 1980	Total 1978	Total 1980
Private road transport	129	186	3 472	5 253	3 601	5 439
Public transport (including taxis)	235	344	482	648	717	992

1. To include costs of:
   i) maintenance of existing infrastructure;
   ii) amortisation, maintenance and operation of vehicles whether paid for by households, administrations or by public transport operators.

Table 7. TRENDS IN THE FINANCING OF PUBLIC TRANSPORT IN THE LONDON AREA (1979-1983)

£ million

	1979	1980	1981	1982	1983
LONDON TRANSPORT[1]					
Operating costs	452	552	622	673	707
Capital expenditure	100	123	147	127	169
Total costs	552	675	769	800	876
Earnings[2] (excluding grants)	378	473	468	542	559
Cost recovery ratio (LT) (%)	68	70	61	68	64
BRITISH RAIL[3]					
Total costs	—	—	—	757	794
Earnings (excluding grants)	—	—	—	448	546
Cost recovery ratio (BR) (%)	—	—	—	59	69
WHOLE SYSTEM					
Total costs	—	—	—	1 557	1 670
Earnings	—	—	—	990	1 105
Cost recovery ratio (%)	—	—	—	64	66

1. Bus and Underground.
2. Includes compensation for concessionary fares.
3. London and South East Railway.

III. THE GENERAL CONTEXT FOR POLICY

London is one of the world's largest cities with a population in the administrative area of Greater London of 6.7 million and a metropolitan area population of 12.1 million (Table 8). The growth of London has been restricted for over 30 years by the enforcement of a "Green Belt" around its periphery. During this time the population within the Green Belt has been falling, and increasing in the rest of the South East Region. For a variety of reasons these population movements have been uneven and a disproportionately high percentage of old, young, low income and ethnic minority groups live within the Green Belt and particurlarly in Inner London.

London's population changes have been associated with, and partly caused by, changes in economic circumstances. Changes in employement have been dominated by a rundown of manufacturing much

Table 8. POPULATION OF LONDON (1961-1981)

Area	1961	1971	1981
Central London[1]	740 705	638 957	507 813
Inner London	2 752 174	2 392 978	1 990 165
Outer London	4 499 564	4 420 411	4 215 187
GREATER LONDON	7 992 443	7 452 346	6 713 165
Outer metropolitan area	4 344 357	5 152 338	5 400 152
Metropolitan area	12 336 800	12 604 684	12 113 317

1. Contains: City of London, Kensington and Chelsea; City of Westminster.
Source: Office of Population Censuses and Surveys.

of which was located in the inner city. The closure and outward migration of factories occasionally punctuated by major reductions in jobs through the closure of installations such as the enclosed docks typify what has happened (Table 9). In central London employment has also been falling but more slowly, because of the buoyancy of service and professional activities. The growth in tourism has meanwhile created jobs which have, to a limited extent, offset the decline in traditional employment; again especially in the central area.

Inadequacies in London's transport system have contributed to the city's changing economic circumstances. Thus while the Central Area, which continues to be well served by rail, has fared relatively well; Inner London where poor road access has become relatively worse than in the outer suburbs and beyond, has suffered particularly badly. Improvements to the outer radials and the construction of a new motorway (the M25) around the edge of the built up area have greatly increased the appeal, especially to growth industries, of the outer suburbs and the attractive towns beyond them. Most of the new developments in the outer parts of the city and beyond are firmly road based and incapable except to a very limited extent of utilising other forms of transport.

The outstripping, by the growth in transport demand, of the ability of city planners and policy makers to adapt and expand the supply of infrastructure is a familiar story; but the disparity between the two phenomena in London is greater than in many other western cities. There are three main reasons why. First, the level of public expenditure allocated to the expansion, adaption and maintenance of urban transport in the UK has, for many years, been low. Moreover, because of the old age of much of the infrastructure, and the necessity to maintain safety standards and the passenger appeal of the Underground, a high proportion of available funds for public transport has been devoted to renewal, repair and maintenance rather than adaption and expansion. Second, significant changes in transport policy have occurred frequently in London leading in turn to changes in direction which have denied London the continuity of policy necessary to achieve major developments[1]. Third, the buildings of London are often of a high standard and greatly valued by Londoners. This has made major changes difficult, especially developments such as new roads running through many districts.

The upshot has been a growing imbalance between the supply and demand for transport, with demand for road use and private transport growing and demand for rail and public transport declining. Investment in automobiles (private, taxi and hire cars) has been large and is of the order of the total cost of operating all forms of public transport in London.

Table 9. ESTIMATES OF EMPLOYMENT IN LONDON AND SOUTH EAST ENGLAND (1961-1981)

	1961	1971	1981
Central area	1 400 000	1 250 000	1 250 000
Inner London	1 250 000	1 100 000	850 000
Outer London	1 700 000	1 730 000	1 550 000
GREATER LONDON	4 350 000	4 080 000	3 650 000

Policy makers have accordingly had to focus on how best to manage this imbalance while carrying on slowly with the development of infrastructure. In both these endeavours a balance has had to be struck between assisting the economy of London and protecting the environment.

The main elements of the transport policy faced by successive administrations are:

— The improvement and maintenance of public transport;
— The selective improvement of the road system;
— The management and selective restraint of traffic; and
— The harmonisation of transport and land use developments.

The importance attached to these elements has varied at different times but, since 1981, priority has been given to the maintenance of public transport. Effort has been devoted to discerning and serving those people most dependent on public transport. Increased emphasis has also been placed on improving conditions for pedestrians and cyclists. Modal split and car occupancy figures are given in Table 10.

Table 10. MODAL SPLIT FOR ADULT (AGED 16 OR MORE) LONDONERS (Autumn 1982)

Percentage

	Weekday peak	Weekday off-peak
Bus	18	22
Underground	11	6
British rail	7	2
Car driver	32	31
Car passenger	8	11
Motorcycle	4	3
Walk	20	26

Source: GLC/LT Panel monitoring.

The most important recent development has been in fares policy. Over the fifteen years up to 1980 fares in London have increased in real terms every year apart from a brief respite in the mid nineteen seventies. A more gradual but consistent reduction in the volume and quality of service took place over the same period.

In October 1981 fares on buses and the Underground were cut by an average of 32 per cent and a system of London-wide zones for bus fares plus a two zone arrangement for the Underground in Central London were introduced. The courts subsequently ruled that this action was illegal and it was necessary to increase fares by 96 per cent. In May 1983, following careful planning, fares were reduced by 24 per cent and zonal fares introduced for both bus and Underground throughout Greater London. A 'Travelcard' that brought together earlier bus and Underground 'season tickets' was introduced at the same time. This change in the fares structure has been unexpectedly successful.

Additional passenger mileage on London Transport's (LT's) services has been almost twice that experienced during earlier fare changes. Hitherto it had been assumed that changes in fares had little effect on road traffic. It is now clear that, given a large change, noticeable effects on road traffic can be expected. Furthermore the effects differ for different types of car use. In London's experience peak period driving to the central area was more influenced than other types of traffic. It was also noted that a simpler fare structure along with Travelcards usable on both bus and Underground has released considerable latent demand for public transport.

The "ups and downs" in fares led to them being about 10 per cent lower, in real terms, in May 1983 than in October 1981, the date of the first change in this series. Furthermore, as part of the policy to improve the appeal of public transport there has been a reversal in the long established trend of service reduction and a marked improvement in bus and Underground reliability. As a result, following thirty years of steady decline in the use of LT's services, carryings increased in 1983 over 1982 and have grown again in 1984 (Table 11).

The GLC's road construction policy is aimed at improving conditions for industrial traffic and public transport and assisting environmental improvement and inner-city renewal schemes. Road projects which could cause heavy local damage to the environment and employment are avoided and in the selection and design of schemes the needs of pedestrians and cyclists are set above the interest of longer distance and through traffic. Moreover road improvement are assessed and planned in conjunction with associated traffic and environmental management measures.

Traffic management has been a feature of transport policy in London for many years but in the last few years there has been a change of emphasis. Greater attention has been given to protecting the environment, assisting public transport and giving a better deal to pedestrians and cyclists. Well established programmes such as the creation of bus lanes (there are now over two hundred in operation in Greater London) have been continued while techniques such as queue relocation, dynamic urban traffic control (SCOOT) and cycle routes have been introduced. At the same time parking regulations have been tightened London-wide lorry restrictions explored.

Table 11. TRAFFIC AND SERVICE STATISTICS FOR LONDON TRANSPORT 1974-1983

	1974	1977	1980	1981	1982	1983
BUS:						
Bus miles run (millions)	175	179	173	174	164	163
Percentage one person operation	38	41	45	48	48	52
Cost/bus mile (pence)	66.2	121.6	197	219	248	257
Revenue/bus mile (pence)	42.6	80.7	118	119	154	161
Bus occupancy (passengers)	17.5	16.2	14.9	14.4	14.2	15.0
Fare (pence/passenger mile)	2.5	5.0	7.9	8.3	10.8	10.8
Bus passenger miles (millions)	3 061	2 901	2 580	2 510	2 330	2 440
TRAIN:						
Train miles run	27	30.0	30.1	30.3	28.7	28.8
Cost/train mile (£)	3.5	5.8	8.9	10.8	12.5	13.4
Revenue/train mile (£)	2.9	4.8	8.4	8.3	9.7	9.9
Train occupancy (passengers)	119	90	88	84	79	94
Fare (pence/passenger mile)	2.4	5.4	9.6	9.9	12.2	10.6
Train passenger miles (millions)	3 210	2 699	2 640	2 540	2 270	2 700

Source: LT Annual Report (1984).

IV. RECENT INITIATIVES

Heavy Lorry Bans

In 1981 the GLC set up independent panel of inquiry into the economic and environmental effects of London-wide lorry bans. The terms of reference were:

"To examine the social, economic and environmental effects of banning heavy lorries within a circular route on or near the administrative boundary of Greater London, the examination to include: The banning of such lorries from the area at all times; the banning of such lorries from the area at night and weekends as an interim or permanent measure; to examine the practicabilities of enforcement of any such ban."

The Inquiry considered a variety of 24 hour bans covering vehicles weighing 7.5, 16.26 and 24.39 tonnes and three different road networks (Table 12). In one network vehicles were banned from all roads within the M25, in another all roads except for a few main radials, and in the third from the whole network except for main radials and certain orbital roads.

The Inquiry relied on two main sources of information, the Greater London Transportation Survey (1981/82) and the Continuing Survey of Road Goods Transport (1981). These provided comprehensive information for weekday operation but for weekends only lorry count information was available. The amalgamation of the two surveys gave data for:

— Vehicle kilometres by goods vehicles by laden weight;
— Vehicle hours by goods vehicles by laden weight;
— Origins and destinations of goods vehicles in London and through traffic;
— Average trip length and drop size; and
— Hourly and weekend taffic.

Heavy lorries are seen as a severe environmental nuisance in London and the Inquiry concluded that current measures were inadequate. Area bans, lorry routes and local traffic management were seen to be only part of what was needed. Responses by operators to lorry bans identified by the Inquiry included switching to smaller vehicles. Some additional break-bulk and transhipment was also predicted, but rearrangement of existing transhipment was seen as being more important. It was also considered that more efficient scheduling, changes in market areas and other innovative practices would reduce the costs of bans.

Vehicle distances and time on the roads before and after bans were used to calculate cost changes. Vehicle operating costs were estimated before and

Table 12. TRENDS IN LONDON'S RESIDENT GOODS VEHICLE FLEET

Size of vehicle – Gross vehicle – Weight (in tonnes)	1962		1972		1981	
7.5 or less	156 075	(82 %)	142 752	(74 %)	171 090	(88 %)
Over 7.5, not over 16.26	26 330	(14 %)	34 968	(18 %)	10 005	(5 %)
Over 16.26, not over 24.39	8 395	(4 %)	9 000[1]	(5 %)	6 438	(3 %)
Over 24.39			5 000[1]	(3 %)	6 608	(4 %)
Total	190 800	(100 %)	192 130	(100 %)	194 141	(100 %)

1. = Estimated

after various lorry bans. These reflected the costs of distribution in London – no costs were included beyond those incurred on the M25 motorway and roads within it. It was agreed that transitional cost would be important if a ban were introduced suddenly, but small if phased over a period of years. Six of the nine bans were estimated to cause changes of plus or minus 1 per cent of the total costs. However these small cost changes relied on a significant increase in the efficiency of fleet utilisation. It would appear that insofar as this was practicable it would mean a reduction in the flexibility and convenience of the service provided and this, in turn, would impose costs on transport users.

The expected environmental benefits may be summarised thus:

— Noise and Vibration – full time bans would reduce noise noticeably on roads affected by vehicles of 16.26 tonnes and significantly more on those affected by lorries of 7.5 tonnes. Benefits would be greater on banned roads if network 2 or 3 were adopted.
— Air Pollution – pollution from all commercial vehicles throughout Greater London would fall by between 8 and 11 per cent. Benefits would be concentrated in the worst affected areas.
— Physical Damage – no firm conclusion was reached about the degree to which heavy lorries damage underground services. Bans on vehicles over 7.5 tonnes would bring significant benefits; bans on lorries over 16.26 tonnes would give benefits with network 1 and disbenefits with networks 2 and 3.
— Congestion – the question of different road space levels for heavy lorries in complex urban conditions and in free flow traffic was considered but not resolved. Overall, the prediction was for little change.

The Inquiry concluded that firms which are planning to move from London for any reason might have their departure hastened by lorry bans but that firms serving the London market would find conditions more attractive than before and move in. It was found that smaller shops would gain a slight advantage in competition with large ones. Overall, it was concluded that shopping centres and industry would not suffer a decline if lorry bans were introduced.

The Inquiry considered the possibility of defining the environmental problems caused by lorries, identifying the vehicle characteristics responsible for them, and banning those vehicles having such characteristics. The purpose of this approach was to maximise environmental gains, while minimising cost to operators. It was found that even with existing technology, significant progress could be made, and that this approach would stimulate further improvement. London alone would provide a sufficiently large market to justify the production of improved vehicles.

The Wood Inquiry did not entirely achieve its aim of resolving conflicts between environmental and commercial interest. However, it made progress in technical questions about the replacement of vehicles following a ban. It was estimated that for blanket bans, the costs would probably be small, but the environmental gains modest. The 'vehicle standards' approach, designed to lead to the production of vehicles suited to London's special environmental and economic conditions, promised a form of control which was progressive and technically and politically feasible.

Following the Inquiry the GLC decided that plans for a night-time and weekend ban on lorries over 16.5 tonnes GVW should be drawn up. Affected firms (of which there are four to five hundred), in addition to being able to switch to smaller vehicles and weekday operation, will be able to use a network of exempt routes and use 'quiet' or 'hushed' vehicles over 16.5 tonnes GVW. The exemption proposed for quieter lorries would progress towards a maximum level of 84 dB(A) (EEC test) for new vehicles by 1988. These exemptions would meet almost all operators' needs while dramatically cutting the night-time and weekend noise of heavy lorry traffic. They offer a balanced treatment of a difficult problem which would secure substantial environmental benefits whilst minimising disruption to London's economy. Implementation is planned for Spring 1985.

The London Electric Vehicle Programme

In 1977 agreement was reached between the GLC and the UK Department of Industry to instigate running trials of electric vehicles. The Department provided the finance and the GLC the staff to monitor the vehicle performance. After an analysis of potential new applications for battery road vehicles it was decided to examine goods vehicles with a 1 to 2 ton payload and three companies were invited to supply vehicles for trials. The firms were Lucas Industries, whose electric drive system was fitted to a Bedford CF chassis, Chloride Technical Ltd. who had modified a Dodge KC60 chassis and Crompton Electricars Ltd, producers of a K85/36 urban delivery van.

The purchase price of the vehicles for their operators was met by a subsidy which covered the difference in cost between a normal diesel van and the more expensive electric ones. Sixty two vehicles participated in the trials, 25 Lucas Bedfords, 25 Chloride Talbots (Dodge) and 12 Cromptons. London, with its congested urban traffic, was chosen as the most suitable location for testing the vehicles which were put into service in 1978 and monitored for 3 years. Monitoring was completed late in 1983 and detailed analysis of the data received is now taking place.

Among factors being analysed are:

- The daily range of the vehicles;
- 'Fuel' consumption; all the vehicle chargers were fitted with Kilowatt hour meters to monitor electricity consumption;
- Reliability; the mean vehicle mileage between failures or breakdowns directly attributable to electric drive or batteries will be analysed;
- Maintenance and part costs; the manufacturers' claim of quicker and simpler servicing and routine maintenance will be examined and the cost of replacement parts analysed;
- The attitudes of transport managers and drivers to electric vehicles are being investigated.

Every operator was requested to monitor at least one internal combustion engined vehicle employed on similar duties to the electric ones in order to allow comparisons to be drawn. Although the detailed analysis has not yet been completed some benefits have flowed from the trials. The manufacturers have been able to give their products extended running trials in ordinary working conditions and design modifications that have improved performance and reliability have resulted.

The trials have also helped to create confidence between government and the electric vehicle industry. Two of the trial manufacturers, Lucas and Chloride, were encouraged to bring their expertise together and to merge their electric vehicle development departments to merge into Lucas-Chloride Electric Vehicle Systems Ltd, a new company, underwritten by the government. This will ensure that all the best of the innovative technology is incorporated in any future vehicles.

The London trials have also encouraged further and larger production of the present generation of electric vehicles. Over 350 vans with Lucas Chloride electric drive systems will be built in 1984. Crompton have also developed a new vehicle, the NP10, which has evolved directly from operating experience in London.

One of the acknowledged disadvantages of electric vehicles is their initial cost which is approximately double that of an internal combustion engined equivalent. Any increase in the manufacturing of electric vehicles is therefore to be welcomed as a step towards mass production which, it is estimated, would reduce the differential to about one and half times.

Interim results are encouraging although it must be remembered that the vehicles are not mass production models based on massive R & D programmes and billions of vehicle miles of operating experience but unconventional low volume prototypes. The electric vehicles were unavailable for 12 per cent of their duty time compared with 9 per cent for comparable internal combustion engined (ICE) vehicles. Intervals between vehicle failures averaged about seven weeks for the more advanced EVs but longer for relatively conventional EVs. ICE vehicles lasted roughly twice as long between failures. Generally the electric vehicles used similar amounts of energy per tonne kilometre as ICEs but more per payload tonne kilometre because of the heavier unladen weight of their on-board batteries.

Fears about inadequate range were dispelled by the trial with some of the advanced vehicles covering 80 kms a day or more if necessary. Freedom of air pollution and quietness of operation are already recognised as environmental benefits of electric vehicles. If their economy in operation can be shown at least to offset their higher initial cost, then electric vehicles should have a role in urban transport.

The London (GLC/LT) Travelcard

In May 1983 a new period ticket (the Travelcard) was introduced on both LT's buses and underground

Figure 1. **TRAVELCARD ZONES IN LONDON**

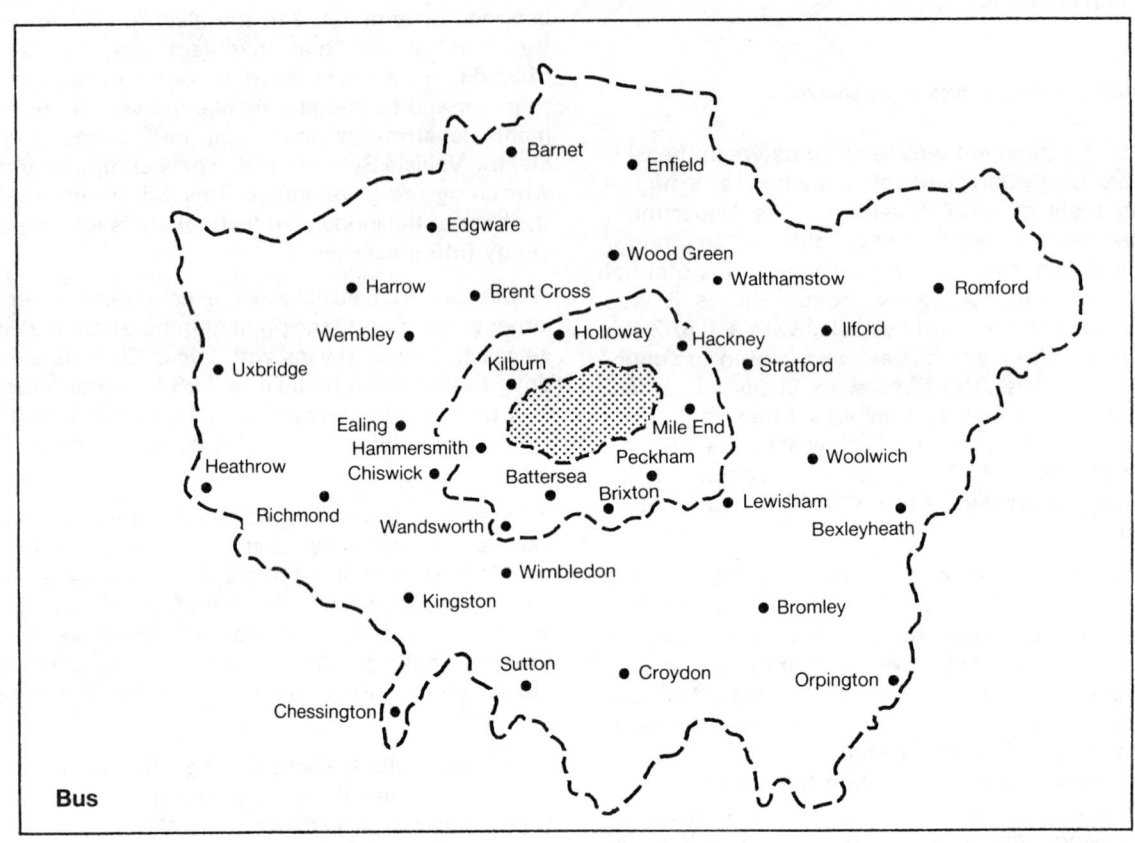

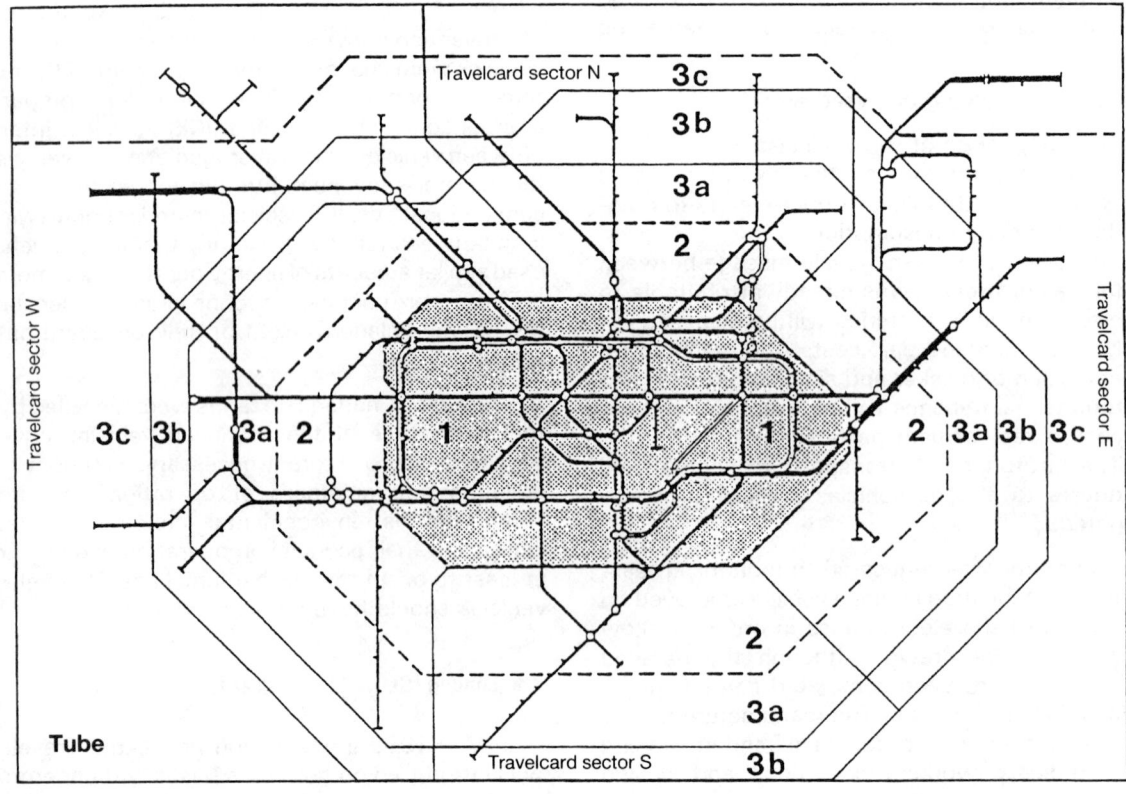

trains. This integrated ticket was very successful and, together with a 25 per cent reduction in fares at the same time, led to a 16 per cent increase in public transport travel.

Use of travelcards is related to a series of annular zones which, together with the central area, cover the whole of Greater London, some 625 square miles (Figure 1). The Travelcards are used in conjunction with a photo-based identity card and are available for periods of 1 week, 3 months and one year. They are valid on all LT bus or underground services excluding such 'specials' as the express bus services to Heathrow Airport.

A number of variations to the system just described have had to be introduced. Travelcards for the central and innermost two annular zones are also valid for travel on buses throughout Greater London. Travel within or to the Central Zone, which covers most of Central London's commercial, retail and entertainment activities, is charged at a higher rate than travel within the suburbs to reflect the attractiveness of the Central Area.

When the GLC was given control of LT in 1970 it acquired a duty to 'promote the provision of integrated economic and efficient transport'. For many years the objective of integration was thwarted by the experience of LT and other UK operators that a scale of finely graduated fares was the most efficient way of collecting revenue (while maintaining as high a level of patronage as possible). Yet an integrated fares system for London required a simple (i.e. coarse) fares scale based on zones. Until recently it was believed that such a simple fares system would have resulted in lower overall levels of patronage, for a given level of revenue, than a finely graduated one.

One of the first steps towards integration came in February 1980 when an experimental flat fare was introduced on the buses in one suburb of London. The main objective was to speed driver-only buses but the results showed that coarser, zonal fares might not lose as much patronage as had been feared. In November 1980 the GLC called for a more widespread simplified and integrated scheme to be investigated. In October 1981 a 'Fares Fair' scheme was introduced which involved a fares reduction of 32 per cent and the conversion of the bus fares to a system of concentric zones. This paved the way for zones for the underground with combined bus and underground tickets (the Travelcards) in May 1983. Further developments are in the pipeline and the GLC has budgeted to extend Travelcards to British Rail's suburban services in January 1985. This will bring common ticketing to all mass transit in London.

Since the introduction of Travelcards the number in use has grown to over 600 000 or about one for every ten residents in Greater London. Travel by bus and underground has increased by 16 per cent, about twice the amount attributable to the reduction in fares while the number of car passengers entering central London during the morning peak has fallen between 10 and 15 per cent. These changes are greater than forecast and have led to a reduction in the estimated cost of the fares package of £23 million in the first year.

A preliminary estimate of the costs and benefits of the Travelcard scheme is set out below (in million pounds):

Benefits	Reduced fares paid by existing passengers	135
	Travel benefit associated with generated trips	11
	Reductions in congestion/accidents etc.	25
	sub-total	171
Costs	Revenue lost as a result of existing passengers' reduced Fares	135
	Revenue gained as a result of new traffic gained	(60)
	sub-total	75
Benefit-cost ratio		2.3

This shows that the scheme was good value for money. The success is due, it is thought, to two main factors. First, many journeys in London call for a combination of bus and Underground stages, something that involved a financial penalty under the old fares system. Second, buyers of Travelcards (who are 30 per cent more numerous than holders of former season tickets) find that journeys in addition to their commuting trips appear free thus making the cards financially attractive.

Future Developments

A number of developments are expected to follow the success of Travelcards. One is further sales of the existing card which GLC market surveys suggest could attract at least 100 000 additional users.

Another development is daily off peak Travelcards valid for the whole of London. These have recently been introduced at a price of £2 for adults and £1 for children and the extension of them to two zones is planned. The inclusion of British Rail in the Travelcard is expected to come into effect in January 1985 and is likely to increase the number of card holders by a further 200 000 three-quarters of them, drawn from existing British Rail season ticket holders and one-quarter of new holders attracted by the combined facility. The British Rail Extension will however be subject to approval by Central Government but assuming it is given the total number of Londoners holding Travelcards and other passes will amount to nearly 1 million. Add to this the million old age pensioners holding free travel passes and well over a third of adult Londoners will be able to travel by bus or the Underground without paying cash fares.

Parking Enforcement – The Wheelclamps Experiment

Increase in car ownership since 1950 have led to dramatic increases in the interference of parked cars with traffic flow and access to buildings. In 1973 when there were 2100 traffic wardens, the highest ever figure, the Metropolitan Police judged the number to be insufficient and stated that 4000 were needed. However, limits on public expenditure led to the imposition of a ceiling of 1500 wardens in 1977-8. This has since been raised to 1800 and the actual number of wardens in post has risen recently from a low of under 1100 to just under 1800.

Notwithstanding this level of enforcement about 2.6 million Fixed Penalty Notices (FPNs), or 'parking tickets', were issued in 1982 in the Metropolitan Police District. Of these about 1.4 million (54 per cent) were paid either with or without a reminder. About 200 000 were unenforceable because they were affixed to diplomatic or foreign vehicles and about 940 000 were unenforceable due to technical or administrative difficulties. The remaining 64 000 were taken to court. Thus only just over 56 per cent of the tickets issued were actually effective. Furthermore, it has been estimated that, because of the shortage of wardens, up to 4 out of every 5 vehicles parked illegally in Central London are not ticketted and the total number of incidents of illegal parking may well exceed 80 million annually. Thus only just over 1 per cent of illegal parkers suffer the consequences of their actions.

It might be thought that the non payers of FPNs are evenly distributed among illegal parkers, but studies indicate otherwise. In one study the police found that out of just over 2700 illegally parked vehicles over 2000 already had FPNs outstanding. Of these the 77 diplomatic vehicles had 428 FPNs of which one had 53. The highest number of outstanding FPNs for a non-diplomatic vehicle was 40. It is thus clear that there is a core of motorists in Central London who persistently park illegally and rely on the low probability of being taken to Court to make the practice worthwhile.

What can be done? Increasing the number of wardens would improve on the 20 per cent of offending vehicles ticketted. Towing away, an expensive and time consuming procedure, is useful but only for vehicles causing gross obstruction. Improved court procedures would, no doubt also help, and the 1982 Transport Act provides for this. Wheelclamps, also allowable initially for a two year period in London under the 1982 Transport Act, are a fresh initiative with several advantages:

— They are effective against all currently unenforceable categories of FPN except diplomats;
— They are effective against persistent offenders many of whom have their tickets cancelled through the Courts' inability to handle their cases;
— To the extent that they affect attitudes to illegal parking they can reduce its actual incidence.

Any vehicle committing an offence may be clamped but, in practice, two categories of offending vehicle are the principle candidates for immobilisation. First, those for which it would be an advantage to interview the owner or keeper or driver. Second, those where a stationary vehicle offence is being committed but where the offence is not of high enough priority to justify removal. Furthermore, there are abundant examples in Central London of places, such as Bond Street, where the incidence of illegal waiting is so high that the removal of all offenders is impractical but where immobilisation of selected vehicles may be expected to have a marked effect.

An experiment with wheelclamps was accordingly set in motion in London in May 1983. An area of about 10 square miles covering parts of Westminster and Kensington and Chelsea was designated as the experimental area. The Metropolitan Police bought 300 clamps for light vehicles plus some special ones for use on lorries and set up seven clamping teams each with a van. The fee for release had been set by the Home Office at £19.50 and in addition the FPN (£10) will generally be payable.

A before and after study is being conducted by the Transport Road Research Laboratory. The before study was carried out in October 1982 and the after study in October 1983. The results are not yet fully analysed but in the first twelve months over 45 000 motorists had their vehicles immobilised and paid £1.3 million in fines and unclamping fees. The inconvenience of being clamped is acting as a deterrent to illegal on-street parking in Central London and it is hoped that the experimental powers will soon be made permanent and their application extended to the whole of London.

It is sometimes claimed that wheelclamps damage vehicles to which they are attached. Nothing to date indicates any such problems. Furthermore drivers in the UK have a responsibility to ensure, as far as is reasonably possible, that their vehicles are fit and safe before driving them. As the presence of a clamp must make a vehicle unfit for use so the driver is responsible to do something about it. He or she is reminded of this by a parking ticket and an additional notice stuck to the windscreen. Thus the driver has to fail to notice these as well as the wheelclamp in order to attempt to drive the vehicle away. The clamp is a large yellow device designed to attract attention. Moreover it is usually fixed to the front driver's side wheel.

It has been suggested that the fixing of a clamp could amount to an obstruction of the highway. Vehicles causing gross obstruction are however towed away, as they were before clamping was allowed. But as there were too many such vehicles to be coped with by towing, the use of clamps has helped to bring this problem within bounds. Vehicles that are the prime candidates for clamping are, furthermore, those with large numbers of outstanding parking tickets and foreign vehicles which, although parking illegally, may not be causing a gross obstruction. For instance they may be overstaying a meter bay.

Public Transport for People with Disabilities

Transport is vital in allowing people to play a full part in society. It is necessary for going to work, shopping and visiting friends or places of entertainment. Many people with disabilities are socially isolated, since they cannot use public transport and have no independent means of travel. In recognising the right of 300 000 Londoners with disabilities to have access to public transport, the Greater London Council has adopted the following propositions:

- People with disabilities have a right of access to public transport equal in availability and price to that of the able-bodied;
- The modification of public transport to accommodate people with disabilities will be promoted and encouraged;
- Limitations exist on the extent to which public transport can be modified in the foreseeable future;
- New public transport systems, such as the Docklands Light Railway should be made accessible to people with disabilities and their needs taken into account at the design stage;
- A door-to-door service involving special services will always be necessary for some disabled people.

The GLC embarked upon the development of alternative services, initially in the form of pilot schemes. These forms of community transport are now being expanded to cover Greater London.

The general aim of Community Transport is to provide for those whose needs are not met by existing public carriers. The 'transport handicapped' may be non car-owners who live in areas poorly served by scheduled bus services or those heavily pregnant women or women travelling alone at night and who live some distance from vehicle stops, for whom travel by public transport presents particular difficulties. Community Transport also provides for those who wish to travel in groups at specific times for specific purposes, such as youth clubs, sports teams, elderly people. One of the more important forms of Community Transport promoted by the GLC is special transport for people with disabilities.

The GLC Taxicard Scheme

'Taxicards' enable people whose disabilities prevent them from using buses and trains to use taxicabs at a price close to that of using public transport. The scheme was launched in 4 January 1983 and limited to 200 members, all of whom had to be resident in the London Borough of Southwark. To qualify for a taxicard, applicants had to be receiving a national mobility allowance or to produce a doctor's certificate stating that they were not able to use public transport. Taxicard holders are able to telephone for taxis without any limit on the number or length of their journeys.

Taxicard users pay a flat fare of £1 plus 15p for each additional passenger. The GLC pays the balance, up to a limit of £6 on the taxi meter, together with gratuities and administration charges. For longer journeys which cost more than £6, the members have to pay the £1 plus any excess of the £6 limit (and the charge for any additional passengers). Taxis are booked, as by other account customers by telephoning one of three radio-controlled taxi companies which, between them, operate around 40 per cent of London's 14 000 cabs.

Taxicards have been welcomed by their users and the representatives of 'disabled' organisations for making a real contribution to the needs of people with disabilities. The scheme has now been extended to cover the whole of London and all seven radio controlled taxi networks are participating. Initially around £2m has been allocated to provide subsidy for the 400 000 trips which are anticipated in the first full year. Up to 50 000 disabled Londoners are expected to be able to use taxicards, a number that is expected to grow following the redesign of the 'London' taxi.

Redesign of the 'London' Taxi

The present taxi is currently being redesigned for production during 1985. The GLC, together with the Departments of Transport and Industry provided a grant to the manufacturer in 1983 of £0.7m each so that the new model could take account of people with disabilities. The new taxi will be able to carry a person confined to wheelchair in safety and comfort. This will entail a sliding partition adjacent to the driver, widened rear doors, raised roof level and lower door sill.

Dial-a-Ride Services

Pending the introduction of the new design of taxi existing public transport services and taxis do not cater for people confined to wheelchairs. Dial-a-Ride services meet this need and in 1982 the Council started to fund three schemes on a 3 year experimental basis.

It subsequently became clear that additional finance would be needed if the Dial-a-Ride operations were to survive. It also became apparent that developments were taking place so rapidly that a three year experiment was too long and that decisions about transport for the disabled in other boroughs and the role of the Council in providing it, should be made sooner.

In December 1983 the GLC therefore agreed to support the development of Dial-a-Ride services throughout London. Services are therefore operational or being set up in 19 Boroughs and total GLC finance is expected to amount to £2m in 1984/85. Applications for further funding are in the pipeline. To assit the development of the new services and their co-ordination, the Council is funding a Federation of London Dial-a-Rides.

Scheduled Public Transport Services

Improvements supported by the Council are also being made to London Transport and British Rail networks. Particular emphasis is being put on access and other facilities at main line railway stations, termini and other interchanges. The running of wheelchair accessible buses on special routes to a scheduled timetable is being considered during the coming year as well as improvements to buses and trains which will assist the ambulant disabled.

Costs

The introduction of the new taxi during the latter half of the 1980s will lead to a review of the relative roles of taxi and dial-a-ride. In the meantime the availability of taxis, at a somewhat higher price, is offset against Dial-a-Ride's lower price, (bus fare equivalent) advance booking requirement (in some cases several days) and skill in handling people with disabilities. Dial-a-Ride also has the potential to carry more than one passenger at a time thereby reducing costs and meeting demand but the complexity of scheduling such trips has so far inhibited progress.

The analysis of the costs of transport for the disabled indicates a hierarchy in terms of value for money. For a relatively small outlay existing public transport services can be improved for large numbers of those with slight disabilities or some sensory handicaps and such improvements benefit all users of public transport. Dial-a-Ride is costly per trip since it does not benefit from the economies of scale of taxi operation but is justified because of the lack of alternative transport for the most severely disabled people. The GLC considers, furthermore, that it has an obligation to provide transport for people with a wide range of disabilities.

In 1984 the revenue support given to London Transport was £190m. Beside this the £4m spent on special transport for people with disabilities is modest. However the cost per trip is many times higher for disabled travellers using special transport than for the ordinary bus passengers. In 1983 the subsidy per passenger journey on London Transport buses was 14p. In 1983 the subsidy per Taxicard trip was £4.03. The subsidy for Dial-a-Ride passenger journeys was even higher but as Dial-a-Rides become more efficient and move beyond the 'development phase' so unit costs will drop.

V. EVALUATION OF RECENT INITIATIVES

Heavy Lorry Bans

The proposal to prohibit heavy vehicle traffic inside London is certainly the most controversial of the five measures presented in this report. It is obviously impossible to evaluate its impact so long as it is only an idea. However, it is likely to have effects touching every aspect of economic activity in the metropolis. It also raises the question of whether to build new roads adapted to the changes in traffic flow. Studies of the measure do not so far seem to have been able to identify all its possible impacts. There is for instance no evidence about the likely effects of lorry bans on urban sprawl. This raises the question whether it is possible to construct new motorways to assist the functioning of London's economic system while avoiding an unwanted dispersion of activities. Judging by the experience of large cities in other countries, those which have built systems of motorways have not avoided tendencies to dispersal.

Electric Vehicles

The introduction of a substantial numbers of electric goods vehicles in London may be expected to bring worthwhile reductions in environmental nuisances. However, this will not occur until such vehicles have attained adequate levels of range and robustness. In the absence of this performance any benefits could be counterbalanced by a loss of mobility and convenience for firms using them. It is therefore desirable to carry out further research aimed at improving the efficiency of electric vehicles. In the meantime the 'London Goes Electric' has shown that there is a substantial range of urban transport applications which electric vans can satisfactorily carry out.

Travelcards

Amongst recent actions affecting transport in London, the introduction of public transport 'travelcards' will probably have the most marked effect on the lives of the region's inhabitants. Contrary to what was done in Paris in 1975, the objectives assigned to 'travelcards' in London seem to have been to increase efficiency, ease of travel and reduce costs. Any reduction in road traffic was counted only as a secondary benefit. In Paris, the principal objective was to increase the integration of different forms of public transport system following an earlier decision to interconnect the networks of the 'Metro' and the National Railways.

Travelcards have been warmly welcomed ty Londoners to an extent exceeding the forecasts of those responsible for them. Not only has the clientele for public transport grown more strongly than foreseen but, in addition, commuter car traffic has been significantly reduced, thereby helping to improve the environment.

Judging by the experience acquired in other cities an innovation of this type is quite slow to 'take-off'. About three years of strong growth in sales may therefore be expected. If to this are added the anticipated benefits of extending travelcards to the suburban services of British Railways, it can be reasonably assumed that the new tariff still has a large market potential.

This potential could be increased still further by modifying the cost of travelcards. At present the difference in cost of travelling in the central and outmost zones is greater for passengers on the Underground than on the buses. The original objective of this arrangement was to off-set some of the beneifts of free 'extra' mobility conferred on holders of travelcards entitled to use the dense network of Underground lines in the City and West End. This price gap does, however, create an opportunity to reduce the cost of certain travelcards and so increase market penetration. Furthermore in pursuing reductions in the cost of public transport fares, recognition needs to be taken of the benefits conferred by public transport on firms of all kinds. One possibility is that they could be called upon to contribute directly to the financing of the running costs of the transit system.

Wheelclamps

Amongst the initiatives described in this report that have, as an objective, the improvement of the environment within inner London, the enforcement of parking regulations by the use of 'Wheelclamps', could be expanded most rapidly. This measure, along with removing vehicles, is one of the most effective in dissuading illegal parking and reducing travel to the centre by car. Wheelclamps however raise delicate legal issues. They also pose the problem of deploying sufficient specialised, qualified and motivated police officers to achieve effectiveness. The threshold of tolerance of motorists could also limit the application of this traffic restraint.

Transport for the Disabled

Transport for the handicapped has reached an advanced stage of development in London. Improved understanding of the specific travel needs of persons unable to use the bus and Underground has led to the introduction of a system of taxis and demand responsive minibuses for which the handicapped pay fares comparable to those of trains and buses. Although the present total cost of transport for the handicapped remains negligible compared with the cost of operating Greater London's conventional public transport system, controlling the costs could become a problem in future (notably for the use of the taxis but also for the use of the demand responsive buses). It will also be necessary to guard against the demand responsive bus services, which are likely to multiply in the coming years, becoming artificially competitive with the existing regular bus services.

VI. CONCLUSIONS

General

The London region is undergoing rapid change. Population is falling, car ownership rising and changes are taking place in the nature and location of employment. The effect of these changes is to create a more dispersed pattern of land uses and trips and a pattern of travel more and more at variance with London's radial rail system.

An outer ring motorway (M25) is being built around Greater London at a distance of about 25 miles from the centre. It promises to contribute to further land use changes and to increased car travel. No comparable roads have been built within the urban area which, in twenty years, has seen a 70 per cent increase in traffic accommodated on an area of roads that has been augmented by only 10 per cent.

Opposition to proposals for new roads within inner London on grounds of air pollution, noise and traffic generation makes it questionable whether new routes can be built without reducing the scale of the works, giving greater compensation to affected property owners and tenants and doing more to reduce environmental nuisances.

Air Pollution

Motor vehicles account for about three quarters of the tonnage of pollutants injected into London's air and are the dominant source of carbon monoxide, nitrogen oxides, hydrocarbons and smoke. Concentrations of ozone exceeding offical GLC guidelines were recorded repeatedly in July 1983 and emissions of smoke from diesels are another cause of concern. Lorry bans, electric vehicles and the improvement of public transport may assist in reducing emissions but stricter control at source over NO_x, HC and smoke from vehicles is needed too.

Irrespective of the extent to which new roads are built, improved techniques for managing traffic and enforcing parking regulations are likely to be needed.

Heavy Goods Vehicles

The noise, smoke and bulk of heavy goods vehicles are considered by many Londoners as serious nuisances. At their present state of technological development electric goods vehicles, despite their environmental advantages, have only a limited role to play in reducing these nuisances. Experiments designed to improve the dependability of such vehicles and to give experience of them by transport managers are nevertheless worthwhile.

In urban areas with large flows of internal HGV traffic, night time and weekend bans on movement, coupled with exemptions for bulk deliveries and certain other classes of goods movement, merit close examination. Giving exemptions to 'hushed' vehicles is a policy that promises to meet transport operators' needs while dramatically cutting night-time and weekend noise levels.

Public Transport

The existence of a concentration of 1.25 million jobs in Central London, the travel needs of the 40 per cent of London households who do not have cars and the desirability of minimising vehicular movements on environmental grounds all point to the importance of expanding the use of public transport services.

London Transport's earnings as a proportion of its total expenditure fell from 70 to 64 per cent between 1980 and 1983. Notwithstanding this, by European Standards, high level of cost recovery, 'Travelcards' introduced in 1983, have been highly successful. They have led car commuting into Central London to fall by ten per cent and causes transit travel to increase by 16 per cent — double the amount forecast on the basis of the 25 per cent cut in fares made concurrently with the introduction of Travelcards.

Estimates of the scope for expanding the use of Travelcards to a wider public indicate that holders of them could amount to about one third of all travellers by 1985.

Wheelclamps

Preliminary results of an experiment with wheelclamps indicate their value at enforcing parking regulations and at recapturing the initiative for traffic authorities when parking tickets have lost their effectiveness.

The influence of wheelclamps on driver behaviour could probably be increased by concentrating on vehicles known to have a record of traffic offences and by being clamped, to pay for all their outstanding tickets.

Transport for the Disabled

About 5 per cent of Londoners are estimated to have physical disabilities affecting their ability to use public transport. The modification of existing bus and railway equipment to assit such individuals will, in most cases, benefit all travellers but may also be extremely expensive. Taxi like services for disabled people are expensive per trip compared with transit travel but offer services suitable for wheelchair users and the very disabled and will be low in cost compared with total transport expenditure.

Providing specialised vehicles and dedicated services for disabled people is more costly than assisting them to take taxis. Developing taxis for use by the general public that can also accommodate wheelchair users accordingly promises to be an aconomical way of providing severely handicapped travellers with mobility. It also avoids stigmatising disabled people by carrying them in distinctive and specialised vehicles.

NOTE

1. In July 1984 the Secretary of State for Transport took over from the Greater London Council responsibility for London Transport.

Chapter 5

LOS ANGELES*

I. INTRODUCTION

Summary

The sheer size and complexity of the Los Angeles region, and the enormity of its travel demands, have forced transportation decision makers to seek transportation system management innovations at many levels. On the whole, the vast Los Angeles region has a surprisingly effective highway and bus system, furnishing relatively safe and reliable daily transportation to its 11.5 million population. Although lacking any rail rapid transit, or even good bus route coverage in certain less densely-developed areas, the region's residents do not express any great dissatisfaction with everyday travel conditions. The recent rate of regional growth in terms of population, employment, family income, and car-ownership — the region will add a population of more than 3 000 000 people, by the year 2 000 — does not suggest that there is any significant transportation problem.

At the same time, the region's air quality is extremely poor. "Smog" alerts are routine, and there is serious concern that, so long as the present growth rate prevails, national ambient air quality standards cannot be met. Because it lacks a rapid transit system, the region consumes more than 5 billion gallons of gasoline annually, and its per capita consumption of motor fuels is one of the highest in the nation. Highway-related noise pollution is a growing problem. Whether or not the transportation system is seen as effective, its environmental, economic, and social costs are very high, indeed.

General Characteristics of the Los Angeles Region

The Los Angeles region is characterised by a unique climate, topography, economy, and lifestyle. For example, its topography includes a narrow coastal strip and lush coastal valleys on the west, foothills and high rugged mountain ranges to the north and south, and inland deserts on the east. The coastal plains and valleys, although constituting a small portion of the region's land area, contain the majority of the population and land devoted to urban activities. The region's metropolitan area is roughly 40 miles wide by 140 miles long, almost three times as large as the Greater London conurbation (Figure 1).

About 11.5 million people inhabited the region in 1980. There are six counties and 158 cities in the region, as well as 64 multi-purpose growth centres of varying size and density. The Los Angeles Central Business District (CBD) stands out as the most highly developed centre, and also forms the nucleus of the regional core — an intensely urbanised area extending from the Los Angeles CBD to the Pacific Ocean, and including ten other major commercial centres.

Highways

The region contains more than 13 000 miles of arterial highways, including almost 2 000 miles of freeways, and an estimated 60 000-70 000 miles of local streets. For lack of regional rapid transit, this vast highway network constitutes the backbone of the region's transportation system, accomodating extensive bus systems as well as trucks and cars. Within heavily built-up Los Angeles County, the freeway system constitutes only one per cent of the total highway/street mileage, but carries 55 per cent of the total vehicle-miles of travel. Nevertheless, congestion continues to be a major problem. Although freeways are generally posted to carry traffic at 55 mph (a national limit imposed after the motor fuel shortages experienced in the 1970s), actual speeds on the freeway system as a whole averaged 37 mph during a typical 1980 morning peak period. On many routes, however, congestion lasted for more than two hours, and crawl speeds were not uncommon.

---

* Case study prepared by Messrs N. Emerson (United States), L. Keefer (United States) and G. Meighörner (Germany).

Figure 1. **THE LOS ANGELES METROPOLITAN AREA AND SOUTHERN CALIFORNIA**

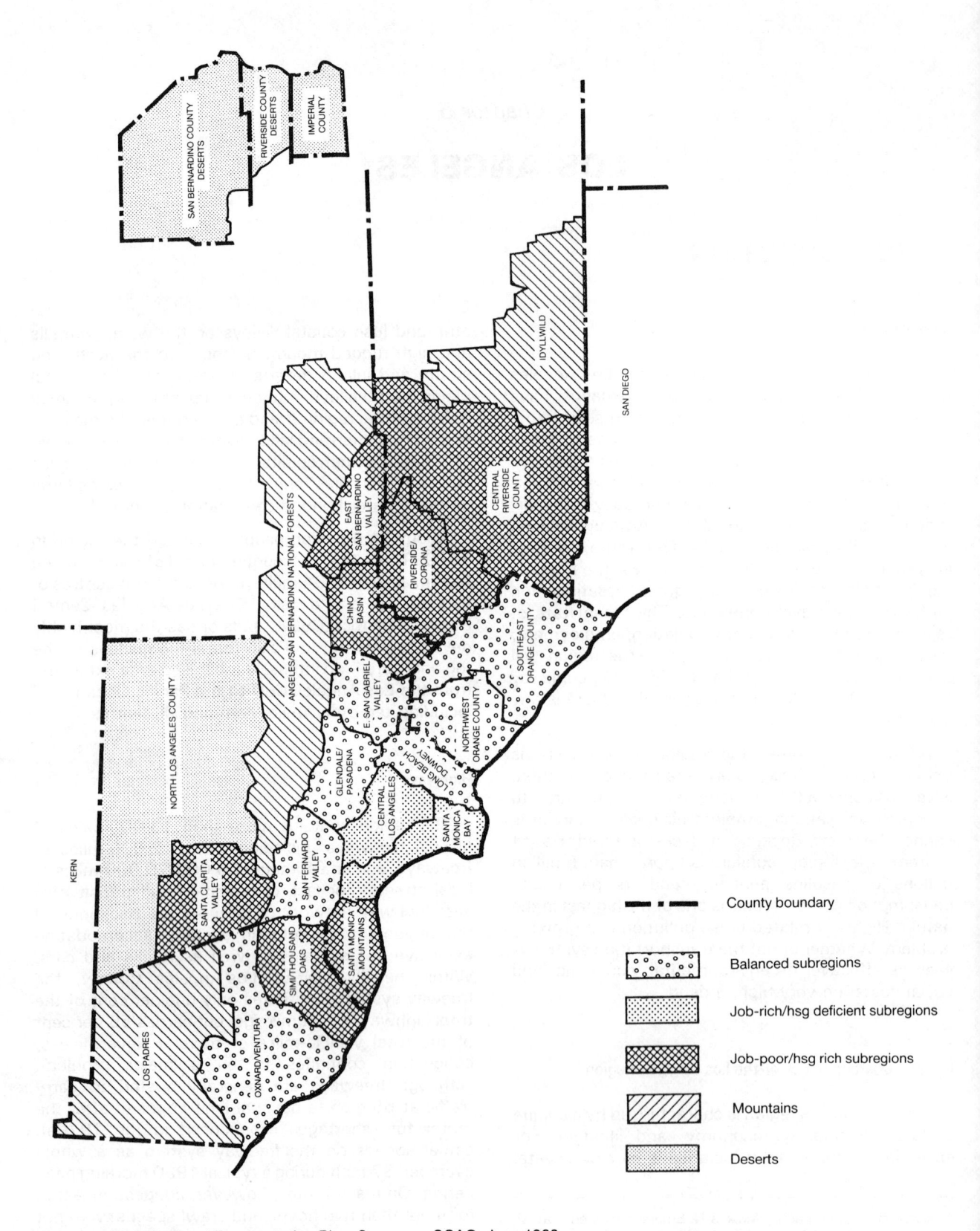

Source: Draft Regional Transportation Plan, Summary, SCAG, June 1983.

Public Transit

In 1980, there were 42 public transit systems in the metropolitan area. Public transit included fixed route, demand responsive, subscription, and charter bus services (there are presently no rail rapid transit services in the region). Approximately 3 467 buses operated on a total of 798 local bus lines. They logged over 135 million revenue miles and carried more than 444 million passengers annually. Over one million riders used the transit systems daily. Another 65 privately operated bus companies provided charter tour, and school bus service. In addition, more than 300 municipal, social service and private sector agencies, operating 5 000 vehicles, offered specialised paratransit services to their client groups. Some 75 taxi companies operated 1 500 taxis in the region and carried over 15 million passengers.

Transportation System Growth Trends

The growth of the highway and bus systems in the Los Angeles region over the past 40 years is illustrated in Figure 2. Population has steadily increased and with it, excepting for the war years, so has total automobile ownership and automobile ownership per capita. That the development of the region's freeway system has paralleled these trends is suggested by the amount of construction that took place during 1950-1955, followed by the even greater growth through the mid-1970s. Figure 2 shows also the growth in the combined bus fleets of the Southern California Rapid Transit District and the Orange County Transit District, the per capita increase in that fleet actually exceeding the per capita increase in automobile ownership during the 1970-1980 period.

Highway and Transit System Funding

About $650 million was spent on the construction and maintenance of the highway system in the Los Angeles region in 1980 (Table 1). The major and increasing source of revenues came from the cities' and counties' general funds (largely property taxes, labelled "conventional local share"). In that year, the cities and counties contributed 51 per cent of the

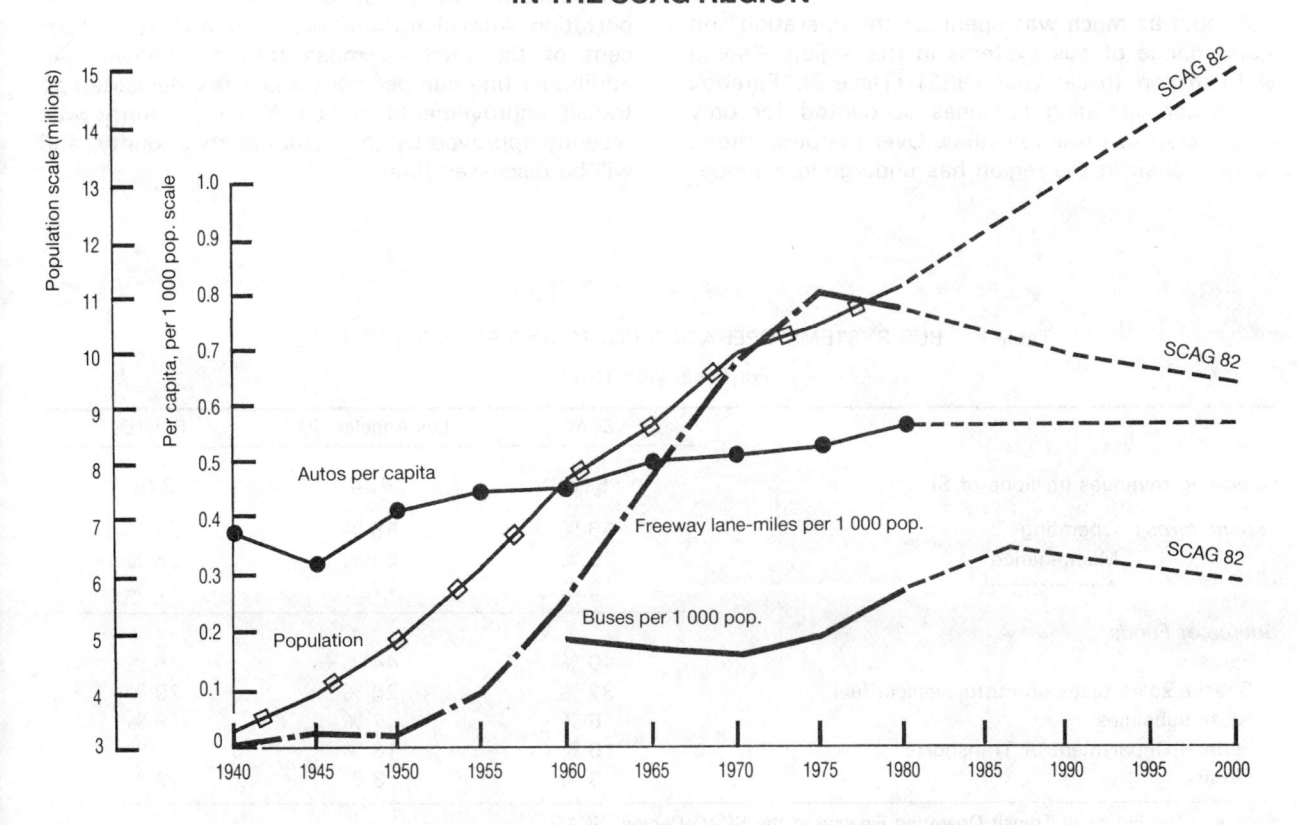

Figure 2. **HISTORICAL TRENDS IN TRANSPORTATION FACILITIES IN THE SCAG REGION**

Source: Draft Regional Transportation Plan, SCAG, June 1983.

Table 1. REVENUES FOR STREETS AND ROADS IN THE SCAG REGION BY SOURCE
In current dollars

	1972	1975	1978	1979	1980	1981
Total amount affected to the road network (millions of dollars)	396	466	595	612	650	752
Sources of revenues (%)						
State of California	44	39	34	33	29	26
Federal Government	2	3	8	11	10	8
Cities and counties (Conventional local share)	39	50	45	43	51	51
Value capture	15	8	13	13	10	15

Source: Financing Streets and Roads in the SCAG Region, SCAG, November 1982.

total highway funding, while the state of California contributed 29 per cent, the US Government contributed 10 per cent, and the private sector through special assessment districts, developer exactions, and developer fees (labelled "value capture" in Table 1) contributed another 10 per cent. The increase in local government revenues has been in response to the declining availability of state and federal revenues.

Almost as much was spent for the operation and maintenance of bus systems in the region — some $517 million (fiscal year 1982) (Table 2). Farebox and other operating revenues accounted for only 43 per cent of these revenues. Over the past fifteen years, transit in the region has undergone a fundamental transformation, growing from a more limited self-supporting system to a larger system requiring over $300 million per year in subsidies. Thus, 32 per cent of bus system revenues derive from a special sales tax on motor vehicle fuel (the Transportation Development Act passed by the California State Legislature in 1971) and another 8 per cent derive from a special State Transportation Act that must be renewed periodically by the Legislature. The US Government, through US DOT's Urban Mass Transportation Administration, provided another 19 per cent of the region's mass transit revenues. An additional one-half per cent sales tax dedicated to transit improvements in Los Angeles County was recently approved by the voters of that county, and will be discussed later.

Table 2. BUS SYSTEMS OPERATING COSTS AND SOURCES OF FUNDS
For fiscal year 1982

	SCAG	Los Angeles CO.	SCRTD
Operating revenues (millions of $)	517	424	378
Expenditures: Operating	56 %	58 %	57 %
Maintenance	25 %	25 %	26 %
Administration	19 %	17 %	17 %
Source of Funds:			
Fares	40 %	44 %	46 %
Special sales taxes on motor vehicle fuel	32 %	28 %	29 %
State subsidies	6 %	7 %	6 %
Federal Department of Transports	19 %	18 %	16 %
Others	3 %	3 %	3 %

Source: The Future of Transit Operating Finance in the SCAG Region, SCAG.

Land Use and Employment Patterns

The Los Angeles region's image as new, sprawling, and lacking features that create a sense of community identity is rapidly becoming outdated. Although its yearly growth was indeed scattered, significant "in-filling" of vacant land has since occured, and the removal of the 13-story limitation on building heights twenty years ago has led to the appearance of true "skyscrapers" — both the Los Angeles CBD and outlying centres containing many new buildings in the fifty to sixty-story range. Surprisingly, the population density of the regional core, at about 4 700 persons per square mile, is the second highest (after New York) of any American city.

The leading employment sectors are manufacturing, with 22 per cent of the regional employment; personal services, with 20 per cent; and retail trade, with 15 per cent. Manufacturing employment is concentrated in the durable goods sector, with aerospace (aircraft, missiles, electronic components, etc.) as the key industry. Other major manufacturing activities include food processing and fabricated metal products. The television and motion picture industry is also notable. Unfortunately, there are significant imbalances between the location of jobs and the population, and this creates considerably more need for long-distance commuting than exists in most metropolitan areas in the United States.

Travel Patterns

In 1980, there were almost 39 000 000 person trips in the region on an average day, only about 1 000 000 of which were made by public transit. About 53 per cent of all trips were for work purposes. Automobiles, trucks and buses produced a staggering 221 million vehicle-miles of travel, about 40 per cent of it during the peak travel periods. The Los Angeles CBD attracts the greatest density of tripmaking in the region, each workday attracting about 690 000 person trips. About two-thirds arrive by automobile, van, or truck; another 27 per cent by bus; and the remainder on foot. Although the region is predominately automobile-oriented, it must be emphasized that the Los Angeles CBD, like many others in the United States, is heavily dependent on public transit for its survival. During the 6am-8am peak period, over 45 per cent of all person trips entering downtown do so by bus. Upon completion of the first, and subsequent, legs of the anticipated rail rapid transit system, the percentage of CBD arrivals by public transit is expressed to rise significantly.

Air Quality

Given the vast dimensions of regional travel, transportation decisions obviously have a wide range of direct and indirect environmental consequences. Perhaps the most notable of these consequences are air and noise pollution and excessive transportation energy consumption. Ringed by mountains rising as high as 10 000 feet, a combination of poor ventilation (light winds and shallow vertical mixing of air currents) and abundant sunshine makes the Los Angeles region an area with a high air pollution potential. That potential is frequently realised. In 1980, monitoring stations centering on Los Angeles County counted 180 days violating the national ozone standard. Of those 180 days, "stage 1" episodes (ozone values exceeding 0.20 parts per million, or ppm), a condition sufficient to cause increased asthma attacks and a decline in lung function for sensitive people) were encouted on 80 days. "Stage 2" episodes (ozone values exceeding 0.35 ppm, a condition enough to cause coughing, chest discomfort and headaches in some people, and a decline in lung function even in the general population) were experienced on seven days. Standards for total suspended particulates (TSP) were exceeded on 70 days, carbon monoxide standards on 40 days, and nitrogen oxide standards on 20 days. No sulphur dioxide violations were reported.

Noise

Outdoor noise levels are highly variable. Levels as low as 30-40 dB(A) occur in the wilderness areas (mountains and deserts) of the Los Angeles region, while the densely urbanised areas create levels as high as 85-90 dB(A). Transportation facilities such as streets and highways, railroads, and airports are among the most important sources of outside noise. Key factors that influence noise levels near highways include distances between noise sources and receptors, traffic volumes, vehicle mix (heavy truck traffic obviously creates higher noise levels), highway configuration below, at, or above grade (e.g., comparing the two locations on the Harbor Freeway, the one below grade results in lower noise levels than the one at-grade), and the presence of noise barriers (such as at the San Diego Freeway). For comparison, the noise level associated with a typical railroad mainline is approximately 70 dB(A) at 100 feet from the track, 65 dB(A) at 215 feet, and 60 dB(A) at 420 feet.

Energy Consumption

In 1979, highway vehicles in the Los Angeles region consumed about 5.6 billion gallons of gasoline and 420 000 000 million gallons of diesel fuel. As shown in Table 3, automobiles accounted for about 75 per cent of the gasoline usage. Perhaps surprisingly, gasoline sales in the state of California and in the Los Angeles region have remained fairly constant

Table 3. TOTAL TRANSPORTATION VEHICLE FUEL CONSUMPTION[1], SCAG REGION, 1979

Million gallons

	Gasoline	Percentage	Diesel Fuel
Automobile	4 193	74.5	Neg.
Light duty truck	799	14.2	
On-road motorcycle	15	.3	
Medium duty truck	145	2.6	
Heavy duty truck	422	7.5	346
Off-road vehicles	48	.9	74
Total	5 622	100	420

1. Does not include buses, airplanes, railroad.
Source: Regional Transportation Energy Analysis, SCAG, March 1982.

Table 4. ENERGY CONSUMPTION BY URBAN TRANSPORT, 1979

Regional totals in million TOE

Mode	Petrol	Diesel Fuel
Private cars and lorries	5 622	420
Publicly-owned and operated buses	2	43
Privately-owned and operated buses	0.5	18
Total	5 624.5	481

throughout the last decade. This is largely the result of a massive swing to smaller, more energy-efficient automobiles (Tables 3 and 4).

II. DESCRIPTION OF INNOVATIONS

Public Transport

Rationalising the Bus System

As streetcars were replaced by buses following World War II, most of the region's private transit operators continued to operate those buses on the traditional streetcar routes. This failed to meet the needs of new urban growth, and led to considerable public dissatisfaction with bus travel, and a continued loss of ridership. In due course, the Southern California Rapid Transit District (SCRTD) was created by the California State Legislature to maintain and expand the existing bus system in Los Angeles County, and ultimately to build an appropriate rapid transit system. After expanding the existing bus system in Los Angeles County (its mandated jurisdiction) through the purchase of additional equipment, SCRTD embarked on a wholesale restructuring of that system. Over a relatively short timespan, accompanied by a massive public information campaign, a north-south, east-west grid system of routes, along with a necessary bus route renumbering scheme, replaced the traditional but often meandering routes. L-shaped and other indirect routes were eliminated in favour of straight-line routes (Figure 3).

At the same time, SCRTD's bus fleet was then among the nation's oldest, much too old for efficient and dependable service. That was largely corrected by an infusion of $261 million in federal funds, the largest single bus improvements commitment in federal transit financing history; with these funds, and matching state and local contributions, over 1 400 new buses were purchased, and seven new maintenance facilities constructed. Today, the average age of RTD's bus fleet is under six years. The new rolling stock includes about 1 200 "advanced design buses" (ADB), 30 double-unit articulated buses, and 22 double-deckers".

Meanwhile, SCRTD construed to make other operational improvements, such as scheduling more express bus routes on Los Angeles freeways. Many of these routes take advantage of high occupancy vehicle (HOV) lanes specifically intended for carpool and vanpool vehicles and buses, and bus by-pass lanes at metered freeway on-ramps (more information about HOV and bus by-pass lanes is presented in the subsequent discussion of freeway system management). SCRTD now operates and 56 "public" express bus routes attracting about 71 000 boardings daily. SCRTD has also inaugurated contra-flow bus lanes on a number of local one-way streets in the central business district as a means of improving upon its scheduled speeds.

This "rationalisation" and improvement of the Los Angeles County bus system was probably the most sweeping of its kind ever undertaken in the United States. It certainly impressed the county voters. In 1980, they approved Proposition A (so-called simply because of its designation as such on the general election ballot), imposing an additional half per cent retail sales tax for transit improvements, by a 54 per cent majority. For the period 1982-1985, proceeds from this "earmarked" tax provide financing for:

Figure 3. **PAST AND PREDICTED BUS OPERATIONS AND RIDERSHIP**

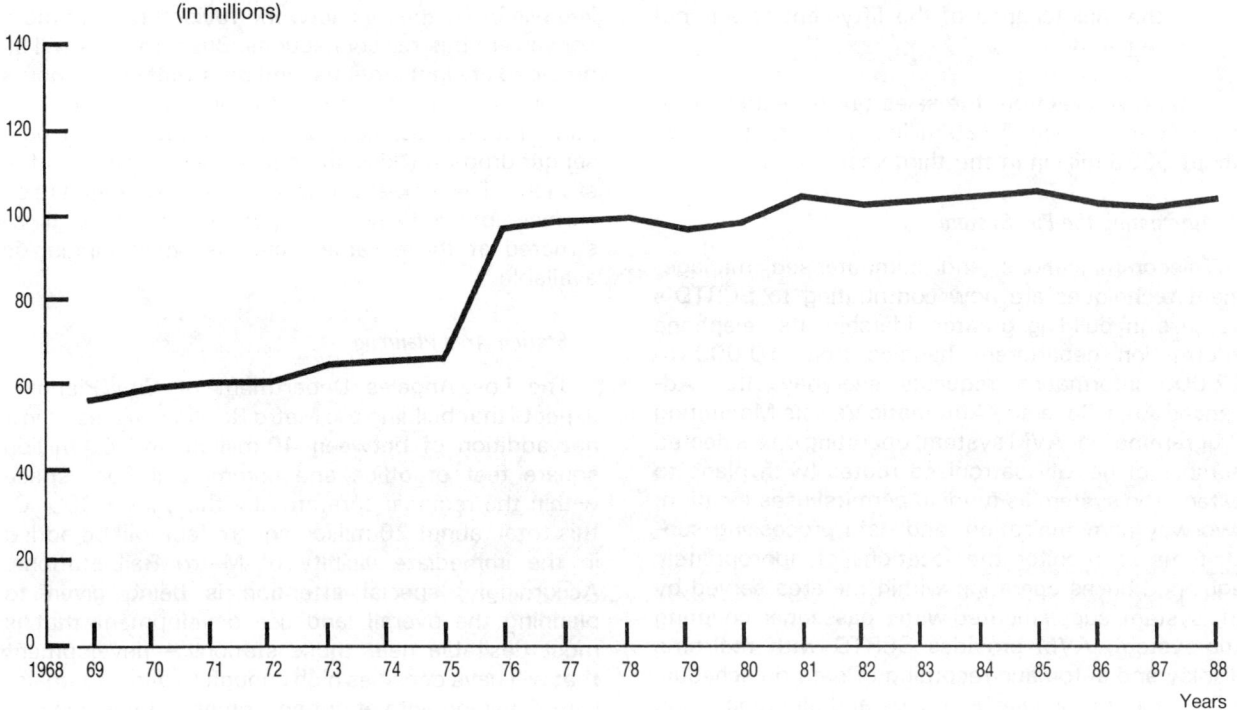

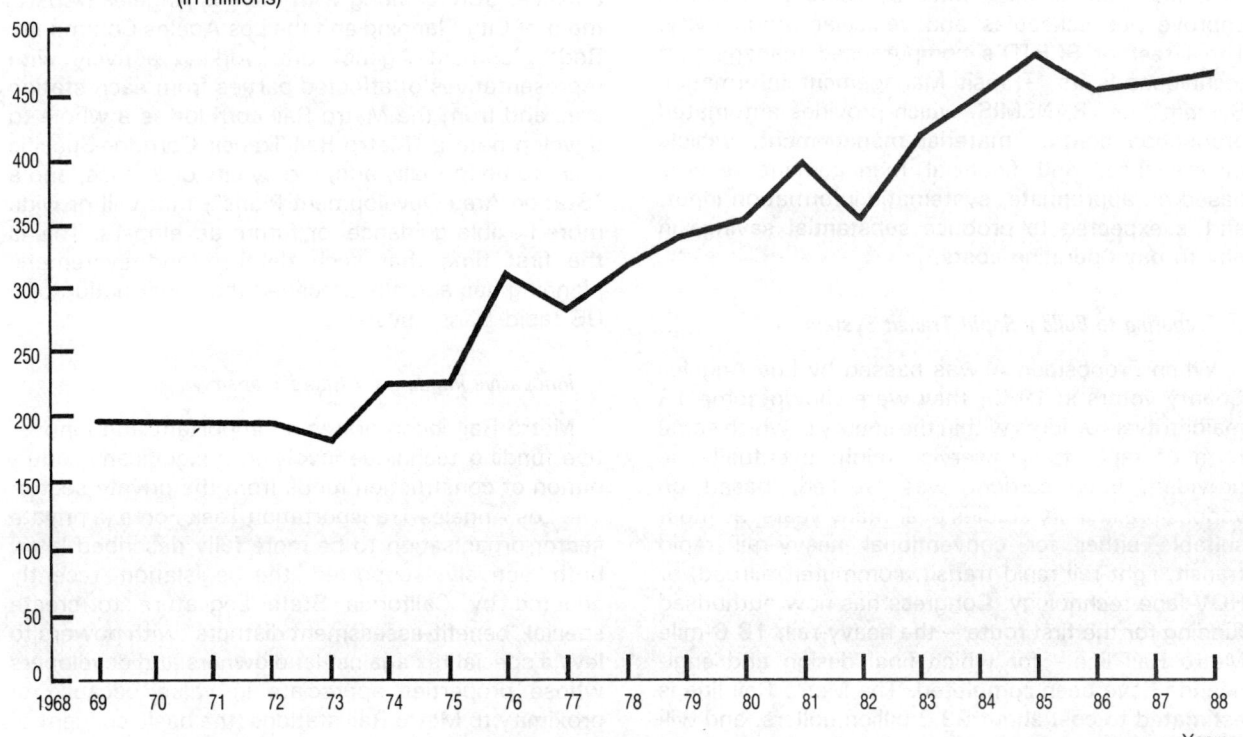

1. Total miles (1969-1988)
   (in millions)

2. Total boardings (1969-1988)
   (in millions)

Note: 1984-1988 data are estimates based on a 1 $ ticket price.

1. Maintaining a fifty-cent bus fare;
2. Improving bus service and making other bus system improvements; and
3. Beginning the construction of a 160-mile rapid transit system within the county (after 1985, the maintenance of the fifty-cent fare is not required.

Total revenues from the sales tax are expected to increase from about $225 million in the first year to about $286 million in the third year.

Automating the Bus System

Telecommnications and computerised management techniques are now contributing to SCRTD's success in building greater ridership. Its telephone information department handles from 10 000 to 12 000 information requests everyday. Its "Advanced Area-Coverage Automatic Vehicle Monitoring Programme", or AVM system, operating on a selected number of heavily patronised routes (with plans to extend the system as funding permits) uses location, two-way communication, and data processing sub-systems to monitor the locations of appropriately equipped buses operating within the area served by the system; supplemented with a passenger-counting sub-system, AVM provides SCRTD with real-time display and automatic recording of data on schedule adherence, passenger boardings and alightings, and passenger loads. This information is used in real-time by dispatchers to control the equipped fleet's operation, and can be used later by route planners to improve bus schedules and vehicular productivity. The latest of SCRTD's computerised management techniques is its "Transit Management Information System", or TRANSMIS, which provides automated production control, material management, vehicle maintenance, and financial management records based on appropriate, systematic information input, and is expected to produce substantial savings in day-to-day operating costs.

Preparing to Build a Rapid Transit System

When Proposition A was passed by Los Angeles County voters in 1980, they were shown some 18 major travel corridors within the county in which some form of rapid transit service might eventually be provided. Each corridor was labelled, based on various preliminary studies over many years, as most suitable either for conventional heavy-rail rapid transit, light-rail rapid transit, commuter railroad, or HOV lane technology. Congress has now authorised funding for the first route – the heavy-rail, 18.6-mile Metro Rail line – for which final design and engineering have been completed. The Metro Rail line is estimated to cost about $3.2 billion dollars, and will have 300 000 daily riders when it opens in 1990. To accommodate this demand, it will be necessary to operate about 120 rail cars, running every three minutes in six-car trains in two directions during the rush hours. With each car having a practical capacity of 170 persons, each train can comfortably carry about 1 000 passengers. Primarily access to the rail line will be by the bus network, revised to offer more convenient bus-rail connections. Bus terminals will be provided at eight stations, and on-street bus turnouts at ten stations. Provisions for auto access include park-and-ride facilities at seven stations, and passenger drop-off ("kiss and ride") areas at seven other stations. The park-and-ride facilities are planned to be surface lots initially, with parking structures constructed at these same locations when funding is available.

Station Area Planning

The Los Angeles Department of City Planning expects that building the Metro Rail line will result in a net addition of between 40 million and 50 million square feet of office and commercial floor space within the regional core area by the year 2 000. Of this total, about 20 million square feet will be added in the immediate vicinity of Metro Rail stations. Accordingly, special attention is being given to planning the overall land use development that is most desirable near those stations – development that will have densities high enough to allow for major joint development yet not so high as to be inimical to existing development (much of which is low-density residential and commercial space). To achieve this balance, SCRTD along with the Los Angeles Department of City Planning and the Los Ageles Community Redevelopment Agency are working actively with representatives of affected parties from each station area and from the Metro Rail corridor as a whole to develop both a "Metro Rail Transit Corridor-Specific Plan, to be formally adopted by city ordinance, and a "Station Area Development Plans", that will provide more flexible guidance for future developers. This is the first time that such detailed and systematic planning has actually preceded the construction of a US rapid transit system.

Innovative Funding for Rapid Transit

Metro Rail incorporates an important and innovative funding technique involving a significant contribution of construction funds from the private sector. The Los Angeles Transportation Task Force (a private sector organisation to be more fully described later) both actively supported the legislation recently enacted by California State Legiature to create special "benefit-assessment districts" with powers to levy a special tax against land owners and developers whose properties appreciate in value because of proximity to Metro Rail stations (the basic concept of value capture) and, in its most impressive accomplishment to-date, pledged $170 million toward completion of Metro Rail. This pledge is basically a guarantee that at least that amount will accrue for construction purposes from benefit assessments through the year 2 000.

By thus oversubscribing the local matching funds required by the Urban Mass Transportation Administration — raising the local share to about 38 per cent instead of the 25 per cent normally needed — the Transportation Task Force was responsible for significantly enhancing the chances for gaining both the initial and additional federal funds needed for constinued rapid transit system development. This $170 million pledge is the largest single private sector funding commitment for a new rapid transit system in US history, and marks the importance attached to good transportation by enlightened Los Angeles businessmen.

Highways

Managing and Refining the Freeway System

The freeway system in the Los Angeles region is probably the most famous in the world. More than 30 000 000 person trips are made daily in the region, and half of them use the freeway system for some part of their length. The most intensively-used part of the system extends over 700 miles in a three-county area, and, with certain exceptions, is all either six or eight lanes wide. The system is almost fully mature, in the sense of having been started largely in the 1940s and 1950s and having most of its present major elements originally completed in the 1960s. Because of certain gaps and missing links, the system is subjected to traffic loads far beyond those originally expected.

The freeway system operate as well as it does largely because of traffic management refinements added long after its original construction. Approximately 40 miles of the system have been re-striped to provide significant additional capacity. Its capacity for carrying people has also been increased by adding HOV lanes such as the El Monte Busway, an exclusive two-lane roadway built in an available median area. This exclusive roadway carries about 42 000 persons daily, half in carpools (required to carry at least three persons per car) and vanpools, and half in express buses. At peak hours, it accommodated 1 200 vehicles an hour, virtually its capacity. Similar exclusive roadways for buses and pool vehicles are intended to be provided by reconstructing some forty miles of the Harbor and Santa Ana Freeways by 1992. Additional existing freeways may be "retrofitted" for HOV lanes as funding becomes available. In addition, a proposal to utilise the median of the Harbor Freeway for an elevated guideway to carry buses and carpools (or even rail rapid transit) is a likely project for the future.

Extensive Ramp Metering

Capacity has also been increased, or some might say, preserved, over what it would otherwise be with an uncontrolled demand, through the widespread application of "ramp metering". Traffic signals at freeway on-ramps limit the number of vehicles enetering the freeways, and help balance demand to capacity by eliminating platooned vehicle entry at freeway on-ramps, and thus reducing upstream shock waves and stoppages on the freeways themselves. Ramp metering is now used at over 600 ramps in the Los Angeles region; another 400 ramps will be equipped with meters in the next five years. At most ramp metering locations, electronic traffic counters have been imbedded in ramp pavements, and information is relayed to controller units in the field, which are being programmed to set the metering rates. These field units are being linked with a joint Caltrans/California Highway Patrol traffic operations centre in Caltrans' Los Angeles office, where traffic flows are monitored and displayed on an electronic wall map. About 50 miles of the freeway system are now so linked. When completed, the center's display will cover over 600 miles of freeways. Eventually it will be possible to over-ride the field units and to set metering rates from the operations centre.

"Balanced Congestion"

Recurrent freeway congestion is also addressed by other traffic system management techniques. These include adding capacity through so-called "spot improvements", such as by adding lanes at bottlenecks related to the original freeway designs, providing auxiliary lanes for weaving maneuvers between on and off ramps, widening ramps, and so forth. These spot improvements are often coupled with ramp metering installations in order to keep the improved traffic conditions from attracting too much new traffic and again causing breakdowns in traffic operations. In deciding where to make spot improvements, a concept of "balanced congestion" is followed: a segment of freeway with notably worse traffic congestion than experienced elsewhere along its length is not brought up to congestion-free standard, but only up to that level of service existing on contiguous segments, even if the latter are themselves somewhat congested. A totally congestion-free system is thus acknowledged to be unaffordable.

Incident Detection

Increasing attention is also paid to non-recurrent traffic congestion that is associated with both planned events, such as maintenance or reconstruction, and unplanned events, such as accidents and other "incidents". Emergency-use telephones have been installed every quarter mile on most of the freeway system in Los Angeles County to help

Table 5. ROAD CASUALTIES[1], 1978-1980

	1978 Killed	1978 Injured	1980 Killed	1980 Injured
Pedestrians	95	883	106	955
Cyclists	1	301	2	358
Motor cyclists	67	2 149	50	2 114
Cars and taxi passengers	396	26 272	349	23 785
Commercial vehicle passengers	1	219	6	249

1. The above accident figures reflect totals from the following:
   - All city streets within Los Angeles city;
   - All county roads within Los Angeles, Orange, and Ventura Counties;
   - All state highways, including freeways, within Los Angeles, Orange and Ventura Counties.

Source: CALTRANS (Letter from District Office for Los Angeles, Orange and Ventura Counties).

motorists seek assistance in the event of breakdowns and minor accidents. An experimental project using closed circuit television cameras along twelve miles of the Santa Monica Freeway has been installed by Caltrans as a further step in incident management. When a possible incident is detected through the use of traffic flow loop-detectors, the incident is further verified through television observation. Use of the cameras enables traffic centre personnel to see any major problem immediately, and to initiate appropriate response activities with a minimum loss of time. Among these responses are the dispatch of emergency equipment to remove disabled vehicles on the roadway, and changeable message traffic signs that direct motorists to pre-planned emergency routes. With so many freeways operating at or near practical capacity during peak periods, Caltrans indicates that it considers prompt attention to even a temporary stoppage absolutely vital to the maintenance of traffic flow (Table 5).

The greatest challenge ever posed by a planned event came during the summer of 1984 when Los Angeles was host city to the Olympic Games. For the most part, they made use of existing facilities scattered throughout four countries in the region. During the three weeks that events took place, some 12 000 athletes were housed at three locations, trained at twenty different sites, and competed at twenty-three other sites. In addition, Olympic officials and workers, members of the news media, and some 75 000 ticket holders were daily travelling to and from the events. Another estimated 1.5 million people were expected to attend non-ticket events during the period. Although various special traffic management schemes have been implemented — such as temporary parking restrictions, signing, signalisation, charter bus service, and park-and-ride facilities — both the freeway and public transit systems have been put to a severe test. Many of the strategies implemented especially for the Olympic Games are expected to have long-term benefits to transportation management in the region.

Table 6. FINANCING OF URBAN TRANSPORT, 1981/82
Regional totals in $ Millions

	Road (1981)	Public transport[1] (1982)
Federal Government	60	98.2 (UMTA)
State or local Government	579	196.4
Users	—[2]	206.8 (Fares)
Others	113[3]	15.5
Total	752	516.9

1. Excludes purchase of new buses.
2. User fees are included in both Federal and Local Government shares.
3. Assessments and fees from landowners and land developers.

Private Sector Contribution

Private Sector – Public Sector Partnerships

In the United States over the last decade, a combination of factors have worked to reduce the availability and purchasing power of federal and state funds for transportation purposes. Gasoline shortages have caused buyers to swing to smaller, more energy-efficient cars, and as a result state and federal gasoline tax receipts dedicated to highway uses (and in some cases to transit uses as well) have generally remained fairly constant, or even declined. At the same time, a new federal administration has in the past few years espoused the philosophy of reduced federal spending for transportation purposes, particularly in the realm of transit operating assistance. With an inflation-driven escalation of construction, operating, and maintenance costs, all levels of government have been hard pressed to afford to meet more than a fraction of the transportation needs that everyone recognises should be met. Increasingly, therefore, public agencies have encouraged private sector help, and the private sector has responded positively not only for altruistic reasons, but also out of an enlightened self-interest.

The Private Sector and Highway Financing

Private sector financing of highway improvements has been significant for many years. One report indicates that, since 1972, funds generated from private revenue sources (that is, value capture through special assessment districts, developer exactions, and developer fees) have totaled $941 million (in 1981 dollars). In 1981 alone, such sources produced $97 million. It is important to understand how this contribution was made, and how private sector attitudes toward helping finance public highway improvements has gradually become more positive. The most important source of private sector highway funding has historically been righ-of-way dedication. Two major types of dedication are most common:

1. through new streets and roads built by the developer of new housing subdivisions, and some commercial or industrial "parks", which after construction are turned over to cities or counties for operation and maintenance; and
2. through existing street and road widening, where an assessment is made against abutting landowners on the principle that street improvements increase the value of that land and the buildings thereon.

What has changed in the last few years is that some private developers are now coming to highway and development agencies requesting improvements that they are willing and sometimes anxious to fund (instead of being required to fund). A number of recent additions to the Caltrans-maintained state highway system provide a good example. They include access ramps from already-built freeways, as well as full interchanges needed to serve specific new private developments. The cost of such freeway additions has ranged up to $30 million for a single project, all in private sector funds. In the past five years that Caltrans has pursued a deliberate policy of encouraging private sector funding, the total private investment in the state highway system in the Los Angeles region has been estimated at $50 million.

Caltrans Private Sector Policy

In 1981, Caltrans further strengthened its commitment to work with the private sector through the establishment of a high-level Office of Transportation Economic Development. Since then, this office has been developing guidelines governing private sector contributions both to highway and transit projects. A newly proposed "Interchange and Overcrossing Policy" would put most simply, see the state paying all of the cost of new interchanges and grade separations, but only up to half the cost of modified interchanges and grade separations, depending upon the levels of observed congestion and accident hazard. A proposed "Policy on Private Sector Participation in Funding Guideway Projects" would limit to a statutory minimum the state funding of fixed right-of-way transit projects in any county that did not develop its own "adequate private sector participation policy", as well as meet a minimum private sector funding percentage. Although both policies have controversial elements, their development and eventual adoption by the California Transportation Commission is strong indication that increased assitance from the private sector is expected in the future.

The Private Sector and Transit Improvements

The private sector in the Los Angeles region has already assited in several ways to improve transit services. Most impressive, as previously reported, is its current pledge through the Los Angeles Transportation Task Force (TTF), to contribute $170 million toward the construction of the Metro Rail project. The TTF is a public-private partnership created in 1981 by the Central City Association of Los Angeles and the Los Angeles Chamber of Commerce. Besides private sector representatives, the TTF is composed of members from the main local and state public agencies involved in transit planning, development, and implementation. Its objectives are:

1. to establish a strategy for central city access and circulation to permit continued economic growth;
2. to assess the implications for the central city of various proposed transit projects; and
3. to engage in legislative advocacy activities aimed at both the US Congress and the California State Legislature.

But other private sector support for public transit has also been important, among them its participation through the "safe harbor leasing" provision of the federal Economic Recovery Act of 1981 in helping SCRTD save a considerable amount of money on the purchase of new buses. Among other things, that law allows a private corporation to purchase rolling stock for subsequent lease to a transit operator; the corporation gains an important tax deduction for depreciation of capital equipment, and the transit operator obtains an advantage in terms of its cash flow situation. SCRTD gained 160 new buses in 1981 through this mechanism, and over 21 other transit companies are currently expected to participate with the private sector in similar efforts.

The Private Sector and Employee Transportation

Private sector participation in transportation matters takes many forms other than through the direct funding of highway and transit projects just described. For example, the El Segundo Employers Association (ESEA), a group representing major aerospace industries concentrated near the Los Angeles International Airport and having upwards of 50 000 total employees, was formed:

1. to work with both highway and transit agencies toward providing better transportation to work for their employees; and
2. to provide employer-based transportation services such as subscription and charter buses, vanpool and carpool programmes, preferential parking for pool vehicles, bicycling facilities, and other alternatives to single-occupant automobile commuting. ESEA has been particularly innovative and willing to experiment, and its various programmes have attracted national attention.

Commuter Computer

Commuter Computer, funded largely through grants from Caltrans, is the largest organisation responsible for the marketing of ridesharing in Southern California (and probably in the entire United States), serving residents of Los Angeles, Orange, Riverside and San Bernardino Counties. Since its founding in 1974, it has worked with some 1 800 companies, and approximately 93 000 people have been assisted in forming carpools. There are currently 300 000 registrants with about 45 000 commuters now ridesharing. Its employer services department forms and maintains the main links with employer-clients, providing ridesharing services (including address matchlists and other statistical data) and promoting the formation of carpools, vanpools, taxipools, and transit usage in all its many forms. Commuter Computer reports that since its inceptin participating employees have used 51 million fewer gallons of gasoline, have reduced aggregate exhaust emissions by some 28 000 tons of various pollutants, and have saved over $150 million in transportation costs.

To make ridesharing more attractive to both employers and employees, the private sector in California recently urged and won the State Legislature's passage of several acts that contained key promotional provisions such as 1) providing deductions from employee's state income tax for costs of commuting to work by transit or rideshare modes, 2) excluding from an employee's gross income for tax purposes any benefits derived from participating in a ridesharing programme, and 3) providing corporate tax credits for employer costs incurred in planning and operating any programme to reduce single-occupant automobile commuting. These legislative acts are among the most far sighted in the United States, and are beginning to be imitated by other states.

Parking Management Programmes

The private sector is also a participant in different aspects of parking management. The City of Los Angeles has, for example, this year passed a zoning ordinance revision that permits reducing the number of parking spaces provided in high-density developments in exchange for developer/landowner commitments to provide certain incentives to ridesharing, transit, and other alternative transportation services. As in several other US cities that have adopted such "parking reduction" ordinances, the concept is that the capital saved by developers in not providing zoning-required, integral parking spaces more than offsets the cost of their contributions toward encouraging other than single-occupant automobile commuting. Considering that the construction cost of a single off-street parking space in downtown Los Angeles is currently about $25 000, many developers undoubtedly will take advantage of the new ordinance and save money by providing the alternative transportation services. A related change in the city code now allows large commercial or industrial employers (100 or more employees) anywhere in the city to provide up to 75 per cent of their required parking at remote locations (instead of on-site so long as high-occupancy vehicle shuttle services between the remote parking lots and their employment sites are continually provided and used.

Environment

Better Transportation for Cleaner Air

As shown in Table 7, according to a 1979 survey by the Southern California Association of Governments (SCAG) mobile sources created slightly more than half of the regional reactive organic gas (ROG) emissions and about two-thirds of the regional

Table 7. TRENDS IN AIR POLLUTANT EMISSIONS IN URBAN AREA, 1976-1982
Regional totals in U.S. Tons/day

Pollutant	Mobile sources 1976	Mobile sources 1982	Stationary sources 1976	Stationary sources 1982	Total 1976	Total 1982
SO_2	40	42	300	158	340	200
NO_x	1 050	700	430	400	1 480	1 100
ROG[1]	1 030	600	670	590	1 700	1 190
CO	9 240	6 700	110	100	9 350	6 800

1. Reactive Organic Gases.

nitrogen dioxide emissions – both of which contribute to the formation of "smog". Mobile sources were responsible for about 80 per cent of the regional carbon monoxide emissions. Thus it is impossible to discuss transportation innovations in the Los Angeles region without reference to the organised attempt to reduce air pollution through transportation actions at both the regional and state levels.

In some instances this attempt was uniquely generated by the Congressional mandate to prepare "transportation control measures". In other instances, the effort was moved ahead by a vigorous California Air Resources Board (ARB) programme to achieve cleaner air in California regardless of federal requirements. Two types of transportation actions can be described under this heading: those that would not have been advanced except for the need for cleaner air, and those would probably have been advanced anyway purely for transportation purposes. Among the SCAG-developed transportation control measures specifically aimed at attaining federal and state air quality standards for ozone, nitrogen dioxide, carbon monoxide, sulphur dioxide, lead, and total suspended particulates, and not previously described, are modified work hours; carpool preferential parking; bicycle and pedestrian facilities; traffic signal synchronization; purchase of low-emission, high fuel-economy vehicles by local governments; conversion to energy-efficient street lighting; and several proposed measures still under scrutiny.

According to SCAG's analysis of progress in implementing these and other transportation control measures in the adopted air quality plan, the Los Angeles region is falling behind in its efforts to achieve clean air standards on schedule (controls on stationary as well as mobile sources having lagged). Although transit ridership is increasing faster than expected, largerly because of Proposition A and the fifty-cent fares in Los Angeles County, the programme to encourage more ridesharing has faltered; this failure has been related to the so-called "oil glut" and lower gasoline prices. If better overall progress toward target emissions reductions progress is not achieved next year, SCAG suggests that additional, more stringent transportation control measures may need to be implemented (Table 8).

Emission Control Devices

Most observers in the United States feel, in any case, that the most effective means of reducing automotive exhaust emissions has been the imposition of emission control devices on the vehicles themselves, and the implementation of inspection and maintenance (I/M) programmes to monitor the effectiveness of that equipment. Standards for new-vehicle emissions, as well as I/M requirements, have been set by the US Environmental Protection Agency (EPA). Since retrofitting of automobiles and light trucks manufactured before 1975 was not required, the vehicle fleet becomes "cleaner" with each passing year as older vehicles are replaced. This fleet turnover is credited for 90-95 per cent of the reduction in total automotive emissions reductions in major metropolitan areas throughout the United States, including Los Angeles.

Motor Vehicle Standards in California

It is particularly noteworthy that California has required somewhat more stringent new-vehicle emission control equipment than has the EPA. The California Air Resources Board ARB's programme includes:

1. Testing and certifying that new vehicles meet unique and stricter California standards;
2. Tests to assure that vehicle owners do not tamper with and alter the required emission control equipment (so that leaded rather than unleaded fuels can be used); and
3. In use vehicle surveillance apart from the owners' responsibility for participation in the I/M programme) to assure that vehicle emission control equipment meets manufacturers' estimated service lives.

Table 8. ESTIMATE OF FUTURE BASIN AIR QUALITY WITH NO FURTHER CONTROL ACTIONS[1]
1987-2000

	Applicable Air Quality Standart[2]	Base period Value[3]	Expected Future Values[3] 1987	Expected Future Values[3] 2000
OZONE				
Maximum 1-hour average concentration	0.10 ppm (S)	0.44 ppm	0.34 ppm	0.35 ppm
2nd-Highest 1-hour average concentration	0.12 ppm (F)	0.41 ppm	0.31 ppm	0.33 ppm
NITROGEN DIOXIDE				
Maximum annual average concentration	0.053 ppm (F)	0.071 ppm	0.055 ppm	0.053 ppm
Maximum 1-hour hverage concentration	0.25 ppm (S)	0.49 ppm	0.38 ppm	0.36 ppm
SULFATE				
Maximum 24-hour average concentration	25 ug/m^3 (S)	45.6 ug/m^3	30.0 ug/m^3	31.3 ug/m^3
CARBON MONOXIDE				
Maximun 1-hour average concentration	–	30 ppm	25 ppm	26 ppm
2nd highest 1-hour average concentration	35 ppm (F)	28 ppm	24 ppm	24 ppm
Maximum 8-hour average concentration	–	24.5 ppm	20.6 ppm	21.0 ppm
2nd highest 8-hour average concentration	9.3 ppm (F)	22.1 ppm	18.6 ppm	19.0 ppm
SULFUR DIOXIDE				
Maximum annual average concentration	0.030 ppm (F)	0.013 ppm	0.009 ppm	0.009 ppm
Maximum 24-hour average concentration	0.14 ppm (F)	0.051 ppm	0.034 ppm	0.035 ppm
Maximum 24-hour average concentration with a violation of a state O_3 or TSP standard	0.050 ppm (S)	0.046 ppm	0.030 ppm	0.032 ppm
TOTAL SUSPENDED PARTICULATE				
Maximum annual geometric mean concentration	75 ug/m^3 (F)	144 ug/m^3	136 ug/m^3	148 ug/m^3
Maximum 24-hour average concentration	100 ug/m^3 (S)	430 ug/m^3	404 ug/m^3	443 ug/m^3
LEAD				
Maximum ¼-year average concentration	1.5 ug/m^3 (F)	3.27 ug/m^3	1.26 ug/m^3	0.90 ug/m^3
Maximum 30-day average concentration	1.5 ug/m^3 (S)	4.28 ug/m^3	1.66 ug/m^3	1.18 ug/m^3

1. Estimate based on forecasts of emissions from mobile and fixed sources, selected.
2. S = State ; F = Federal.
3. The figure shown is the most probable value taking into account the range of values (minimum and maximum) possible.
Source: SCAG (1982), Air Quality Management Plan.

California's I/M Programme

The inspection and maintenance (I/M) programme, which will only become operational in early 1984 following prolonged controversy, requires the state's testing every two years to see that all passengers and light trucks are not polluting excessively. Vehicles that fail the test must be repaired and re-inspected before they are allowed to use California roads. Fees for the new programme will be set by licensed inspection stations. Competitive forces are expected to limit the fee to about $10. Adjustments or repairs to reduce emissions from vehicles that fail the test are limited by law to $50, with allowance for subsequent increases due to inflation. Despite the delay in implementing I/M, California remains one of the few states to move ahead with such a programme, and it is considered critical to reducing air pollution in the Los Angeles region.

California's Vapor Recovery Programme

California is also the national leader in implementing vapor recovery systems at gasoline filling stations, only two other states having similar systems. Special nozzles on the pump hoses collect vapors that would otherwise escape into the air during fill-ups, and route them through vacuum-assisted pipes back into underground storage tanks where thy are condensed back into gasoline. In 1980, such systems were credited by ARB as having prevented the loss of 49 million gallons into the air, and for reducing statewide hydrocarbon emissions by 420 tons a day, almost 15 per cent of the total hydrocarbon emissions from all sources.

Highway Noise Abatement Programme

The Caltrans "Community Noise Abatement Programme" is a retrofit programme to construct soundwalls along existing freeways. Residential areas immediately adjacent to freeways and experiencing noise levels of 67 decibels (Leq) are eligible if the residences were built prior to 1974 and prior to when the freeway was open to traffic, or prior to the completion of major reconstruction or widening that resulted in more than a 3 dB(A) increase in freeway noise levels. Noise abatement measures for development adjacent to freeways after 1974 are the responsibility of local governments. Soundwalls cost from $1 million to $1.5 million per mile to construct, and funding problems do not permit Caltrans to build them as fast as they are requested by eligible developments. Even so, about 30 miles of variously designed soundwalls have been built since 1976. As part of its retrofit programme, Caltrans will also provide air conditioning and insulation to control noise levels in public schools adjacent to freeways if the classroom noise level exceeds 50 decibels on the "A" scale. All new freeways incorporate noise abatement measures in their design (in accordance with US DOT's Federal Highway Administration regulations of 1976) wherever predicted noise levels would exceed 67 decibels (Leq).

III. EVALUATION AND CONCLUSIONS

The Transportation System Today

Despite its complexity and sheer size, the Los Angeles region has a surprisingly good transportation system, and its 11.5 million residents are famed for having a "mobile" lifestyle in which they measure travel in units of time rather than distance. People and goods move about with a relative degree of freedom. Although there is traffic congestion on many freeways during peak travel periods, there has never been an instance of "grid-lock", as in some other US cities, where traffic on signalised city streets comes to a dead stop.

Traffic engineers credit this mobility largely to the exitence of a mature, high-capacity freeway network serving the major concentrations of urban activity. Because this freeway network carries a disproportionate share of all travel in the region — in Los Angeles County, more than half — it has also been the focus of extensive transportation system management measures designed to yield its greatest possible carrying capacity. In contrast to many older US cities, the region is also blessed with an abundance of six-lane and eight-lane arterial streets, many with interconnected, progressive signalisation schemes. In short, the region has an excellently maintained and managed highway system, often considered a model for imitation by highway officials elsewhere in the US.

The Transportation System Tomorrow

Nevertheless, given the phenomenal and sustained growth experienced by the region, and the general expectation that new highway construction will be extremely limited; it appears to be only a matter of time before the accompanying growth in travel demand finally outstrips the ability of transportation agencies significantly to increase the capacity of the existing freeway and arterial street systems through transportation system management measures. Although there are plans to retrofit certain freeways with additional vehicles, funding limitations preclude such retrofitting on all freeways.

Given the expected population growth, there are few real choices for maintaining present levels of freeway service: a massive construction programme to add still more freeway mileage is ruled out not only by cost considerations, but also on environmental grounds — rights-of-way that would satisfactorily minimise environmental impacts are simply no longer available. Nor is there any real chance that the basic pattern of land use activities, particularly the locations of jobs in relationship with housing, can be re-arranged in order to reduce the amount of needed travel. The real enemy is the single-occupant automobile, and it is only by striking directly to reduce the general public's reliance on that mode of travel that good freeway operations can be maintained for very much longer.

Not surprisingly, then, many groups within the region, including the private sector, have increasingly stressed the need for improving the public transit system. The Los Angeles all-bus system, second largest in the US, has already been extensively modernised, and county voters in 1980 approved a new sales tax, with part of the proceeds dedicated to the eventual construction of a 160 mile, mixed-technology rapid transit system. Construction of the Wilshire Boulevard "starter line", or Metro Rail, is expected to begin shortly, and a Los Angeles Long-Beach light rail line is in preliminary design. Funding limitations, unfortunately, preclude constructing more than these first two routes until after the year 2 000.

For this reason, the two new rapid transit routes may be too little, too late. They are expected to carry perhaps 400 000 to 500 000 trips a day, out of a projected year 2 000 regional total tripmaking by all transportation modes of 50 000 000 trips a day, or about one per cent. Although they will immeasurably strengthen the Los Angeles central business district, and confer an important "organising" effect on land use development in the corridors served, the Metro Rail and Los Angeles-Long Beach light rail lines will not solve the region's longer-term transportation demands as projected by the Southern California Association of Governments.

One of the underlying problems not yet fully resolved in the region is that of fragmented transportation planning and implementation authority. With a variety of separate highway and transit agencies having a variety of separate responsibilities, it has apparently been extremely difficult in years past to agree on a common course of action to improve and maintain transportation services. Cooperation and coordination certainly exist, but only within the context of the sometimes conflicting legislation that established the various agencies. This situation is particularly galling where environmental objectives — supported by some of the strongest environmental laws in the world — cannot be met because of institutional constraints that proscribe against more rational transportation system development.

The issue of funding, however, still more basically transcends all questions of planning and technology (such as a proposed mixed-mode rapid transit system that will require transfers between modes to complete many trips). SCAG sees a total shortfall in public agency funding of at least $14 billion out of the estimated $61 billion needed for highway and transit system improvements through the year 2 000. Although the Los Angeles region in general, and the Los Angeles Transportation Task Force in particular, can take great pride in having committed $170 million in private sector funding toward construction of the Metro Rail line, there is serious question about how far the private sector can go in making up this shortfall. Aside from mounting traffic congestion toward the end of this century, one of the projected results of this funding shortfall is a continuation of poor ambient air quality. As shown in Table 8, SCAG shows all federal air quality standards will be violated not only in 1987, but as far ahead as the year 2 000.

Recent Transportation Progress

While the future is thus somewhat clouded, what has been accomplished during the last decade with respect to implementing innovative transportation programmes and policies clearly demands respect. After several unsuccessful attempts, an additional sales tax was finally passed by Los Angeles County voters to build a rapid transit line and to make other transit system improvements. At the same time, few new freeways have been built, and almost total emphasis has shifted to getting the most out of the existing system through transportation management techniques. As radical as some of these management techniques may have seemed to many southern Californians, they have in fact been accepted without serious question, and compliance on the part of the average driver has been excellent. Also at work has been the growing recognition by the private sector that it must participate more vigorously in transportation decision-making — that its own best interests are directly served by promoting, and even financially supporting, a better transit system and various forms of ridesharing. These developments, both individually and collectively, have made the Los Angeles region an exciting arena for transportation innovation.

How much have recent transportation improvements contributed to cleaner air and less noise pollution? This question is really impossible to answer definitively. However, it is certainly possible to infer that air quality, for example, would now be much worse had not a strong effort been made to substitute more transit and ridesharing for the single-occupant automobile travel mode. In fact, according to SCAG reports, the concentration of certain pollutants in certain subparts of the region has measurable declined during the last decade. Even if such progress cannot be credited to more enlightened transportation programmes and policies (it is actually more a result of emission control systems on new vehicles and controls on stationary source emissions), the general public recognition that automotive exhausts are a major source of air pollution that can be reduced through changes in transportation programmes and policies is clearly an important step forward.

The situation with respect to noise pollution is not so optimistic. Although, because of the region's spread-out character, transportation-related noise (other than near major airports) does not yet seem to be the more serious problem that it is in New York, Chicago, and other major US cities, noise is beginning to emerge as a serious environmental concern. Noise near some freeways has reached levels that have required, under federal highway noise regulations, the California Department of Transportation to inaugurate a noise barrier freeway retrofit programme in the vicinity of sensitive receptors, which programme the

Department says would be accelerated were more funds available. At the same time, most of the Department's new highway construction projects (limited though they be) include appropriate noise abatement measures. Little if anything else is being done, however, systematically to identify and take steps to solve highway related noise problems on other than state owned freeways, and they are generally ignored by other transportation agencies.

Prospects

Despite the problems just pointed out, the Los Angeles region has nevertheless made impressive strides toward achieving more effective transportation and a better living environment. Its most striking attributes are:

1. A willingness to experiment and to seek necessary changes;
2. A recognition that both the private sector and the general public need to be positively involved in the planning and implementation of transportation innovations; and
3. A growing understanding that the historic reliance on automobile transportation can only lead to continued environmental, social, and economic troubles.

This combination of characteristics should enable the region to continue to find and implement even more exciting transportation innovations in the future.

Chapter 6

MUNICH*

I. THE URBAN AND REGIONAL CONTEXT

The City of Munich is the capital of the Free State of Bavaria and one of the Federal Republic of Germany's leading economic and cultural centres. It constitutes the core of a relatively prosperous region covering over 5 000 km² and containing 2.3 million inhabitants. The pattern of settlement reflects Munich's predominant role in the region's economic and social affairs: it is mono-centric and highly concentrated. Some 1.3 million people live within the city boundaries, a densely built-up area covering only 310 km² (Table 1). Another 1 million people live in about 170 small towns and villages scattered throughout the region. Half of these live in an inner belt of about 900 km² which together with the city constitutes the Greater Munich urban agglomeration. The remaining 0.5 million live in an outer belt covering 3 800 km². Many of the communities in this predominantly rural area are small, remotely located villages.

Munich's inherent attractiveness as a business location and as a place to live, work and study in has been enhanced by the city's administrative status as State Capital, by its strategic location at the centre of the region's road and rail networks, and by the absence of any major competing urban centres nearby (Figure 1). But despite these important advantages, long-term trends towards the dispersion of population, employment and travel have been threatening to weaken the city's position in the region, while the still heavy concentration of population and economic activities in the inner city have created problems of overcrowding, traffic congestion and pollution. This case study shows how the city's elected representatives and planners have responded to these challenges since 1965. They have resolutely pursued a policy designed to achieve a better balance between public and private transport and thereby improve overall travel conditions and environmental quality in Munich and its region.

---

* Case study prepared by Messrs. G. Meighörner (Germany), H. Doleschal (Germany), S. Falk (Sweden), and B. Pearce (United-Kingdom).

Trends in Population and Housing

The population of the Munich region has grown continuously over the past three decades. The city's population, however, has been declining in relative terms since 1961 and in absolute terms since 1971 (Table 2). The growth in the area outside the city has largely been due to people immigrating from other parts of the country and from households in Munich's centre and ageing inner city districts moving out to settle in the suburbs.

Changes have also occurred in the structure of the city's population. The proportion of employed persons has declined while those of students and elderly persons have increased. Average household size has fallen from 2.3 persons in 1960 to 1.9 in 1980, but there are more large and single-parent families. The proportion of registered foreigners has increased from 6.9 per cent in 1960 to 17 per cent in 1980. Growth has been highest amongst those immigrant workers and their families who are difficult to integrate into German society.

Despite the decline in population the demand for housing in Munich has increased. The number of households increased from 591 000 in 1970 to 648 000 in 1980. Although average household size has decreased, the average living space required per inhabitant rose by 20 per cent over the same period. The supply of housing within the city has simply not been able to keep pace. In the inner city and certain neighbouring districts, the construction of new housing has been hampered by the lack of available space and the high land prices generated by intense competition from other non-residential activities. The outcome is that Munich is now faced with a severe housing shortage and rents for flats, especially in new apartment buildings, are very expensive. The rise in local costs of living increases the pressure on low-income households to move out of the city.

These changes have also accentuated differences in the social composition of the city's various districts.

Figure 1. **THE CITY OF MUNICH AND NEIGHBOURING SUB-CENTERS**

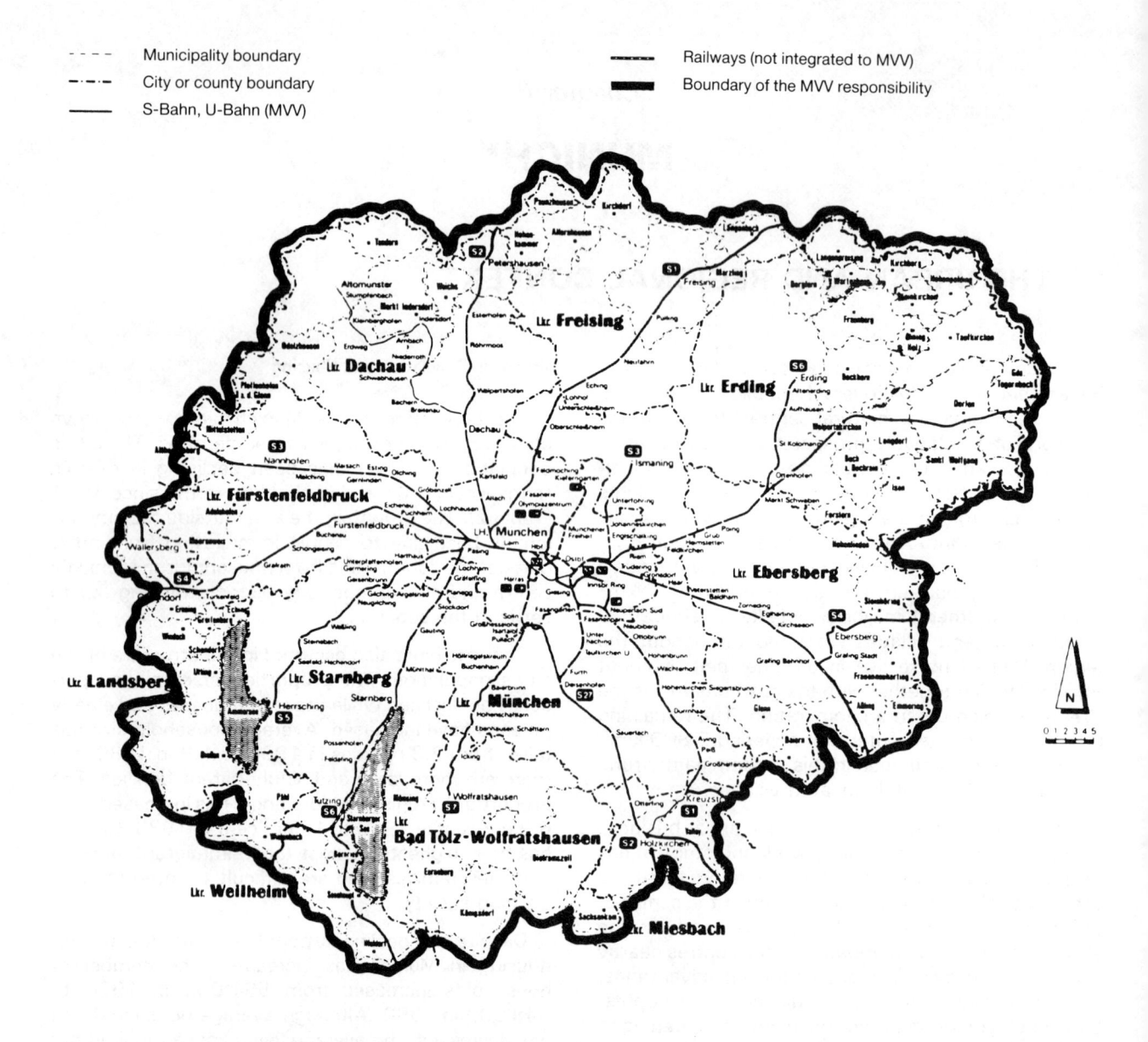

- - - - Municipality boundary
- · - · - City or county boundary
———— S-Bahn, U-Bahn (MVV)

▬▬▬ Railways (not integrated to MVV)
▬▬▬ Boundary of the MVV responsibility

Table 1. THE BUILT-UP AREA IN THE CITY OF MUNICH, 1950-1982

	1950	1960	1970	1980	1982
Total (km^2)	311.55	310.01	310.55	310.39	310.39
of which:					
Covered by buildings (%)	(23.9)	(31.6)	(37.3)	(38.5)	(38.9)
Covered by transport facilities [1] (%)	(11.7)	(11.6)	(13.5)	(13.7)	(13.8)

1. Includes airports, railways, and roads, streets and squares.
Source: 1983 City Development Plan.

Table 2. POPULATION IN THE CITY OF MUNICH AND THE REGION, 1950-1982

	1950	1961	1970	1980	1982
City of Munich	931 937	1 085 014	1 311 978	1 298 941	1 287 080
Surrounding area	572 046	629 323	800 432	1 001 142	1 021 636
The Munich region	1 403 983	1 714 337	2 112 410	2 300 083	2 308 716
The city's share of the total (%)	(59.3)	(63.3)	(62.1)	(56.5)	(55.7)

Source: 1983 City Development Plan.

The city centre's steadily decreasing population has become a mixture of extremes: of students and retired people, of large families and single persons, and of low-income and highly affluent households. The older, densely built-up districts surrounding the city centre mainly contain lower-income households. These are followed by newer, low-density districts extending across the city boundaries and mainly containing higher-income households.

Trends in Economic Structure and Employment

The total number of persons employed in the City of Munich peaked in 1970 and has remained relatively stable since then (Table 3). This contrasts with employment trends in the area outside the city where the number of jobs has carried on growing. Marked changes have also occurred in the structure of employment. There has been a definite shift towards tertiary activities and these now account for about 62 per cent of the jobs available in the city. Of the 51 500 firms currently located in Munich, 19 400 are in the personal services sector, 12 400 are in commerce and trade, and 12 300 (including about 500 industrial enterprises) are in the crafts sector. There has also been a tendancy towards a relatively high concentration of industrial jobs in a few firms. The ten largest industrial companies currently employ over 104 000 workers or 35 per cent of the total in this sector.

These trends towards greater specialisation and concentration are not without risks. Serious financial or growth problems in major industrial enterprises, increasingly high rationalisation in industry and the crafts sector, and the growing use of modern information technologies in the tertiary sector have increased the risks of structurally linked collapses in employment. A recent forecast of employment in Munich indicated a decrease of 20 000 jobs by 1990. Thus a gradual rise in the unemployment rate to the level of other cities seems likely. The comparative percentage figures for the end of 1982 were as follows: Munich 5.9; Berlin (West) 10.2; Cologne 11.5; Saarbrucken 13.4; Dortmund 13.7.

Munich's economy has become highly dependant on exports. In 1982 foreign countries accounted for 40 per cent of the turnover of industrial and crafts firms with 20 or more employees in Munich. The volumes of exports are largely governed by external and often unpredictable factors, adding to the uncertainty about the city's economic prospects.

The commercial and other service activities are mainly located in the city centre and adjacent districts to the north. About 85 per cent of the available floorspace in the city centre is occupied by government administrations and public services, head offices and branches of business and financial institutions, department stores and shops, hotels, restaurants and entertainment establishments. The industrial activities are mainly located on the outskirts, with heavy

Table 3. EMPLOYMENT BY MAJOR SECTORS IN THE CITY OF MUNICH, 1961-1982

	1961	1970	1975	1980	1982
All sectors	670 500	784 000	770 000	788 150	771 000
of which:					
Primary sector (%)	0.5	0.4	0.3	0.3	0.2
Secondary sector (%)	43.0	46.0	39.7	39.7	38.2
Tertiary sector (%)	56.5	53.6	60.0	60.0	61.6

Source: 1983 City Development Plan.

concentrations in the vicinity of major transport arteries. The strong competition between commercial activities for advantageous locations in the city centre and the impossibility of industrial expansion due to lack of space has prompted a number of firms to move out of the city to cheaper more spacious sites.

Trends in Road Traffic and Environmental Conditions

The total number of cars, vans and lorries in Munich has trebled since 1960, from about 181 000 to over 540 000 at the present time. Similar growth has occurred in the rest of the region. The highest growth has occurred in the number of private cars and vans (Table 4). Up until 1972 the number of private cars and vans in Munich exceeded those registered in the surrounding area. Since then, Munich's share of the total for the region has fallen, from just over 50 per cent to 45 per cent by 1982. This trend is consistent with the decentralisation of population and has resulted in increased commuter flows to the city.

Space is scarce in Munich. About 10 per cent of the city area is occuppied by roads and parking facilities. Traffic volumes on the road network have increased enormously in recent decades with the growth in private car ownership and road haulage activities. About 3 million trips are made daily within the city and private motor vehicles account for 35 per cent, i.e. almost 54 per cent of all motorised trips. The supply of road space has only marginally increased over the past decade and traffic jams regularly occur during peak periods on certain parts of the network. The routes worst affected are the city's partially completed ring-roads, the main radial access routes to the city and the bridges over the River Isar, which constitutes a north-south barrier just to the east of the city centre. At weekends and holiday periods, traffic jams frequently occur on the southern and eastern sectors of the middle ring-road as Munich's citizens head for the nearby recreational areas to the south of the city. Within the inner city traffic problems arise on the service network as commuters search for places to park their cars in densely built-up districts with little or no off-street parking facilities.

Road traffic is also a major contributor to environmental problems in the Munich region. Cars and lorries emit thousands of tons of pollutants into the atmosphere every year. As pollution from industrial and domestic sources has diminished, road traffic's share of the total has increased and it is now the major source of air pollution in Munich and the region (Table 5). Heavy lorries, cars and motorcycles are also the predominent source of noise disturbance in the city and high noise levels disturb thousands of people living along the city's extensive main road network. Although the number of persons killed and injured on the roads has significantly decreased over

Table 4. PRIVATE CARS AND VANS IN MUNICH AND THE REGION, 1960-1982

	1950	1961	1970	1980	1982
City of Munich	250 000	340 000	360 000	450 000	472 000
Surrounding area	180 000	310 000	465 000	625 000	578 000
The Munich region	430 000	650 000	825 000	1 075 000	1 050 000

Source: MVV, 1983 City Development Plan.

Table 5. ESTIMATED YEARLY EMISSIONS OF POLLUTANTS BY SOURCE, 1974/75
tons/year

Emission source	SO_2	Partic.	CO	NO_x	HC	Total
Industry	2 800	330	–	2 450	–	5 580
Energy production	10 200	650	–	2 900	–	13 750
Domestic heating	5 000	2 700	22 300	5 800	–	35 800
Air traffic	70	25	850	200	600	1 745
Road traffic	2 400	4 800	109 000	25 400	14 000	155 600
Total	20 470	8 505	132 150	36 750	14 600	212 475
Road traffic's share (%)	11.7	56.4	82.5	69.1	95.9	73.2

Source: 1983 City Development Plan.

Table 6. ROAD ACCIDENTS IN THE CITY OF MUNICH, 1970-1982

	1970	1974	1982
Total accidents	45 946	29 470	45 214
Total injuries	12 118	9 822	9 584
Of which:			
Pedestrians (%)	—	(16.2)	(14.0)
Cyclists (%)	—	(8.8)	(17.1)
Total deaths	246	193	119
Of which:			
Pedestrians (%)	—	(47.1)	(52.0)
Cyclists (%)	—	(7.3)	(11.8)

Source: Local Experts.

the past decade, there has been a disturbing rise in accidents among users of two-wheeled vehicles in recent years (Table 6). Heavy lorry traffic is mainly concentrated on certain radial routes and the middle ring-road but they must pass through residential areas to the north and west of the city for lack of adequate connections with the federal motorways arriving there.

Transport Policy in the Munich Region

Transport planning in Munich has been closely tied to urban development policy. The idea of a better balanced transport system allowing both for the increase in essential road traffic while improving local public transport was first formulated in the city's 1963 General Urban Policy Plan. This idea was further elaborated in a tributary document, the General Transport Plan, which was approved by the City Council in 1965. Both of these plans have been revised every ten years or so. As circumstances have changed, the emphasis in planning policy has clearly been tipped in favour of public transport.

The 1963 General Urban Policy Plan

The General Urban Policy Plan approved by the City Council in 1963 set out the urban development strategy to be pursued up until 1990. The overriding goal of this strategy was to maintain and reinforce Munich's predominant role in the regional economy by increasing its attractiveness as a location for firms and households. The attainment of this goal largely depended on the fulfillment of two related and potentially conflicting pre-conditions:

i) The improvement of accessibility to the city centre for essential business traffic, workers, shoppers and tourists; and

ii) The prevention of traffic congestion and environmental deterioration from reaching unacceptable levels in the city centre and neighbouring districts.

Although it was considered that local transport problems in Munich had reached a crucial phase and that urgent action was required, the City government also considered that these problems could only be solved by a comprehensive and coherent long-term transport policy. This policy was formulated two years later in the 1965 General Transport Plan.

The 1965 General Transport Plan

This plan, which sought to establish a new balance between private and public transport, rested on a number of basic premises:

i) Good road links were essential to serve local business activity and personal travel requirements, both within Munich and also between the city, the region and the rest of the country, and should therefore be developed;

ii) The build-up of excessive levels of road traffic in the city centre was undesirable and should be controlled, especially as physical and financial constraints meant that likely future demands for road and parking space in the city centre could never be fully accomodated;

iii) A high quality public transport system could be an attractive alternative to the private car for local travel and should be developed and promoted to enable it to become the dominant means of passenger travel to and from the city centre;

iv) Different modes and networks could complement one another and maximum advantage should be taken of this.

Although the 1965 General Transport Plan set separate objectives for private and public transport, the various actions it proposed were conceived as mutually supporting elements of one comprehensive strategy.

Actions Concerning Public Transport

A fundamental objective of the 1965 plan was to completely reverse the prevailing trends in modal choice by raising public transport's share of the daily motorised trip market from 50 per cent in 1965 to 65 per cent by 1990. This implied a substantial improvement in the speed and quality of service then offered by public transport. It meant, for example, reducing the duration of home-to-work trips on public transport by a factor of two or three. In common with federal transport policy at that time, Munich's planners accorded priority to high-speed, high-capacity modes operating on exclusive rights of way.

The plan therefore provided for:

- The creation of a region-wide rapid rail network comprising a suburban rapid railway (S-Bahn) designed to provide a fast rail link between the suburbs and the city centre, and a new underground railway (U-Bahn) designed to serve the centre and fringe areas of the city;
- The protection and improvement of the existing tram and bus systems.

The policy objective concerning the complementarity of different modes and networks produced perhaps the most innovative aspects of the plan:

- The creation of a public transport authority responsible for developing and administering an integrated system of service planning and promotion for all local public transport services operated by municipal, regional and federal transport undertakings in the Munich region;
- The restructuring of the tram and bus networks to avoid the duplication of services and to provide feeder services to suburban rapid rail and city underground stations;
- The provision of adequate facilities for the speedy and convenient transfer of passengers between the S-Bahn and U-Bahn, and the tram and bus networks; and
- The creation of park-and-ride facilities at suburban rapid rail and underground stations to enable motorists to transfer to public transport rather than drive on into the city centre.

The 1965 plan sought to improve the flow of road traffic in Munich, by:

- Concentrating it on a network of high capacity urban expressways and main roads; and
- Relieving the city centre of undesirable through traffic and excessive terminating traffic.

Both these objectives implied extensive new road building to improve Munich's main road network and its connections with equivalent regional and federal networks. The principal elements of the main road network envisaged for 1990 included three new orbital routes allowing local, regional and long-distance traffic to by-pass the city and its centre. These ring-roads were to be completed by a series of radial roads and interconnections with several state roads and six federal motorways which terminated on the fringe of the inner city.

The second objective also entailed the implementation of a strict policy of parking control and traffic management to discourage non-essential use of private vehicles in the city centre (particularly by commuters) and in combination with measures to improve conditions for pedestrians and public transport users (e.g. the extensive pedestrianisation of Munich's historic centre, the "old town", and traffic priority measures for trams and buses).

II. THE DEVELOPMENT OF THE REGIONAL RAPID RAILWAY (S-BAHN/U-BAHN)

The 1965 General Transport Plan provided for the reorganisation and modernisation of existing public transport networks in the Munich region as well as the construction of a new underground railway to serve the capital city. The planned development of the suburban rapid rail (S-Bahn) and the municipal underground (U-Bahn) systems required a major and carefully phased investment programme spread over three decades. The most intensive network development phase occurred between 1966 and 1972 and coincided with preparations in Munich to host the 1972 Olympic Games.

The Development of the Suburban Rapid Railway (S-Bahn)

The Initial Situation

Prior to May 1972 the Federal Railways operated local commuter services on thirteen lines between Munich's two mainline terminals (Munich Central and Munich East) and suburban stations lying between 40 to 45 kms from the city centre. These local services were provided for workers and schoolchildren travelling daily to and from Munich, but they had two major shortcomings:

i) The railway network was divided into two sectors with lines terminating at stations on opposite sites of the city centre: eight at Munich Central to the West and five at Munich East. Inaddition, the line installation and operations (track density, electrification, signalling and rolling stock) were not standardised. Through services between the western and eastern sectors were therefore impossible and travellers arriving at the city were faced with extra walking or transfers to trams or buses before reaching their final destination;

ii) The Federal Railway's inter-city passenger services and goods trains had priority over the local commuter services on shared line sections. It was therefore impossible to run local rail services to a fixed timetable or guarantee regular frequency.

The planned suburban rapid railway (S-Bahn) was designed to overcome both problems by improving network connectivity and the reliability of services throughout the region. The basic idea behind this project was to create a new, homogeneous rail network from existing lines so that, in theory, every train could go from one point on the network to any other point. For this to be possible it was first necessary to:

a) Interconnect the existing lines selected for the new S-Bahn network; and
b) Standardise the tracks, stations, rolling stock and equipment to be used on it.

The First Phase of S-Bahn Development

The suburban rapid railway is being developed in two phases. The first phase lasted from 1965 to 1972 and entailed the following operations:

i) The construction of a 4.2 km long tunnel and six S-Bahn stations underneath the city centre, between Munich Central and Munich East, to house a trunk line connecting six of the western lines to the five eastern lines;

ii) The overhaul of track electrification and signalling on twelve lines (one of the western lines retained its independent operating status);

iii) The introduction of new rolling stock comprising 3-wagon motive-units which can be operated as short (3-wagon), normal (6-wagon) and long (9-wagon) trains according to demand;

iv) The modification of platforms on all stations and, where necessary, the renovation or relocation of stations themselves.

The outer terminals of the Federal Railway's former local service were maintained for the new suburban rapid railway giving a total network length of about 412 kms.

Work was begun on the tunnel in 1965 and the S-Bahn system came into service on 28th May 1972

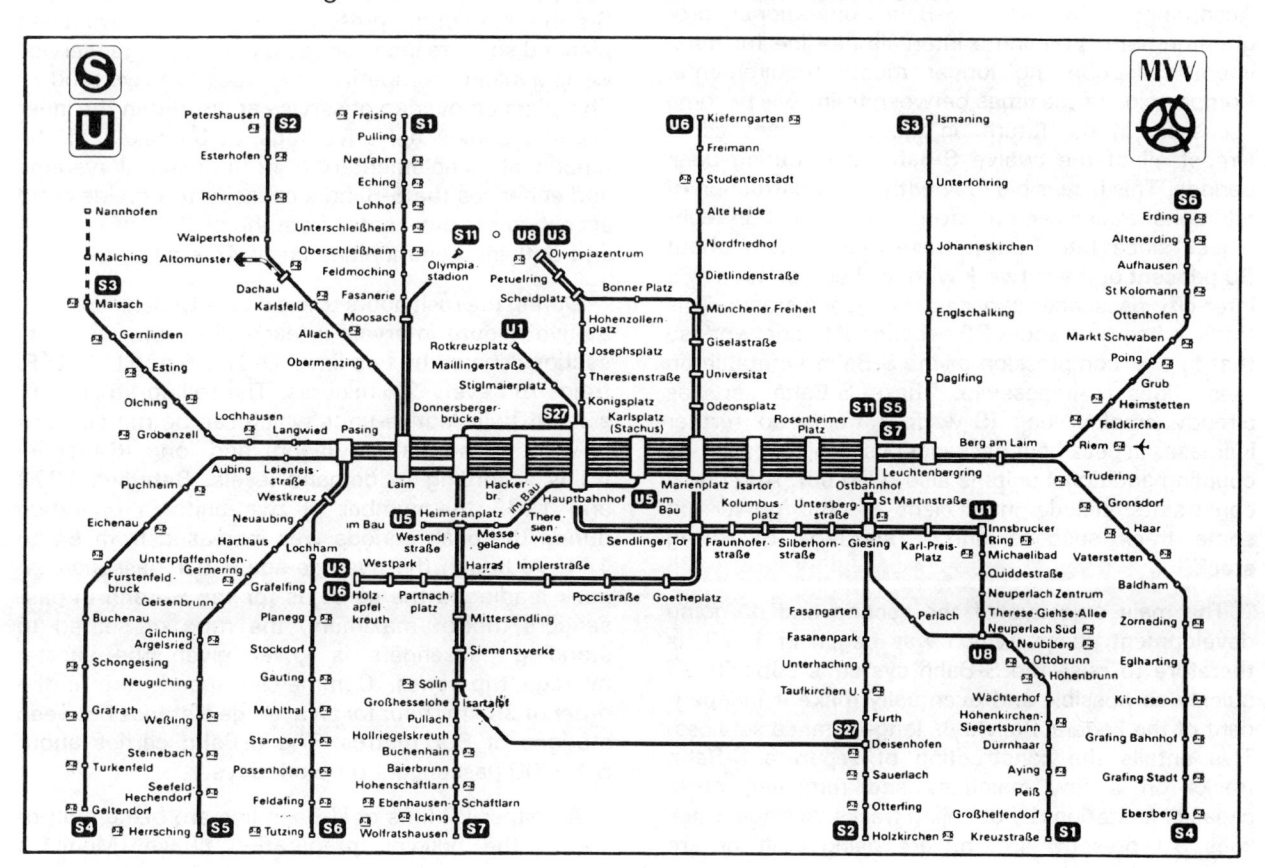

Figure 2. **THE S-BAHN AND U-BAHN NETWORK**

with the operation of seven regular through services passing underneath the city centre (Figure 2). On the outer sections of the S-Bahn network, trains basically run every 20 minutes in each direction at peak hours and every 40 minutes at other periods. The funnelling of the through services into the tunnel gives service frequencies of from two to six minutes on the trunk line depending on the time of day. The S-Bahn carries about 600 000 passengers on work days, almost half of them on the trunk line section. Thus, although the principal function of the S-Bahn is to serve the suburbs by assuring a fast link with the city centre, the trunk line also functions as an underground line assuring fast east-west services across the city. The S-Bahn stations constructed in the tunnel serve the most important activity centres in the heart of Munich. They are equipped with good passenger interchange facilities for convenient transfer to underground (U-Bahn) lines and to surface tram and bus services.

The Second Phase of S-Bahn Development

The quality of service offered by the S-Bahn system represented a clear improvement over the local commuter services it replaced. The big increase in traffic that occurred easily exceeded forecasts and the number of triple motive-units operated during peak hours was raised from 109 in 1973 to 169 in 1983, mainly by increasing the length of trains (Table 7). Accordingly, the initial S-Bahn operational programme, with 20 minute intervals per line for maximum utilisation, no longer meets requirements. Compression of the times between trains will become necessary in the future on the outer branches of almost all of the twelve S-Bahn lines during peak periods. This is feasible only with the construction of additional tracks and improved train protection techniques since the S-Bahn trains must share about 90 per cent of the network with the Federal Railway's inter-city passenger services and goods trains. This traffic is heavy on about 30 per cent of the network so that further compression of the S-Bahn timetable on these lines is impossible. These S-Bahn services already employ long (9-wagon) trains so further increases in peak hour passenger carrying capacity by coupling additional units is also ruled out. Additional constraints include insufficient electric power on some track sections and a shortage of rolling stock.

The main objective of the second and on-going development phase, which was begun in 1973, is therefore to raise the S-Bahn system's capacity as quickly as possible and eventually make it independent of the Federal Railway's long-distance services. This entails the constructing of separate S-Bahn tracks on a few mainlines and removing other capacity limitations by doubling tracks on single track lines. At present, 47 kms are being built or are planned. The extension of standardised operational and technical installations is also planned. The only actual prolongation of the suburban rapid railway concerns the city's new airport, Munich II. The S-Bahn already services Munich's existing airport on the eastern outskirts of the city at Riem. It is planned to connect the new airport, which is still in the planning stage, to the city centre by extending an S-Bahn line (S-3) about 19 kms to the north of the city.

The Development of the Underground Railway (U-Bahn)

Work was started on Munich's new underground railway in 1965 and the first section was opened to service on the 19th October 1971. It was 8.7 km long and jointly used by two U-Bahn lines (U-3 and U-6). The connection of a 3.5 km branch of line U-3 with the Olympic Complex was completed in early 1972. By 1983 some 37 kms of track had been built and 41 kms of line opened to service (Table 7). All four U-Bahn lines now in operation (U-1, U-3, U-6 and U-8) run underneath the city centre on a basically north-south axis (Figure 2). The U-Bahn's principal role is to serve local travel needs within a radius of about four to five kilometres of the city centre and, within this zone, to collect and distribute S-Bahn users making intermediate and longer distance trips. The points of intersection between the U-Bahn and S-Bahn networks, present and future, have been planned so as to increase rapid rail passenger service while avoiding competition between the two modes. This planned overlap of service areas within the inner suburbs allows full advantage to be taken of the functional complementary of both rapid rail systems and enhances the S-Bahn's capacity to provide rapid access to the city centre from the centres of suburbs and outlying communities up to 45 kms away.

During the rush hours trains on the U-Bahn lines run at five minute intervals in each direction. On route sections shared by two lines (U-3/U-6 and U-3/U-8) trains run every 2½ minutes. The rolling stock consists of twin motive-units which can be run as short (2-wagon), normal (4-wagon) and long (6-wagon) trains according to demand levels. Between 1973 and 1983 the number of twin-units in operation during the peak periods was increased from 54 to 141. Although the average space per passenger at peak loading is the same as for the S-Bahn (4 passengers/metre^2 maximum) the ratio of seated to standing passengers is lower given the shorter average trip times. Commercial speeds are in the order of 35 km/hour for an average distance between stations of 800 metres. The U-Bahn carries about 600 000 passengers on work days.

A further 22 kms of U-Bahn line are being built or are in the building preparation phase. Munich's

Table 7. THE MVV'S UNIFIED TRANSPORT SYSTEM, 1973-1983

Mode/year		Network length (km) Track/road route		Stations and stops	Lines	Rolling stock
S-Bahn	1973	410		125	8	109[1]
	1983	412	428	144	8	169
U-Bahn	1973	9	14		2	54[2]
	1983	37	41	43	4	150
Tramway	1973	112				631
	1983	84	158	171	11	174
City bus	1973	281			62	418
	1983	373	528	727	72	509
Regional bus	1973	618				129
	1983	2 263	2 551	1 446	159	285

1. 3-wagon trains.
2. 2-wagon trains.
Source: MVV.

City Council has already approved the construction of about another 15 kms. Thus by the middle of the 90s a total of 78 kms of underground line will have been built. The new 1983 Urban Development Plan provides for the construction of about 20 kms more. Thus in the final phase the U-Bahn network will comprise about 90 to 100 kms of line. As Munich's underground network is extended, further connections will be made with the suburban rapid railway. This will enable passengers to transfer to and from the U-Bahn and S-Bahn on the outskirts of the city and so relieve pressure on the trunk line section of the S-Bahn in the city centre. The ultimate objective of this co-ordinated development of lines and network junctions is to produce a single highly efficient regional rapid railway system.

The Financing of the Regional Rapid Railway (S-Bahn/U-Bahn)

Since the start of works, both the suburban rapid railway and the Munich underground have been financed through a specially created enterprise — the Munich Tunnel Company. The Federal Republic of Germany, the Free State of Bavaria, the City of Munich and the Federal Railways are partners in this financing company and allocate funds through it to finance construction programmes.

The Suburban Rapid Railway (S-Bahn)

The funding of the first S-Bahn development phase (1966-1972) was agreed on in September 1965, with the Federal Government, the Bavarian Government, the Federal Railways and the City of Munich sharing the costs according to a given scale or on the basis of fixed sums. The funding of the second and on-going development phase is based on the agreement of May/June 1975, between the Federal Government, the Bavarian Government and the Federal Railways. Funds in aid are available from the Federal Government under the provisions of the 1971 Local Transport Funding Act which earmarks a part of the revenues from the mineral oil tax for investment in local transport development (the revenues thus earmarked are now allocated on an equal basis between local road and public transport projects). Additional aid from the state level is made available under the provisions of Bavaria's Local Transport Programme. Thus, under the 1975 funding agreement the Federal Government finances 60 per cent of the eligible costs, while the Free State of Bavaria meets 26⅔ per cent of the costs and the Federal Railways 13⅓ per cent. The financing operation is implemented through the Munich Tunnel Company and the Federal Railways are responsible for carrying out the works. Under both of the above-mentioned agreements, about DM 1 200 million were invested in the suburban rapid railway up until the end of 1982. In addition, outside the scope of these agreements, the Federal Railways has carried out works costing about DM 200 million and provided rolling stock costing about DM 550 million.

Munich's Underground Railway (U-Bahn)

The Federal Government and the Free State of Bavaria also contribute to the funding of Munich's underground railway under the auspices of the Munich

Tunnel Company. The Federal Government provides aid covering about 60 per cent of the eligible costs under the Local Transport Funding Act while the Free State of Bavaria provides aid covering 20 per cent of the eligible costs under the Bavarian Local Transport Programme. Eligible costs include the construction or extension of infrastructure and supporting facilities, with a series of exceptions such as planning and rolling stock, and amount to about 90 per cent of the actual costs. Thus, all in all, the Federal Government in theory meets 54 per cent and the Free State of Bavaria 18 per cent of the total costs of construction works on the Munich underground railway. The City of Munich meets the remaining 28 per cent out of its own budget.

The cost of the U-Bahn, including stations, has ranged between DM 20 to 80 million per km, with the higher figure occurring in the city centre. About DM 3 100 million have been spent to date on building Munich's underground railway since the start of works in 1965. The annual volume of works now stands at about DM 300 million and it will be necessary to keep to this value in view of the financial requirements in the next few years. The new 1983 Urban Development Plan envisages an investment programme of DM 1 569 million over the 5 year-period 1983 to 1987. The progress in construction works of three to four kilometres of new U-Bahn line a year which can thereby be achieved is necessary in order to guarantee the high standard of public passenger transport in the Munich region in the long term and exploit the building trade's capacity on a continuous basis.

As the U-Bahn's funding partners cannot provide the stated percentage of investment subsidy for every annual sum, the volume of construction works which can be carried out is based on assistance programmes which are drawn up and renewed every year by the Federal Government and states, with the available resources and each project that has been announced being taken into account. These programmes are discussed under the direction of the Federal Ministry of Transport within a Federal State working group and therefore negotiated on a practical basis. It is to be stressed that continuity in the provision of funds from the Federal Government and the Free State of Bavaria has never been seriously threatened since the beginning of the planning process in 1966. Neither at the Federal Government and State level nor at local government level has serious difficulties arisen concerning the subsidy of DM 230 million per annum, despite the shift in political power and economic difficulties. This has been to the advantage of planning provisions and building trade activity. This continuity can be assumed for the future although circumsances may temporarily require special short-term funding arrangements to guarantee continued works on the U-Bahn system.

Recent Funding Initiatives to Ensure Scheduled Development of the U-Bahn

As petrol has been consumed more sparingly, Federal resources from the oil tax have not been sufficient since 1980 to subsidise the underground construction works to the full 60 per cent of eligible costs, as had been the case until then. But to be able to go on with the works underway according to schedule, the City of Munich has decided to engage in credit funding operations. These outside funds, amounting to about DM 50 million a year, are to be paid off as soon as possible from future Federal financial aid. However, a commitment by the Federal Government for this purpose has not yet been given. It has simply been guaranteed that the pre-financed projects remain eligible for grants. In such situations formal authorisation is required in each case for the so-called premature start of works. The interest on the temporary funding of the Federal resources that are lacking must be paid by the City of Munich. Without the additional efforts by the city, the yearly volume of works, based on reduced Federal aid, would go down to about DM 100 million per annum, since the supplementary funds put up by the other partners would have to be withdrawn according to the automatic provisions in the funding scheme. The results would be a stop to allocations that would last for years and a slowdown in current operations. The disadvantages from the viewpoint of transport, energy, environment and employment policy of such a check on the development of Munich's underground has prevailed on Munich's City Council and the funding partners in the Munich Tunnel Company to assume the additional costs of funding the credits as long as this burden appears compatible with Munich's other commitments.

III. THE INTEGRATION AND RATIONALISATION OF PUBLIC TRANSPORT

Up to 1972 public transport in and around Munich was provided by a variety of public and private undertakings each functioning independently of one another:

- The City of Munich's Municipal Transport Division operated dense tram and urban bus networks;
- The Federal Railways operated extensive commuter rail and regional bus networks;
- The Federal Post Office operated a regional bus network; and
- About forty small private companies operated rural bus lines throughout the region.

Each of these operators had their own networks, timetables and fare system. This arrangement was hardly adequate to the task of providing effective local public transport in the Munich region and by the early 1960's the private car was steadily capturing a growing share of the daily travel market. This growth in private car traffic not only undermined public transport operators finances but brought associated problems of traffic congestion, environmental pollution, and urban sprawl. It was in response to this increasingly unsatisfactory situation that the 1965 General Transport Plan called for the creation of a regional transport authority to organise an unified and efficient public transport system offering an attractive alternative to the private car.

The Munich Transport and Fares Authority (MVV)

In 1971, the Federal Republic of Germany, the Free State of Bavaria, the City of Munich and the Federal Railways signed an agreement to create a regional transport authority. The Munich Transport and Fares Authority – or "MVV" (Müchner Verkehrs und Tarifverbund GmbH) – is legally constituted as a limited liability company. It is jointly owned and managed by the City of Munich and the Federal Railways (each has a 50 per cent share in the intial capital). The two partners, as public transport operators, are responsible for assuring the supply of public transport. They own their respective transport facilities and rolling stock, conduct business affairs, and bear their costs. The MVV, for its part is responsible for organising the supply of public transport in accordance with demand, planning this supply economically, and providing the City of Munich's Municipal Transport Division and the Federal Railways with the necessary operating instructions.

The authority is governed by three executive bodies; the Board of Directors; the Administrative Board, and the Managerial Board. The Administrative Board comprises two representatives each from the City of Munich (the Lord Mayor and the Director of the Underground Railway Department), the Free State of Bavaria (Bavarian Ministries of Transport and of Finance), the Federal Republic of Germany (Federal Ministries of Transport and of Finance) and the Federal Railways. The Managerial Board has two equally powerful directors, one of them appointed by the City of Munich and the other by the Federal Railways.

The MVV, with a small staff of 85 persons, works on the principle that it should only carry out those tasks which the operating companies cannot perform better or more quickly themselves. According to its articles of incorporation, therefore, the MVV has responsibilities in the following areas:

- General transport planning (including service integration and rationalisation) and on-going development of the unified system (including new lines, stations and park-and-ride facilities);
- The directors of transport service provision, including the coordination of timetables and dispatching systems for all the different operators, with constant rescheduling in response to fluctuating demand;
- Commercial and marketing strategy, including the establishment of uniform fares and the organisation of ticket sales;
- Business management, including the estimation of required levels of investment and profitability, budgeting and accounting, and the distribution of fares receipts and subsidies between operators in return for transport services rendered.

The MVV also carries out basic research, traffic surveys, and all press relations and publicity work for the operating companies. The latter, however, are responsible for the planning and implementation of construction works, maintenance and repair, operations, inspection, and legal and insurance matters on their respective networks.

The Reorganisation of Tram and Bus Networks

Since the inauguration of its unified transport and common fares system on the 28th May 1972, the MVV has allotted the larger part of the traffic to the regional rapid rail system (S-Bahn/U-Bahn). One of the MVV's basic planning principles, as set forth in its articles of incorporation, is the avoidance of parallel service and duplication. No two points linked by the S-Bahn or the U-Bahn should also be linked by a parallel tram or bus route. Thus, as the regional rapid railway has been developed and extended, tramway and bus lines have been restructured to meet the new situation. Within the City of Munich both trams and buses now mainly fulfill a complementary role by providing feeder services to the suburban rapid railway and city underground stations. The overall improvement in public transport services means that passengers are not disadvantaged by this restructuring process.

The Rationalisation of the Municipal Tramway System

The 1965 General Transport Plan called for increases in the efficiency of Munich's tramway system. One-man-operation was adopted and special programmes designed to increase commercial speeds and service reliability were implemented. These included physically segregating extensive lengths of tramline on the inner network from other road traffic, temporarily reserving traffic lanes for exclusive tramway use by means of overhead reversible traffic lights, and installing special traffic signals at junctions to accord priority to trams. By 1975 just over half of the inner tramway network had been protected in this manner enabling average speeds to be raised from about 15 km/hour to 20 km/hour. Further protection measures became increasingly difficult to implement without making conditions intolerable for motorists. The temporary reservation of traffic lanes and priority at junctions were eventually abandoned as it proved more effective to limit private traffic in general, especially on roads with a high proportion of tram and bus traffic. To begin with, these improvement programmes were accompanied by the continuing growth of tramway ridership. But in 1978 tramway ridership peaked and has been falling ever since. Present ridership on work days amounts to about 380 000 passengers.

The opening of the S-Bahn and, particularly, the U-Bahn systems had a marked effect on tramway ridership. The decrease in traffic on tram lines running parallel to the underground was so heavy that the economic basis for their continued operation was removed. Thus over the past ten years the municipal tram network has been reduced in length from 112 to 84 km. Tramway operation will be further reduced as new sections of the U-Bahn are opened in order to avoid the duplication of routes and services. A discussion is at present in progress in Munich on whether the tramway should be kept or not in those parts of the city that cannot be served by the suburban rapid railway or underground systems in the future. The energy requirements per passenger is an important factor and new technologies are being carefully observed. Another aspect which may be of importance in the final decision concerns the environmental consequences of scrapping the tramway system. The results of a recent evaluation study indicate that the replacement of the trams by conventional buses could increase the emissions of air pollutants by public road passenger transport modes by about 10 per cent (Table 8). However, the emissions from an all-bus system would still remain a tiny proportion of the total volume of air pollutants expected to be emitted by road traffic around the years 1990 and 2000 (i.e. about 0.6 per cent).

The present procedure of gradual network reduction in step with development of the U-Bahn cannot go on indefinitely. It has been estimated that a minimal network length of about 55-58 km is necessary for viable tramway operation, i.e. within a cost coverage range comparable to that of the city's other public transport modes. Moreover, a decision will have to be taken in 1986 on whether or not to renew the tramway systems increasingly obsolecent rolling stock. The public's nostalgic attachment to the tramway and local shopkeepers' fears concerning possible losses of custom (tram stops are situated

Table 8. ESTIMATED YEARLY EMISSIONS OF POLLUTANTS (MINIMAL VALUES) BY SURFACE PASSENGER TRANSPORT IN THE CITY OF MUNICH, 1995-2000

tons

	Private cars (around 1995)	Mixed bus and tram (around 2000)	Bus only (around 2000)
Carbon monoxide	71 535.8	237.0	262.1
Hydrocarbons	6 925.5	28.4	31.4
Nitrogen oxides	3 296.9	110.6	122.3
Sulfur oxides	178.3	39.5	43.7
Lead deposits	31.1	14.0	15.5

Source: MVV.

every 500 metres along routes whereas underground stations are situated about every 800 metres) will undoubtedly weigh in the balance for or against its complete closure. The City Council will come to a final decision following the completion of the necessary studies and a comprehensive public debate on the recently published "Public Transport Policy for the Year 2000".

The Reorganisation of Municipal and Regional Bus Networks

The municipal bus system also benefitted from priority measures, such as the reservation of traffic lanes for buses, called for in the 1965 General Transport Plan. The city bus system has also been reorganised and expanded (Table 7). Buses usually serve all sectors situated between the suburban rapid rail and underground lines, providing feeder services to S-Bahn and U-Bahn stations inside the city. Several lines also connect lower-density areas on the outskirts to the centre. Approximately one-third of the fleet belongs to private transport companies operating city services under franchise to the Municipal Transport Division.

The regional bus networks have also undergone considerable reorganisation and expansion since the MVV was established. Initially, only the regional bus networks operated by public undertakings – the Federal Railways and the Federal Post Office – could be incorporated into MVV's unified transport and common fares system. These two networks together accounted for about 60 per cent of the regional bus routes. In 1977 these networks were taken over by the Upper Bavarian Transport Company, a private undertaking jointly owned by the Federal Railways and the Federal Post Office.

The MVV considered that it could only completely fulfill its function of providing effective and attractive public transport in the region if the remaining rural bus lines were integrated into its system of unified transport and common fares. In 1973 it developed an overall plan to improve the rural bus services by:

– Improving connections throughout the entire region;
– Making county seats more accessible to the areas around them; and
– Establishing better connections between the State Capital of Munich and the surrounding countryside, above all by linking the rural bus lines to the S-Bahn network.

The Free State of Bavaria is responsible for controlling public transport in the eight counties surrounding the City of Munich, but any changes in service provision requires the cooperation of all of the county and district authorities concerned. At the present time over 260 separate municipalities and rural districts lie within the MVV service area, as defined by the terminals of the suburban rapid railway network. The first of the MVV's planned regional bus networks was put into operation in Munich County to the south of the State Capital on 28th May 1978. Other counties followed suit and by 1982 seven of them were incorporated in the unified transport andcommon fares system. An attempt has since been made in one county to include all school bus transport, which is financed by the state, in the regular public transport network and thereby make such transport services available to everybody.

The integration of all bus routes within the MVV's service area made it possible for the counties, with the active support of the Free State of Bavaria, to participate in the financing and organisation of bus services and thereby foster regional planning and structural development through transport policy. The county authorities now determine the services to be offered and establish the timetables, routes and bus stops in cooperation with the MVV. The reorganisation of the regional bus networks greatly facilitates access to the suburban rapid railway. Those people who do not live directly adjacent to S-Bahn lines now have access to an integrated transport network, a coordinated service timetable and a common fare system. Despite the takeover of planning, scheduling and pricing activities by the counties and MVV, it was possible to maintain the commercial and legal independence of the Upper Bavarian Transport Company and 42 much smaller private transport operators.

The MVV'S Common Fare System

The common fares system developed by the MVV applies to the entire service area and to all of the local public transport modes operated within it. To facilitate fare payment the MVV divided its service area into concentric fare zones radiating out from the centre of Munich. The two innermost zones – the central zone and the ring zone – cover the extended city area. The fare price consists of a basic charge for two zones, plus a fixed amount for each additional zone. Tickets are sold individually for single trips and in multiple ticket strips for several trips. Every strip ticket represents the fare for one zone. These tickets may be purchased from automatic vending machines in S-Bahn or U-Bahn stations and at bus and tram stops, or from tram and bus drivers. Lower fares are accorded for "short trips" not exceeding two fare stages indicated on schematic route maps. Two 24-hour tickets are also available, primarily for tourists. One of them covers the urban agglomeration while the other is valid for travel throughout the entire MVV service area. In 1981 the MVV also introduced two family travel passes, one for average sized families and the other for large families, to permit more convenient travel by parents and children.

Weekly, monthly and yearly travel passes are also available for regular travellers. Fares also depend on the number of zones crossed but for economic reasons a more complex zoning system consisting of a multitude of small zones is used. The bearer must buy a special stamp corresponding to the value of the pass and affix it to his pass. These stamps can be bought at 350 sales points, most of which are located in privately owned shops. The passes are valid all day long every day of the calendar week, month or year. All of MVV's tickets are available at two prices: full price for adults and reduced price for children between the ages of 5 and 15.

As soon as the MVV was established in 1972, all ordinary ticket operations for the entire unified transport system were mechanised and the sale of stamps for travel passes was franchised to private agencies working on a commission basis. The MVV's common fares system goes hand-in-hand with the self-service "honour system" of ticket cancelling. Once purchased MVV tickets must be cancelled by the passengers themselves before they are valid for travel. Automatic cancelling machines are located at station entrances or on platforms in the S-Bahn and U-Bahn systems and inside trams and buses. Ticket inspection is carried out on a random basis in stations and on-board vehicles.

The mechanisation and privatisation of ticket sales, and the "honour system" of ticket cancelling offered substantial economic advantages. Their introduction allowed total ticket sales and inspection staff employed in the different transport undertakings to be reduced by about one-third. The potential for further savings from these measures has now been exhausted and present economy efforts are concentrated on further reductions of inspection staff through the introduction of television surveillance systems.

The Costs of the Unified Transport and Common Fares System

Munich's Unified Transport and Fares Authority is required, under its articles of incorporation, to ensure that its services are provided at least cost. In 1982 the total operating costs of MVV public transport services amounted to about DM 800 million. The revenue from its passenger transport services consists of fares collected "on the network" and various compensatory payments for services rendered. These include contributions by the Federal, State, and county governments and certain third parties for the carriage of the seriously disabled, and compensatory payments from the Bavarian Government for the free transport of schoolchildren in accordance with Bavarian legislation. When these financial contributions were taken into account, total revenues amounted to DM 410 million in 1982, leaving a deficit of DM 350 million. This deficit works out at DM 0.66 per passenger carried.

The costs of operating public transport services must be met by the various operating companies participating in the unified transport and common fares system. The MVV distributes all ticket receipts to the transport operators according to a prearranged formula based on the number of kilometres of service provided. The deficit is therefore met contractually by the Federal Government and the City of Munich. Since 1978, seven of the eight counties surrounding Munich have taken over the financial responsibility for regional bus transport, with the support of the Free State of Bavaria, for specified proportions of the costs. They meet the operating costs by refunding appropriate sums to private bus companies for services rendered under the contractual agreements and receive an equivalent proportion of the fare revenue from these bus routes from MVV's common fares system.

Table 9. MVV'S OPERATING COSTS, REVENUES AND DEFICITS, 1972-1982
DM millions

	1972[1]	1973	1974	1975	1979	1980	1981	1982[2]
Costs	278	493	559	578	658	705	757	801
Farebox revenue	129	213	236	246	317	395	412	413
Compensation	–	5	6	5	37	36	39	40
Deficit	149	275	317	327	305	274	306	348
Rate of cost coverage[3]	46.4	43.6	42.7	42.93	51.0	59.2	57.4	54.3

1. For the period 28th May to 31st December only.
2. Estimates.
3. Calculated as follows: $\dfrac{\text{Farebox revenue}}{(\text{costs} - \text{compensation})} \times 100$.

Source: MVV.

The average rate of cost coverage for all MVV transport services has steadily increased from around 44 per cent in 1973 to about 54 per cent in 1982 (Table 9). This has been achieved through a consistent policy of adapting services to meet demand, through continuing efforts by each operator to reduce its personnel, and the steadily increasing number of passengers using the system. The present rates of cost coverage are accepted by all political committees concerned, since the promotion of public transport has always been considered as a social service and not directly comparable with economic activities that cover their costs. The popularity and acceptance of the MVV's unified transport and common fares system are due, among other things, to the positive effects of this basic concept.

Since the establishment of MVV, a fares policy has been maintained that benefits those inhabitants passengers living in the peripheral areas of the region. This has resulted, of course, in heavy settlement of the rural districts in the neighbouring counties, which can only be checked with difficulty through a restrictive land use and settlement policy implemented by state and regional bodies. In the meantime land prices have risen accordingly, so that the financial incentive for living in outer areas is being reduced. But the MVV's operating deficit is greatly marked by the losses made on services provided in the outer region. Since 1972 there have only been three price increases in fares (May 1974, January 1976 and January 1980). The pricing system recently reviewed in 1983 and a new fare increase of 8.3 per cent is due to be introduced early in 1984. As from 1984 fares policy will have to be directed towards a more balanced financial situation, which will have effects on the population trends and the modal split in the entire region.

IV. ROAD DEVELOPMENT POLICY

The road development policy advocated in the 1965 General Transport Plan entailed a major investment programme designed to "harmonize" and improve the city's main road network. Key features of the planned network were the construction of three high-capacity orbital routes permitting through traffic to by-pass the city and the improvement of connections with the Federal and state road networks. The completion of this ambitious road building programme was effectively prevented by widespread public criticism of the tremendous costs and damaging effects of road construction works in the city centre. In response to this reaction the Lord Mayor of Munich called for a radical change in road development policy. In the revised General Transport Plan of 1975 the objective of increasing capacity to match traffic demands was firmly rejected and the budget for new road construction was reduced by about 80 per cent. These drastic cuts mainly affected planned outlays on new expressways inside and on the fringes of the inner city.

The Shift in Road Development Policy

The City Council decided that no new major roads or intersections would be built in the area demarcated by the city underground's service area. Some aspects of the former strategy to harmonize the existing road network were to be newly evaluated and financed on a small scale only, mainly along the outskirts of the U-Bahn service area. Harmonization meant modest and locally fixed projects to complete missing parts of existing roads and streets and to increase traffic flow at some intersections and streets without producing structural damage, especially in residential areas. There was to be no hidden promotion of previous large-scale network programmes now officially abandoned. Beyond the U-Bahn service area and in sectors of the city lacking access to the suburban rapid rail (S-Bahn) system, the deficient parts of the road network were to be developed on a larger scale, especially along certain tangential arteries. Finally, the revised General Transport Plan also provided for measures to protect the residents of existing buildings situated along busy main roads from traffic noise disturbance and pollution. This meant insulating windows on the exposed façades of buildings and, as far as possible, making appropriate changes to the interior layout of apartments. Thus the development of new main roads could now depend, among other things, on the feasibility of protecting adjacent populations from noise disturbances along the proposed route. Since 1975 the city's road development programme has been characterised by four types of action:

i) The abandonment of major new road construction in the inner city area, and even a reduction in the capacity of certain completed sections;

ii) Continued but mainly gradient-free development of the middle ring-road;

iii) Active pursuit of planning further works on the outer ring-road; and

iv) The extension of traffic noise abatement actions to protect populations living and working along main roads.

The Development of Major By-passes

Work was first begun on constructing the inner ring-road around the "old town", about 0.6 kilometres from the city centre and on the middle ring-road, about three to four kilometres from the city centre. Additionally, in order to free the urban road network of long-distance through traffic, work was begun on a stretch of the outer ring-road in the east, about 15 to 16 kms from the city centre, in order to link up two federal motorways converging on Munich from Nuremberg and Salzburg.

The Inner Ring-Road

The construction of an eight-lane expressway around the old town to divert through traffic away from the city centre and to distribute local traffic in the inner city was planned in stages. First, a 1.2 km section forming a by-pass on the north side of the old town; then a 0.6 km section on the south side to complete the loop around the old town. The intersections on the ring-road were to be made large enough to take even more traffic on a second level at a later, final stage. No adverse impacts occurred during the development of the western portion of the northern by-pass as a wide arterial already existed there. The northeastern portion of the by-pass, however, was newly created, in part by demolishing residences. This led to considerable damage to the social structure of the neighbourhoods through which the road passed due to separating effects.

The northern section of the ring-road was completed at the end of the 1960's. Its resemblance to an urban motorway and its damaging effects in splitting up established residential districts was subsequently criticised. Also, the volumes of traffic actually using the new road fell significantly short of its design capacity. On the basis of this experience and, above all, as a result of the public's change in attitude to major new road projects in Munich, the planned construction of the southern section and the creation of a second level were dropped from the revised General Transport Plan in 1975. Soon afterwards, in 1978, it was decided to reduce the width of the completed northeast section from eight to an average of six lanes by converting the outer lanes to on-street parking (with vehicles parked end-on to the kerbside). This operation cost almost DM 5 million.

The Middle Ring-Road

Following the abandonment of further development on the inner ring-road and the slow progress with the outer ring-road after 1975, the middle ring-road became the main route for relieving the city centre of undesirable through traffic. As it will not be able to effectively perform this function until it is completed, road construction investment funds have been mostly used for gradient free development of the middle ring-road and for improving its connections with major radial routes. The path of the middle ring-road intersects with the U-Bahn network at two points on the fringes of the inner city. The road construction works at these sites were therefore coordinated with the development of the U-Bahn and profited from the fact that there is no local opposition to the extension of the city's underground system. On other sections, however, it was necessary to intensify accompanying measures to reduce the ring-roads impact on the environment, notably due to traffic noise. These measures illustrate the increasing importance attached to environmental considerations in road construction and improvements in Munich.

The Outer Ring-Road

The outer ring-road is essentially a "motorway box" designed to link-up, in a first phase, two federal motorways converging on Munich from Nuremberg and Salzburg. Its main purpose is to alleviate traffic congestion and undesirable environmental impacts in Munich by concentrating holiday traffic, which can hardly be controlled owing to its extreme peak volumes, and heavy lorry through-traffic on to a route completely by-passing the city. The completion of the eastern section of the outer ring-road in 1975 coincided with the wave of public opposition to new road building. Since then a firm stand has been taken against completing the southern section of the ring-road in order to protect the local tourist area there. However, the residential areas in the northern and western sectors of the city are still exposed to undiminished and unacceptable levels of congestion and pollution as there is no junction between the ring-road and the motorways arriving there. The city government considers the construction of the northern and western sections of the outer ring-road as being extremely urgent. With an efficient and attractive works programme, the existing eastern and new northern and western sections should keep existing pressure on the City of Munich from through traffic to a minimum by acting like a protective bell. A majority of the population and the environmentalists basically accept these plans. However, quite different views arise on questions of detail. For example, as in the case of a small area of woodland with special biotope characteristics which is affected by the works on the north-western section. Thus this project, which is being sponsored by the Federal Government and not by the City of Munich, will still take years to implement.

Noise Abatement Measures along Main Roads in Munich

The public's sensitivity to and resentment of noise disturbance has greatly increased over the past decade, Road traffic is undoubtedly one of the major sources of noise disturbance in Munich. The highest outdoor noise levels are to be measured along Munich's main road network. It is 493 kms long and represents abut 23 per cent of the total road network length of 2 184 kms. About 110 kms of main road lie within the densely built-up area demarcated by the middle ring-road. Noise levels of between 71 to 75 dB(A) along the major part of the main roads here (Table 10).

In addition to planning controls on the design and location of new buildings, the City authorities have undertaken a number of measures to protect people living along busy roads from traffic noise nuisance. These include resurfacing cobbled streets, constructing noise barriers, insulating windows on exposed building façades and tunnelling new sections of the middle ring-road.

Resurfacing Cobbled Streets

Munich still has extensive stretches of cobbled streets and for years the City authorities have been implementing a programme to resurface them with asphalt in order to reduce traffic noise. These operations are carried out in response to demands from the public. The authorities note the petitions by district commitees, citizen groups and similar bodies and place them in an appropriate position on a priority action list. Existing traffic volumes, housing density and the presence of local residents especially sensitive to noise are all taken into account in the appraisal of needs. So far 15 of the resurfacing operations booked for in 1981 have been implemented, mainly in areas on the city outskirts. A further 65 operations are to be carried out in 1983. The City of Munich finances the greater part of these operations but subsidies covering about 25 per cent of eligible costs and a low-interest loan covering 41 and $^2/_3$ per cent of the eligible costs are available as grants under the Bavarian Loan Programme. Road resurfacing reduces traffic noise but can increase accident risks by speeding up traffic. It also results in higher road maintenance costs.

Construction of Noise Barriers Along Main Roads

The construction of noise barriers (walls and embankments) as part of road building projects has been carried out for about the past ten years. The rebuilding of a 4 km long road (Ingolstadter Strasse) was among the first projects to incorporate the construction of noise barriers to protect local residents. These protection measures are comparatively costly. About DM 6 million of the DM 35 million being spent on the development of another 2 km long main road (Von-Kahr Strasse) are required for noise protection barriers and related measures, while the replacement of an existing noise barrier built along the middle ring-road in 1972 (Candid-Hangauffahrt) cost DM 1.65 million. In cases of road improvement where buildings already exist, the City meets the costs of purchasing the property required and erecting the noise barriers when these are technically justified. However, in the case of new building development projects the investors must meet the costs of noise protection measures. In future, whenever it is technically feasible and can be required for reasons of urban development, garages will be built into the noise-prevention embankments, for example along green spaces on parts of the road. However, funds will have to be provided by the road work sponsors on an even greater scale to enable new connecting roads to be built. Apart from significantly reducing noise levels, these measures are also designed to achieve structural improvements and increase traffic safety along the road concerned.

Table 10. ESTIMATED PROPORTIONS OF THE MAIN ROAD NETWORK EXPOSED TO DIFFERENT AVERAGE DAILY NOISE LEVELS

	Average daily noise levels in dB(A)					
	62 or less	63-65	65-70	71-75	76 or more	Total
Within and including the middle ring road	–	–	16	83	1	100
For the city as a whole	1	5	36	51	7	100

Source: 1983 City Development Plan.

Insulating Building Façades Against Road Traffic Noise

Since 1975 the City authorities have been conducting annual programmes in support of noise-insulating glazing on particularly busy roads. The choice of eligible premises is partly made by means of traffic noise measurements, but today very largely on the basis of estimates. The following noise limits, outdoor noise levels measured a one metre distance from exposed façades, were set for the inclusion of premises in such programmes:

1975:	80db(A)
1976:	79db(A)
1977:	76db(A)
since 1978:	75db(A)

The decisive factor is the noise disturbance value which is derived from the traffic volume to be expected in the long term, the distance between the building and the edge of the road and the combination of traffic (private cars, lorries and trams).

The costs of these programmes are shared equally by the City of Munich and by the Bavarian State Ministry for Regional Development and the Environment. Subsidies totalling DM 15.5 million were granted in the years 1978 to 1982. Aid was thereby provided for the installation of 17 572 noise-insulating windows and doors on 609 sites with 5 298 flats. With an average of 2.36 occupants per flat in the urban area, it was thus possible to make life in their flats bearable again for about 6 375 people in 1978 and 1979 alone, including the residents of three homes for elderly people.

Programmes for noise-insulating windows are greatly appreciated by the population involved. This is illustrated by the general insistence on being included in them. The programmes which were carried out for the first time by the state capital of Munich in conjunction with the Free State of Bavaria in Federal territory are now being implemented in a series of other towns in and outside Bavaria.

These measures are voluntary and have no legal basis, since the Noise Protection Act adopted by the West German Federal Parliament on 6th March 1980 could not be implemented following an objection by the Federal Council and an unsuccessful attempt at mediation. Consequently, at present there are no legally binding maximum values for which noise-insulation precautions would have to be taken by the road works sponsors when such values are exceeded. Pending the adoption of national regulations since mid-1983, the Federal Minister of Transport has therefore established guidelines for protection against noise on highways for which the Federal Government is responsible and has requested the states as the competent authority to use them as approved technical rules. As an interim measure for state, county and local roads, an order was issued through the Bavarian State Ministry of the Interior on 20th December 1982 on the procedure to be adopted until further notice regarding road construction projects.

The inclusion of adequate noise-abatement provisions in road development schemes is an increasingly important pre-condition for project approval, but the necessary measures can significantly increase construction costs. For example, the noise protection provisions for a big road project in Northwest Munich were not deemed sufficient by the appropriate administrative court. The route and the planned number of lanes were confirmed, but the inclusion of all noise protection provisions in the building plan was demanded, as were a reduction in the road width by eliminating the provision for an avenue, a reduction in parking space and reduced widths for median strips, cycleways and footpaths. The costs for the anticipated passive noise protection will thus amount to about 25 per cent of the construction costs.

Tunnelling New Sections of Main Road to Reduce Local Noise Disturbance

The eastern section of the middle ring-road was completed first. Following the completion of the eastern section of the motorway box in 1975, which brought some relief on the completed eastern section of the middle ring-road, work was then begun on further sections to the south and west. The path of the middle ring-road runs through densely populated parts of urban redevelopment areas at a number of sites on the edge of the inner city, particularly along the western edge. It was feared that grade-free construction here would lead to further structural decline because of the resulting isolation of certain blocks of buildings. There was also the problem of increased noise disturbance for residents living along the ring-road. The elimination of housing along the ring-road and its replacement with less noise-sensitive industrial establishments was not a feasible solution since there was no corresponding demand from industry on the scale required and established residential areas would have been seriously damaged by the necessary demolition works. The only feasible solution remaining was to put the new sections of the middle ring-road underground.

The first underground section in the northwest (the Landshuter Allee) is already completed and further tunnelling of long sections of the middle ring-road is underway on sections in the west (the Trappentreustrasse) and south (the Brudermuhl Tunnel). The tunnel sections in the west and south replace either planned or temporary bridges and are considerably more expensive. In certain cases construction costs are five times greater than the costs of demolishing houses and building new accomodation elsewhere. The Trappentreustrasse project costs over DM 500 million and will be covered by funds provided under the Local Transport Funding Act. The Brudermuhl project costs DM 166 million, half of which will be met by the City of Munich. Taken together, these two projects tie up almost all of the state's subsidy resources for a decade. The construction works impose temporary but non-negligible inconvenience on local residents. However, these much more expensive tunnel solutions are accepted by the local populations who consider that these expenditures result in an improvement in the general environmental situation and are not merely intended to speed up traffic.

The Financing of Road Development and Improvement

Road transport legislation in the Federal Republic of Germany defines various classes of road: federal, state, county and municipal connecting and local roads. This classification implicitly indicates the highway authorities responsible for financing the construction and maintenance of particular roads. In towns of more than 80 000 inhabitants, the municipality is usually responsible for actually building every class of road within its boundaries but without necessarily having to pay for the works on non-municipal roads. The road network in Munich is composed of various classes of road ranging from local access roads to motorways. In practice both federal and state road building programmes in the city are mainly carried out by the municipality using the funds provided by these higher levels of government. Regional road funding facilities do not exist. As a rule, roads in Munich are financed out of general budgetary resources, credits, allocations from the administration's budget, general reserves, grants and so on. Various possibilities exist for obtaining aid for purely municipal road projects but the amounts accorded and maximum limits on funding of individual projects are subject to increasing constraints.

Possibilities of Aid

As for local public transport investment projects, the Local Transport Funding Act provides for federal support of approved local road investment projects. According to the current rule, funds-in-aid may be provided for a maximum of 60 per cent of the eligible costs. However, this aid is difficult to obtain for projects costing over DM 5 million and the total annual grant is limited to about DM 20 million. Owing to the current cost-intensive operations underway in Munich, this means that the city's possibility of aid for new road building projects is already blocked for many years to come (up until the early 1990's). Thus money for new road building projects must be found elsewhere. In the case of projects with eligible costs of over DM 12.9 million, a further possibility of Federal Government aid exists under the Financial Redistribution Act. It is paid out of funds from the Motor Vehicle Tax (hardship fund) and generally amounts to 10 per cent of eligible costs. According to the Railway Crossings Act, a municipality as a rule has to meet only a third of the project costs (including the purchase of land) for eliminating level crossings. The remaining two-thirds are shared between the Federal Government and the Federal Railways. But the possibility of a subsidy for the municipality's share exists under either Article 71 of this Act or the Local Transport Financing Act. According to the provisions of the Federal Construction Works Act, financial contributions can be demanded from residents in new housing estates to build roads for purely access purposes. The municipality's share, as specified in this Act, must cover at least 10 per cent of the eligible costs. As a rule the City of Munich prefinances the contributions to be collected from residents out of the city budget. The Free State of Bavaria accords a grant to the City of Munich paid out of the funds collected by means of the state motor vehicle tax which currently amounts to about DM 38 million. This aid is not earmarked for new construction works but can also be used for road maintenance, the provision of equipment, winter operations and so on. Finally, limited financial aid is also available for the provision of cycleways and footpaths under the Free State of Bavaria's Leisure and Recreation Programme.

Table 11. EXPENDITURES ON ROADS IN THE CITY OF MUNICH, 1975-1982

DM million

	1975	1982
Development costs	54.6	140.5
Operating costs	100.7	154.6
Total expenditure	155.3	295.1[1]

1. Total is made up of funds from the following sources:
   Federal Government = DM 3.6 million.
   State Government = DM 78.4 million.
   City Government = DM 200.1 million.
   Contributions made by residents under the Federal
   Construction Works Act = DM 13.0 million.

Source: Local Experts.

Since 1975, road development and improvement in the city have become more dependent on aid from Federal and State governments (Table 11). The road investment programme outlined in the new 1983 Urban Development Plan envisages expenditure totalling DM 520 million spread over the five-year period from 1983 to 1987 inclusive. However, the annual expenditures are planned to decrease by almost half from DM 155 million in 1983 to DM 81 million in 1987.

The Development of Munich's Cycleway Network

Traditionally bicycle use in Munich was associated almost entirely with leisure activities. In recent years, however, its use for work and shopping trips has greatly increased and it is now frequently used as a means of travelling to and from suburban rapid rail stations. Unfortunately this increase in cycling is also associated with a big increase in the number of accidents involving cyclists. Munich's transport authorities have therefore taken a number of measures to increase the safety of cyclists and facilitate bicycle use in and around Munich. In particular, the City of Munich has endeavoured to extend the existing cycleway network as rapidly as possible. As a result, the cycleway network was extended by 150 km (from 530 to 680 km) over the past four years at a cost of DM 22.8 million. It is planned to double the length of the cycleway network in the coming years. Most of the cycleways are funded out of the resources for general road construction programmes. For example in 1982, these resources amounted to almost five times an annual package cycleway grant of DM 1.5 million accorded by the Free State of Bavaria. In general, new cycleways are built in connection with general road construction programmes such as new roads, section conversions, and renovation works following the construction of underground railways and so on. In future, cycleways will be marked on pavements/sidewalks only in exceptional circumstances.

V. TRAFFIC MANAGEMENT POLICY

Pedestrianisation in the City Centre

The General Urban Policy Plan of 1963 provided for the creation of an extensive pedestrian area in Munich's "old town". The first planning approach formulated in 1966 stated two preconditions for the success of this scheme:

i) A highly attractive system of combined suburban rapid rail and underground services had to guarantee easy access to the city centre; and
ii) The old town had to be protected from non-essential traffic by the construction of an inner ring-road and a system of traffic cells and supporting restrictions.

The implementation of the pedestrianisation scheme was thus highly dependant on the development of the new regional rapid trailway (S-Bahn/U-Bahn) system and the city's main road development programme.

The construction works on the S-Bahn tunnel and stations underneath the city centre provided a perfect opportunity for creating the first big pedestrian zone. The planned S-Bahn system was universally accepted by the public and although the tunnelling took many years and disrupted traffic on several major roads, the parallel development of the pedestrianisation scheme encountered no marked opposition from local shopkeepers or motorists. Most drivers accepted all the traffic diversions occasioned by the work and by the time the tunnel was completed they had become used to the new traffic situation. The initial pedestrian precinct covered an area of about 50 000 metres^2. It was constructed in 1971 and opened to the public in 1972. The first phase of development cost about DM 12 million and was mostly financed by the city government itself, as it was not certain at the outset of the planning phase if the scheme would be a success. As it turned out, the scheme was highly successful. It benefitted local shopkeepers and was very popular with shoppers and tourists.

The next development phase (Theatingerstrasse) which had been planned from the outset, only required the modification of the ground layout and traffic regulations. However, strong opposition had to be overcome from a few shop-owners, who feared that the pedestrian precincts would have a negative effect on their regular customers, before it could be implemented. The City Council therefore decided not to start the works until half of the project costs could be covered by collecting financial contributions from every house-owner and shop-owner in this prestigious shopping area. Only then would the City meet the other half. This "voluntary" collection of funds succeeded through the efforts of a respected representative of the local retail business community who over a period of three years, went from door-to-door collecting and even gave energetic reminders to absent foreign property owners. Only then in 1976 were the streets and squares renovated. This second phase cost DM 2.5 million and brought the total pedestrianised area up to 85 000 metres^2. This procedure, which can only be successful by bringing in a mediator from the business community, has since been used in many other towns. The fears expressed by the opponents of the second phase have not been confirmed.

Gradual Traffic Restraint Following the Amendment of the 1971 Federal Highway Act

For years traffic planning and management in Munich, and in the Federal Republic of Germany as a whole, has been subject to two sets of legal constraints:

i) The principle of "Anliegerverkehr" or "right-of-way", which safeguards the freedom of residents and business owners to enter or leave their property. German jurisprudence traditionally allows a fairly broad interpretation and application of this legal concept. Thus whenever new traffic control measures are implemented, accessibility to all adjacent premises must be maintained or satisfactory compensation made to affected parties;

ii) The provisions of the 1971 Federal Highway Act, whereby traffic planning and management policy was governed by the principle of assuring the "safety and smoothness" of traffic, with "smoothness" being taken to mean the guarantee of adequate road capacity at all times.

The legal basis for traffic restraint was only recently established following the amendment of the Federal Highway Act. Under this amendment, which came into force on the 1st August 1980, the principle of "safety and smoothness" was replaced by the principle of "safety and organisation". The new legislation no longer obliges road transport authorities to guarantee unlimited road capacity. Instead, it now enables them to limit or ban the use of certain roads in order to protect the residents from vehicle noise and exhaust emissions and, if need be, divert traffic in order to improve public safety or support orderly urban development. The administrative orders required for traffic diversion can now be issued by the same department. Also, in contrast to the previous principle, which only allowed the use of road engineering, the new principle allows combinations of road engineering and traffic regulations to be used. The possibilities for traffic restraint include the modification of the road network or the traffic control system to check through traffic, such as the creation of one-way streets, cul-de-sacs, and "traffic restraint areas". The newly defined traffic restraint areas resemble "play streets" and the "town yards" or "woonerven" first developed by Dutch traffic planners.

Traffic Restraint Areas in Munich

The creation of traffic restraint areas involves redesigning short stretches of road to give motorists the impression that its residential function has precedence over its traffic function. This is best achieved by reconstructing the road's cross-section to remove all physical distinctions between the carriageway and the pavement. The impact of this structural downgrading of the road as a thoroughfare may be reinforced by additional modifications to road entries and on-street parking spaces, and by street markings, street furniture, and traffic signs indicating that special traffic regulations are in force.

In Munich such schemes are usually implemented only when roads are scheduled to be built or improved. The City's Public Works Department generally offers local resident's associations the choice of a conventional operation, with carriageways separated from the pavement by a kerb, or a uniformly paved road cross-section, and then leaves them to decide. Most decisions are only just obtained, often after heated discussions at meetings. This can create strained relations between neighbours who have to go on living together in the same street. Most objections to town yards are either based on the argument that the usual safety standards for pedestrians will be eliminated if the pavements are not clearly separated from the carriageway, or that the risks of disturbance and damage to parked cars will increase if children are allowed to play in the street. Older residents are especially concerned about these aspects and tend to be more negative in their attitudes. Younger residents, particularly parents with young chidren, tend to be more in favour of such schemes.

The implementation procedure requires that traffic using the road must first be reduced to a low level, not more than 60 to 200 vehicles/hour in the daily peak depending on the circumstances, before the reconstruction works may take place. For this reason most schemes have been located in quiet streets such as cul-de-sacs or short streets linking two parallel roads. The maximum authorised speed on restrained road sections is limited to 5 km/hour. As motorists cannot be expected to drive so slowly for more than about one or two minutes, the length of the restrained road sections is also limited and must not exceed 200 metres for cul-de-sacs or 350 metres for link roads. The police would not be able to enforce the speed restrictions otherwise. Once the decision has been taken to create a town yard and the traffic flow criteria have been attained, the road is redesigned and the necessary construction works completed. There is no trial period.

So far about twenty town yards have been completed or are being constructed in established and newly built-up areas on the city outskirts. The Federal Construction Works Act stipulates that up to 90 per cent of the costs of providing access roads in new housing estates must be met by the residents themselves, with the municipality paying the remainder. This offers the possibility of implementing more schemes than if the City of Munich had to fund them in their entirety. The original fear that the costs of constructing town yards and installing associated street furniture would be higher than for conventional road designs has not been confirmed. With economical design the costs are about the same. The city's Public Works Department made allowance for costs of about DM 100-120/metre^2 but the actual costs vary between DM 90-105/metre^2. These figures may be compared with the DM 200-400/metre^2 required for the construction of pedestrian areas in the city centre. Total costs for most schemes implemented, therefore, range between DM 300 000-550 000.

In densely built-up inner city areas, similar schemes to convert stretches of road into town yards or play streets, where parks and playgrounds are lacking, will remain an exception even in the future, as the funds available are very limited. They are almost exclusively used when scheduled renovation works on carriageways following extensive excavation works (for example, on account of construction of new U-Bahn lines or the extension of pipelines for the city's district heating system) create an exploitable opportunity. The creation of town yards usually results in fewer on-street parking spaces being available. In districts where on-street parking is the only solution available to local car owners the district committees generally prefer more, rather than less, parking space and usually oppose such schemes. Somewhat different traffic management plans are therefore being developed for the peripheral areas of the inner city.

Traffic Restraint in an Outer District of the Inner City

An undesirable consequence of the pedestrian areas and traffic restrictions implemented in Munich's historic core was to increase the pressure of through traffic by non-residents in neighbouring districts in the inner city. The entry into service of radial U-Bahn lines within the middle ring-road provided an attractive incentive to such motorists to switch from their cars to rapid rail transport. It also laid the basis for restraining non-local through traffic within this area. Comprehensive traffic management plans to check through traffic are therefore being drawn up and implemented in the peripheral areas of the inner city. In their initial phase these schemes largely rely on traffic reassignment measures such as one-way street systems. Each scheme is tested for a certain period before the final rebuilding works on the road are carried out. The scheme implemented in one area is an inner city district bordering directly on the old town (Gartnerplatz Viertel) offers an interesting example.

Following the opening of radial underground line (U-1/U-8) the City's Department of Public Works decided to restrict non-local through traffic on a main road (Corneliusstrasse) running parallel to the new U-Bahn line during a trial period. The scheme, which included one-way streets and supporting road engineering to correct kerbs and so on, was tested for nine months before the City Council approved the department's recommendation to make it permanent. As a result of the traffic restraint measures it was observed that:

i) About half the original through traffic has been transferred to public transport modes and the other half onto nearby main roads;
ii) The diversion of traffic to neighbouring districts, which had been feared, has not occurred; and
iii) There has been a general increase in pedestrian traffic.

While the desired objective was immediately achieved on the main road network, certain traffic adjustments were still required on the secondary service network to stop unwanted traffic from pushing its way into the area. After these had been made, it was possible, in 1983, to start preparations for modification works on the roads and areas where traffic had been restrained. The provisions now being implemented are intended to reduce the area for moving vehicles by considerably widening pavements (for example, on the Gartnerplatz itself) and by increasing the number of parking spaces on-street.

The reactions of local political representatives and inhabitants to this scheme were very complex. Opinions were divided at local residents' meetings. Majorities were in favour of schemes with one-way loops as a means of keeping out all non-local traffic. Other groups strongly rejected these suggestions owing to the lack or increased difficulty of access to the district. The breakthrough did not come until after five yers of planning work and public discussion, when the Department of Public Works made a recommendation to the City Council whereby through traffic was made considerably more difficult, but accessibility was not reduced to any great extent and almost all available connections were maintained for local public transport. This procedure for traffic restraint, which will be used in other inner city districts or is scheduled in conjunction with the introduction of new U-Bahn lines, shows that street space can be won back for pedestrians along major thoroughfares without making access more difficult for local residents.

VI. PARKING POLICY

The primary aim of parking policy set out in the 1965 General Transport Plan was to protect the inner city, and notably the "old town" from excessive terminating traffic and long-term parkers. The demand for parking in the city centre was to be strictly controlled but in a selective and sensitive manner so as to avoid harming local commerce or imposing undue inconvenience on local residents. Thus, on the one hand, motorists commuting to workplaces in the centre were to be encouraged to park on the outskirts of the city, or even further afield, rather than in the city centre. On the other hand, adequate parking opportunities were to be provided for delivery services and shoppers. A secondary policy aim was to provide adequate parking facilities for residents. Priority was to be given to populations living in densely built-up districts on the periphery of the inner city where off-street parking facilities were scarce or considered too expensive, and where competition from non-residents (commuters, shoppers, tourists, etc.) for on-street spaces was strong. This aim has assumed increasing importance in recent years as residents in these area districts have continued to move out to the suburbs. Both policy aims were to be achieved by concerted action employing a range of complementary measures designed to control the number, distribution and use of parking spaces.

The Control of On-street Parking

Parking Bans and Charges

Parking bans were progressively introduced in Munich. This allowed the number of long-term parking spaces in the city centre to be reduced from 17 000 to about 12 000. At present about 15 per cent of Munich's streets (out of a total of about 6 500 for the city as a whole) are subject to such bans. In most cases parking is prohibited at specified times only. Short-term parking on the remaining on-street spaces has been encouraged by charging time-related parking fees or by establishing blue disc control zones. Although these measures have affectively encouraged short-term parking in the old town, the enforcement of parking controls remains a difficult task in other parts of the inner city. The illegal parking and unloading of vehicles continues not only in controlled or metered spaces but also on sidewalks and cycleways. The police have sufficient personnel to make conditions as tolerable as possible.

Any attempt to substantially reduce such behaviour by Munich motorists would require a stricter enforcement effort in combination with higher parking fees and fines for over-long or illegal parking. Such a course of action is now possible but there seems to be a general reluctance to adopt it because of the lack of parking facilities for residents in densely built-up inner city areas. Indeed, there is some evidence of parking controls being relaxed:

- Up to 1980 the maximum fee that could be charged by local authorities for on-street parking spaces was limited by Federal law to DM 0.1 per one-half hour. This legislation has recently been replaced and communities may now charge up to DM 1 per hour. Little incentive exists in Munich for raising parking fees however because the money thus collected does not go to the municipality but to the Free State of Bavaria;
- The blue disc parking scheme allows motorists to park for a maximum of three hours at a given space in a controlled zone. This period can be adapted to local demand but so far shorter periods have not been introduced in the city centre;
- Since 1981 parking bans have been either completely lifted or shortened in 135 streets and the need for continuing such restrictions is currently being reviewed. The authorities now show the greatest moderation when new parking bans are to be established. This relaxation of parking controls sometimes results in conflicting interests in the peripheral areas of the inner city where the traditional objectives have been to further reduce parking by non-residents and protect the environment. Each case is treated according to the specific local context and the conflicting objectives are weighted accordingly before the final decision is taken.

Parking Licence Schemes

Over the years the restrictions of long-term parking in the inner city has led to a deterioration in parking conditions for residents and local businesses in the older residential neighbourhoods bordering the old town and in similar areas located around the inner city. Many of these districts have become rundown and are in need of rehabilitation. As the original mainly middle-class residents have moved out to the suburbs, less affluent and often immigrant populations have taken their place. The rehabilitation of these neighbourhoods is considered essential for maintaining the vitality of the city, particularly of the central commercial area in the old town. The provision of adequate parking opportunities for residents is considered one means of stemming further population loss and urban decay. For a long time relief measures were delayed because suitable sites for new parking facilities were scarce.

Conditions were especially critical in the Lehel district to the northeast of the old town. This district covers an area of 115 hectares and largely consists of older apartment buildings. The 15 000 esidents with their 4 800 cars must share the 3 455 on-street parking places with about 15 000 commuters and their cars. On the 21st January 1981, after years of preliminary work, the Munich City Council decided to accord special parking rights for residents by introducing a system of parking licences. The aim of this project was to improve living conditions in Lehel and to check the move of the resident population away from this area. The necessary regulations for a twelve-month experiment were introduced on the 1st May 1981. The scheme involved reserving about 1 920 on-street parking spaces for use by holders of parking licences or special permits. During the twelve-month trial period the authorities issued some 3 200 cences to residents and another 1 0E0 special permits to businessmen, self-employed workers, etc. The parking licences and permits take the form of a card which must be displayed on the windshield of vehicles when parked on the reserved spaces. A nominal fee of DM 40 was charged to cover the administrative costs for each licence issued. The experiment proved to be quite successful and it was transformed into a permanent arrangement in 1982. The City authorities are thereby endeavouring to obtain long-term experience with such schemes since the possibilities of extending them to other areas is by no means certain.

Extensive feasibility studies were also initiated in other districts on the edge of the inner city. These included a large redevelopment area, parts of the city near the Beer Festival ("Oktoberfest") site, and two smaller areas. The study findings were negative, however, because of the problems involved in implementing parking licence systems over extensive areas. The areas examined contained numerous commercial activities, it was impossible to demarcate boundaries, and the resultant limitations on general mobility were unacceptable. Such feasibility studies are now directed at individual streets or high-rise buildings which suffer greatly from parking by outsiders, even though they consist almost exclusively of residential facilities.

The Development of Off-street Parking

The Control of Parking Garages in the City Centre

Local building codes still in force in Munich impose high minimum requirements concerning the number of parking places to be provided in new apartment buildings and office blocks. Complying with these provisions places a heavy demand on planning programmes for administrative and office buildings and generates high additional costs for an investor. In Munich for example, each parking space in large underground garages currently costs up to DM 45 000. There are cases, however, in which compliance is not possible for technical reasons and is not desirable because it conflicts with general urban development policy. This is the case in Munich's old town where limitations on parking space are established policy. Consequently, the City Council has for long authorised the administration to waive the obligation to meet the prescribed 100 per cent parking provisions for new buildings there. The number of parking spaces actually constructed depends on the site location and the type of use. In the extreme case, only 10 per cent of the specified number of spaces in the planning provisions has to be met. The remaining 90 per cent has to be discharged. This means that for every parking space avoided the investor has to pay an exemption fee of DM 7 500 into a special municipal fund. The money collected is earmarked for investments in public off-street facilities at sites outside of the restricted area (demarcated by the inner ring-road). Understandably, this possibility is frequently used and the funds collected in this way at present amount to some DM 35 million.

The Construction of Parking Garages Outside the City Centre

The exemption procedure described has enabled the municipality to provide new multi-storey or underground carparks in districts bordering on the city centre as well as more remote park-and-ride facilities at stations along the suburban rapid railway network. Once built, the municipality concedes the operation of these parking facilities to appropriate private undertakings. Motorists only use the parking spaces in the district garages if the rent is kept within reasonable limits. Thus the authorities are sometimes obliged to subsidise these operations. The provision of district car parks is increasingly expensive but the exemption fee has remained stable since its introduction. It is therefore intended to revise the exemption fee in the near future. This is all the more necessary because of the City Council's decision in the autumn of 1982 to provide a further 19 district parking garages. As no general budget funds can be provided at present or in the next few years, this two-stage programme is to be financed out of the parking exemption funds. In the first stage, eight facilities will be financed out of the exemption funds collected up until the end of 1983. In the second phase, the eleven facilities planned will be financed out of the exemption fees accruing from 1983 onwards. According to this programme the municipality will provide the building plot but the view is that responsibility for building and operating the facilities should be transferred, as far as possible, to private undertakings. As none of the facilities planned are designed to be operated at a profit, the city will provide a non-refundable subsidy out of the parking exemption fund.

Where suitable surface sites for multi-storey carparks are lacking they have been built below ground, usually under squares or parks. This is often advantageous for traffic and environmental reasons. A good example is the planned replacement of a large surface carpark for about 2 000 vehicles on the Konigsplatz by an underground garage beneath a neighbouring greenspace in front of the Alte Pinakothek. Both sites are world famous and structurally speaking extremely sensitive. Constructing a multi-storey carpark above ground was eliminated because, among other reasons, the ramps required could not be built without seriously affecting the quality of the site. Functional and technical difficulties in providing the ramps on the new site also had to be accepted, so as not to impair the characteristics of the area in front of the famous gallery and keep the additional utilisation of the adjacent streets to a minimum. Only a few years ago the majority of political representatives and citizens would have been opposed to such an extension of parking facilities in the inner city because of the likelihood of increases in road traffic and its negative effects. However, due to the serious shortage of parking space and the great importance attached to keeping residents in the inner city, the new programme for multi-storey carparks, which are mostly intended for the inhabitants of the districts concerned, is now being accepted.

The Development of the Park-and-Ride System

Prior to the creation of the suburban rapid railway, there were about 2 600 parking places available at the Federal Railways' stations. The demand for parking spaces at railway stations increased rapidly after 1972 following the introduction of the S-Bahn and the integration of public transport networks and services throughout the region. A growing number of motorists preferred to ride public transport to the city centre rather than drive all the way only to waste time searching for a parking space. The Munich Transport

Figure 3. **THE LOCATION OF MVV PARK-AND-RIDE FACILITIES**

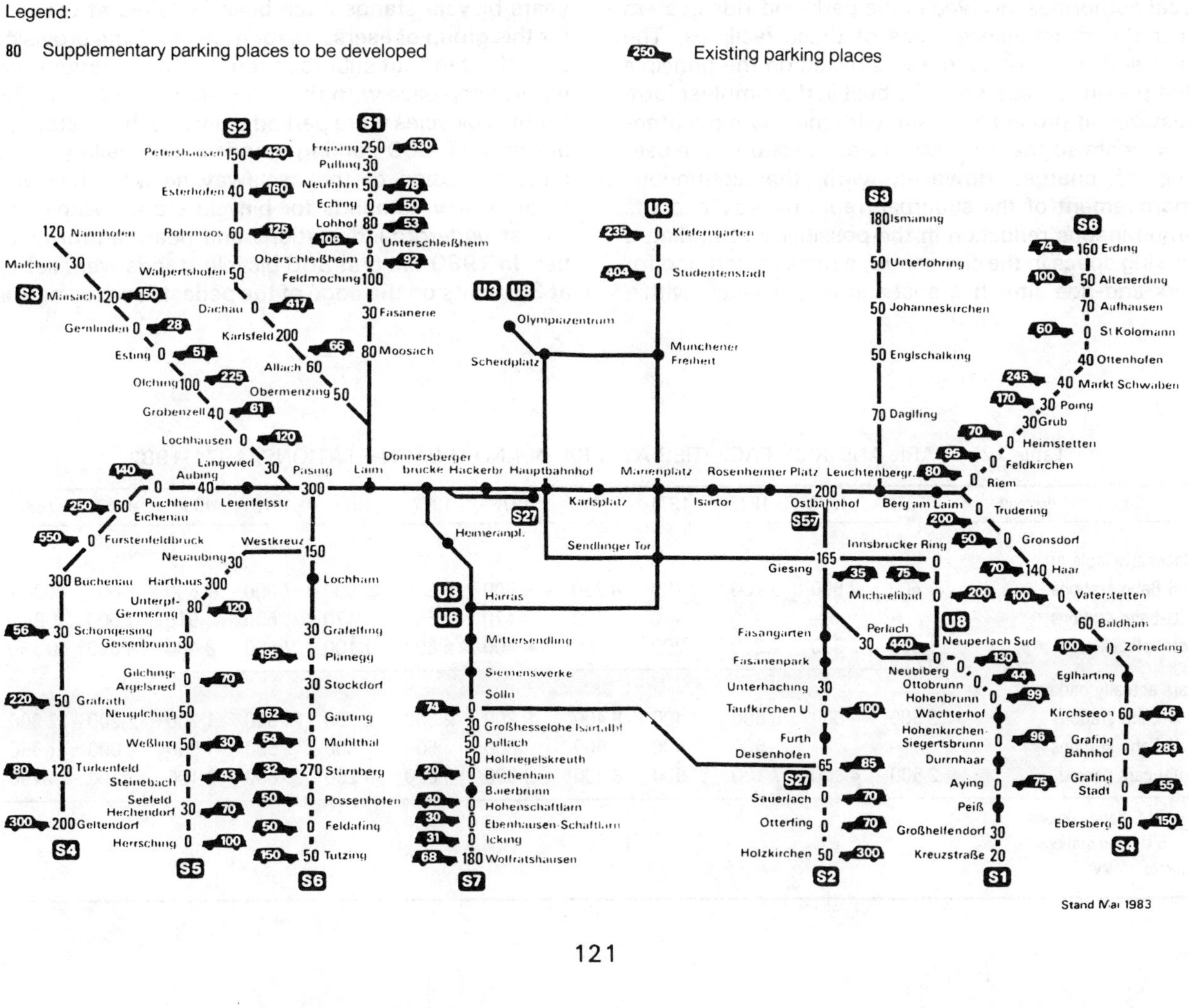

Legend:
80 Supplementary parking places to be developed
250 Existing parking places

Stand Mai 1983

and Fares Authority decided to encourage this practice by providing parking spaces free of charge for its customers at S-Bahn and U-Bahn stations. Most of the park-and-ride facilities contain spaces for between 150 to 300 cars (Figure 3). The number of places available was increased from 4 200 in 1973 to 10 200 in 1983. According to studies by the MVV the considerable gap between demand and supply has scarcely narrowed despite the extension of the park-and-ride system (Table 12). At present the number of cars actually parked is still 40 per cent higher than the number of parking spaces available. This means that many rapid transit users have to park their cars on-street in the neighbourhood of the S-Bahn or U-Bahn stations, sometimes causing local parking problems. Further rapid development is therefore envisaged to serve this latent demand and another 8 665 spaces ($1/3$ at 14 U-Bahn stations and $2/3$ at 67 S-Bahn stations) are in construction or planned for 1990.

The park-and-ride programme is a cooperative effort between the Federal German Government, the Free State of Bavaria, the City of Munich and the Federal Railways and agreement has been reached among the signatories for the financing of the programme. A fundamental prerequisite for the extension of these parking lots, however, is that the local authorities involved in the park-and-ride system bear the maintenance costs of these facilities. The park-and-ride programme was based on the principle that parking places would be built in the simplest form possible, at ground level and with minimum maintenance costs so that they could be provided to the user free of charge. However, with the continuous improvement of the suburban rapid railway and the simultaneous reduction in the possibility of finding a parking space in the city centre, a marked demand for park-and-ride sites has arisen in recent years within the city boundaries. The urban park-and-ride system is therefore being developed to meet this local demand, as far as possible, outside the middle ring-road in order to reduce illegal parking outside houses and shops in the streets around S-Bahn and U-Bahn stations.

Since sufficiently large and cheap sites are rarely available in the urban area, heavy investment is required for building multi-storey and underground carparks. The first underground park-and-ride garage was built on the Innsbruck ring-road some years ago. Two new park-and-ride facilities were recently built on the U-6 underground line in the northern part of the city. One of them, a six-storey garage located at the Kieferngarten underground station, provides spaces for 235 cars cost about DM 4.6 million. It is open daily from 04:15 to 01:45. No fees are charged. Motorists parking there must simply be in possession of a valid public transport ticket on leaving the garage. Anybody caught without a ticket must pay a fine of DM 40 as well as being liable to prosecution.

In addition to motorists, park-and-ride traffic includes several thousand users who travel to stations on two-wheelers. Most of them are young passengers and this traffic is naturally very dependent on the weather conditions. Nevertheless, in recent years bicycle stands have been installed at stations for this group of users. As for motorists, the provision of cycle stands at suburban rapid railway stations can hardly keep pace with the growing demand. In 1983, 19 500 bicycles were parked daily at S-Bahn stations but only 16 600 parking stands were available. The rapid extension of the cycleway network has also created new demands for bicycle stands within the city at underground stations and near central facilities. In 1980 the first 355 bicycle stands were set up at 36 points on the edge of the pedestrian precincts in

Table 12. PARK-AND-RIDE FACILITIES AT S-BAHN AND U-BAHN STATIONS, 1971-1982

Supply and demand	1971	1972	1973	1974	1975	1976	1977	1978	1979	1980	1981	1982
Places available at:												
S-Bahn stations	1 500	2 500	3 700	4 250	4 250	5 900	6 300	6 600	7 000	7 500	8 000	8 800[1]
U-Bahn stations	–	–	470	470	470	470	500	470	600	910	1 000	1 354[2]
Total places	1 500	2 500	4 200	4 700	4 700	6 400	6 800	7 100	7 600	8 400	9 000	10 150
Cars actually parked at:												
S-Bahn stations	2 500	4 800	6 600	7 400	8 400	8 800	9 300	9 700	10 300	11 200	12 200	12 900
U-Bahn stations	–	–	500	500	500	500	500	500	500	900	1 000	1 350
Total cars parked	2 500	4 800	7 100	7 900	8 900	9 300	9 800	10 200	10 800	12 100	13 200	14 250

1. 71 S-Bahn stations.
2. 5 U-Bahn stations.

Source: MVV.

the old town. At the beginning of 1983 the City of Munich was responsible for the upkeep of about 3 500 bicycle stands. Of these, about 2 200 are at S-Bahn stations and about 750 at underground, tram and bus stops. The installation target for the years 1983-84 is for over 2 000 new bicycle stands costing more than DM 100 000.

Special Coach and Lorry Parks

Munich's importance as a tourist attraction and regional economic centre poses increasing problems of providing sufficient parking space for coaches and lorries. Temporary parking site solutions for visiting tourist coaches are planned to be replaced by a programme designed to provide more appropriate facilities. An initial project plans for the construction of a large garage with service facilities located underneath businesses premises four kilometres from the town centre, near the intersection of an urban motorway, the suburban rapid rail and city underground railways.

The need to set up lorry parking areas became greater in 1980 with the change in road traffic regulations, for there is now a parking ban on lorries of over 7.5 tonnes in dwelling areas at night, as well as on Sundays and holidays. According to the objectives of the new 1983 Urban Development Plan, lorry parking lots are therefore to be provided on the most access roads on the edge of the city. In the meantime an existing parking area to the north of Munich has been extended. The construction of a parking area to the south of Munich, which was originally planned for the short-term, is at present not possible owing to access problems and massive opposition from nearby residents.

VII. ENVIRONMENTAL PROTECTION POLICY

About 1.5 million litres of motor fuel are consumed daily in Munich producing substantial amounts of atmospheric pollutants. Motorists must submit their vehicles to mandatory technical inspections every two years at designated sites. Road-worthy cars complying with the technical and environmental standards are indicated by means of a coloured mark on their number plate. A voluntary vehicles inspection scheme is organised twice a year at 15 garages to enable motorists to check the pollution emission levels of their cars. Motorists must pay DM 30 to have their cars tested. Despite these controls, road transport accounts for almost three-quarters of the total emissions of air pollutants in the Munich region.

The City authorities realise that the main emphasis of all action to improve air quality must lie in the improvement of vehicle technologies and that other action conditions can never achieve anything but limited success. Nevertheless, actions can be developed and implemented at the local level, where such phenomena have their worst effects, to complement the broader actions being pursued at the national and international level. The City of Munich has therefore created a small Department for Environmental Protection. This newly created body has been charged with developing and implementing a local action programme to stimulate public support for environmental protection policies with the active cooperation of motor vehicle manufacturers and oil companies.

Experiments Concerning the Municipal Vehicle Park

The municipal vehicle park comprises about 4 000 motorised and non-self-propelled vehicles, such as construction plant and machinery. Although this fleet accounts for only a tiny proportion of the total number of vehicles on the road in Munich, the City authorities have embarked on a series of innovative measures to promote the use of less polluting vehicle technologies and fuels. The main objective is to make the advantages of such technologies better known to the public by using the municipal fleet in a pioneering experimental role. These measures include the promotion of lead-free petrol and liquified petroleum gas (LPG) and the use of cleaner, quieter lorries and buses.

Initiatives to Reduce Vehicle Exhaust Emissions

About 15 to 16 per cent of all vehicles in Munich are diesel fueled and market forecasts have suggested that this proportion could rise to as high as 25 per cent by the year 2000. The City authorities are preparing a field study for the standard installation of smoke filters on diesel vehicle engines. A local motor vehicle manufacturer (Mercedes) is giving a diesel lorry to the municipal vehicle parc for this purpose in the first half of 1984 and there is the promise of financial participation by the Bavarian State Ministry for Development and the Environment and from the Federal Department of the Environment.

By the end of 1983 about 40 vehicles in the municipal fleet were equipped with exhaust catalysers. The more widespread use of exhaust catalysers is at present hampered by vehicle registration procedures in the Federal Republic of Germany. Every new car constructed in the country must have a licence of homologisation proving that it is built in conformity to the technical and environmental standards in force. But if a vehicle is fitted with an exhaust catalyser it loses its license and a new one takes about a year to acquire. This procedure acts as a disincentive to vehicle owners. Of course, the retro-fitting of exhaust catalysers does not offer a completely satisfactory means of reducing vehicle emissions of air pollutants, since to be truly effective vehicles must be run on lead-free fuel. Otherwise the lead additives in petrol can damage exhaust catalysers and greatly reduce their performance once a certain number of vehicle-kilometres have been travelled. It is for this reason that the City authorities have decided to promote the use of lead-free fuels.

Initiatives to Promote the Use of Lead-Free Fuel

The City of Munich plans to operate older vehicles meeting certain technical requirements on lead-free petrol. At present about 300 vehicles are being inspected for this purpose and about 20 of them are already in service. In connection with this experiment, the first lead-free petrol pump for the lead-free operation of vehicles belonging to government vehicles (i.e. both the city and state departments) was recently set up in Munich at a cost of DM 5 000. More can be installed if required. Negotiations between the City authorities and oil companies are also underway with the objective of operating lead-free filling stations on a commercial basis and thereby enabling environmentally conscious citizens in Munich to drive lead-free. The first commercial filling station offering lead-free and ordinary petrol came into operation in 1984. The German Automobile Association's magazine already lists those makes of car models which can be run on lead-free fuel.

The Lord Mayor of Munich's decision to promote the use of lead-free petrol at the local level was taken without any guarantee of support from the State or Federal Governments. The Munich initiative has brought a big response from other communities, resulting in an on-going exchange of experience and information among, for example, members of the Associations of Bavarian and of Federal Towns. As a result, about 20 local authorities in the Free State of Bavaria have decided to install lead-free petrol pumps to service their own small vehicle fleets. These local activities also counted a great deal in the decision of the Federal Government on 21st July 1983 to set up the legislative basis for the introduction of lead-free petrol throughout the Federal Republic of German as from 1st January 1986. The Federal Government is also prepared to go it alone on this move, although it considers that its main objective is to inspire a united procedure in Common Market Countries. A Common Market resolution was therefore already set up in June 1983 at a session of the Council of Ministers of the Environment with the support of the United Kingdom, Denmark and The Netherlands. This resolution requested the Commission of the European Communities to submit proposals for the introduction of lead-free petrol by 15th April 1984. On July 4th, the Federal Minister of the Interior said the Cabinet had reaffirmed its decision to introduce lead-free petrol from 1st January 1986, and approved several parallel actions to maintain the Federal Government's momentum in its attempt to reduce vehicle exhaust emissions. These actions included:

i) The sale of lead-free petrol at all of the 272 service stations operated along the Federal motorway network (lead-free ordinary petrol by the end of 1985 and lead-free super by the end of 1986) by offering financial incentives to station owners;

ii) Exemption from the motor vehicle tax from between five to seven years for motorists buying new cars of up to 2.5 litres displacement for vehicles purchased between 1st July 1985 and 31st December 1989 and running on lead-free fuel; and

iii) The possibility of direct subsidies of up to DM 1 500 for each car using lead-free fuel purchased in 1986 with the amount of the subsidy falling to DM 800 for cars bought in 1989.

Liquified petroleum gas (LPG) is also a lead-free fuel. At present there are about ten petrol stations in Munich selling LPG. The installation of the equipment required for running vehicles on LPG is still being left to private initiative. However, the City authorities are examining the feasibility of a field study using vehicles in the municipal fleet converted for operation on LPG. The intention here is to use LPG on three buses fitted with catalytic converters to eliminate NO_x emissions.

Initiatives to Reduce Vehicle Noise Emissions

The City authorities have decided to change vehicle procurement guidelines so that utility vehicles in the municipal fleet with a useful load of three to six tonnes can gradually be replaced by quieter vehicles equipped with encapsulated engines. Engine encapsulation will make it possible to bring down the noise level from the currently permissable 86 dB(A) to 79 to 80 dB(A), i.e. by about 6 dB(A). A decrease of 10 dB(A), for example, cuts the noise volume by half. This action is already underway. Another local motor vehicle manufacturer (MAN) will present a quiet lorry to the Lord Mayor of Munich for this purpose. In addition, about 60 buses will be fitted with engine casings to reduce noise emissions in the near future.

VIII. EVALUATION OF ACTIONS TO IMPROVE TRANSPORT AND THE ENVIRONMENT IN THE MUNICH REGION

Actions Concerning Public Transport

The 1965 General Transport Plan to develop and promote local public transport in Munich and its region has been resolutely pursued for almost two decades. Even after the change in political majorities on the City Council in 1978, the consistent objective is to be noted in its resolutions to improve living conditions in the city through the continued development of public transport modes. One of the most striking aspects of this policy is that incentives and not compulsory measures are used to encourage motorists to switch from their cars to public transport. The justification for this plan, which had been set and maintained on a long-term basis, has been confirmed by the growth of ridership and the increased share of the tracel market by public transport. This success, which so far has been achieved at politically acceptable cost, has been based on:

- The constant development of the city underground railway and the continuous adaptation of surface tram and bus networks to this extension;
- The improvement of capacity and service quality on the suburban rapid railway through the development of lines and the introduction of more trains;
- The development of the park-and-ride system for car drivers and cyclists and the provision of feeder bus services throughout the region;
- The unified route network, the co-ordinated timetables, the common tariff and common sales promotion and advertising implemented by the Munich Transport and Fares Authority;
- The good working relationships between the various government bodies and transport undertakings co-operating in the development of local public transport in the city and region.

This co-operative effort brought four decisive improvements for the citizens of Munich and its region:

- An integrated transport network comprising S-Bahn, U-Bahn, tram and bus lines;
- A co-ordinated service timetable with more frequent services, shorter vehicle transfer times and greater passenger carrying potential;
- A common fare system with unified prices and tickets for all means of local public transport; and
- A basic plan for future growth, allowing for integrated and economical expansion of the entire system.

The popularity of this multi-modal system is largely explained by the expansion of the service area, the improvement in service quality and the easier access and convenience of use. The opening of the S-Bahn network resulted in a considerable shortening of journey times for travellers and more than tripled the catchment area for local public transport from about 500 km^2 in 1971 to about 1 600 km^2 when the trunk line under the city centre came into service. The number of train departures per day was also almost doubled. In 1971 the number of local commuter trains per day totalled 290 (about 200 from Munich Central and about 90 from Munich East). In 1972 this figure had been increased to 573. The integration of the S-Bahn network with the city's U-Bahn, tram and bus networks and with the regional bus networks has brought further improvements. The ratio between seated and standing passengers is high and has been calculated so that most passengers have a chance of obtaining a seat for trips lasting more than 30 minutes. The numerous transfer facilities, the standardised ticketing and the self-service open nature of the system has made it easier to use the different modes, both individually and in combination.

Trends in Public Transport Ridership and Modal Split

Between 1973 and 1982 the total number of passengers carried on the services provided by the Munich Transport Fares Authority increased by 30 per cent, from 358 million to 464 million (Table 13). This growth rate would be considerably higher if the period before 1972 were taken as a basis for comparison. Yearly growth rates of up to four per cent from the previous year's traffic are no longer recorded because of the general recession. However, the number of passengers has remained stable, in contrast to other German towns.

The city's underground system has assumed an increasingly important role in the local public transport system. Daily ridership has steadily increased from about 230 000 passengers in 1973 to 600 000 in 1982. Ridership has also increased on the suburban rapid railway. In 1971 the Federal Railway's local commuter services carried about 160 000 passengers per working day. Following the connection and improvement of the network and the introduction of S-Bahn through services, it was anticipated that this figure would rise to 210 000 passengers. The actual results far exceeded this figure. During the first complete year of S-Bahn operation, 1973, ridership

Table 13. TOTAL RIDERSHIP OF MVV PUBLIC TRANSPORT SERVICES, 1972-1983

	1972[1]	1973	1974	1975	1976	1977	1978	1979	1980	1981	1982
Millions of passengers	203	358	395	405	412	417	432	449	451	463	464
Index 1973 = 100	57	100	110	113	115	116	121	125	126	129	130

1. Only for the period 28th May to 31st December 1972.
Source: MVV.

totalled 430 000 and the S-Bahn system currently carries about 600 000 passengers per day.

This growth in public transport ridership cannot be attributed to any one particular mode, but rather to the overall improvement in service efficiency and quality. Over 1.6 million passenger journeys are made daily on MVV services. If these are separated out statistically by mode this gives a grand total of almost 622 million passenger trips a year (Table 14). This represents an increase of almost 50 per cent compared to the days before the MVV. These figures show that more and more people are making better use of the MVV's integrated services in that greater numbers are using feeder tram and bus modes and then changing to the S-Bahn and U-Bahn.

Data for 1977 indicated that Munich's inhabitants made about 1.9 trips per day on average and total volume of daily traffic in the city amounted to about 3 million trips. About 36 per cent of these trips were made on foot or on two-wheels, 35 per cent were made in private cars and 29 per cent by public transport. Thus private cars accounted for almost 55 per cent of motorised trips made in the city. Already though, the objective of establishing public transport as the main mode for travelling to and from the city centre has been attained (Table 15). The steady growth in public transport ridership has had a significant effect on modal split (Table 16). Public transport's share of motorised trips was around 37 per cent in 1970, 45 per cent in 1977, and is somewhere between 45 and 50 per cent today. Forecasts indicate a continuing expansion of public transport's share to around 53 per cent in the period 1990-2000.

Areas Requiring Further Action

The development and promotion of local public transport, especially of the underground system, should increase incentives to transfer from the car to public modes. The aim behind investment is to get most people in Munich to travel by public transport. Accordingly, this form of transport has precedence over individual motor traffic in the densely populated areas within the middle ring-road and in town district centres. However, the outer areas of the town must continue to be served by car.

The further reinforcement of public local transport is a pre-requisite for a reduction in environmental pollution from vehicles and at the same time for guaranteeing the town's transport capability on a long-term basis. Bus and tram networks are to be adapted to the suburbs rapid rail system on an optimum basis and the possibilities of transferring from private to public transport should be consolidated so that public local transport will become even more attractive. The very success of this policy has

Table 14. TOTAL MVV PASSENGER TRIPS BY SEPARATED OUR MODES, 1973-1982

Millions of trips

	1973	1974	1976	1978	1980	1982
S-Bahn	131.7	145.5	153.9	163.1	177.3	185.2
U-Bahn	59.6	68.8	93.3	97.4	115.1	155.4
Tramway	155.9	171.9	165.3	172.1	160.1	141.1
City buses	102.5	110.5	131.1	137.5	142.8	148.1
Regional buses	7.4	7.7	9.6	11.7	18.6	18.5
Total trips	457.1	504.4	553.2	581.8	613.9	621.3
Index 1973 = 100	100	110	121	127	134	136

Source: MVV.

Table 15. MODAL SPLIT (MOTORISED TRIPS) BY TYPE OF TRAFFIC, 1977-1980

Type of traffic	1977 survey Public transport	1977 survey Private transport	1980 estimate Public transport	1980 estimate Private transport	Millions of trips made daily
	Percentage				
"Old town" traffic [1]	67	33	70	30	0.4
Internal traffic [2]	45	55	46	54	1.9
City traffic [3]	43	57	44	56	2.5
City-regional traffic [4]	34	66	35	65	0.6
Work related traffic with a destination in the "Old town"	73	27	75	25	
All work related traffic	44	56	46	54	

1. Trips with an origin or destination in the "Old town".
2. All trips with an origin and destination in Munich.
3. Internal traffic plus trips with either an origin or destination in Munich.
4. Trips with an origin inside Munich and a destination outside the city.
Source: 1983 City Development Plan.

generated new tasks. For example, it is now necessary not only to reinforce the capacity of the suburban rapid railway on the outer branches of the network but also their power supply. Another task is to make the decision-takers realise the effects of pricing policy on the structure of the town and region.

Actions Concerning Private Transport

Private cars continue to play an important role in local travel and road haulage and delivery services are essential for the efficient functioning of the city's economy. The road transport policy pursued in Munich since 1975 has achieved some success in fulfilling the initial objectives:

i) Work has continued on completing the middle and outer ring-roads and these routes serve to keep a certain amount of long-distance through-traffic away from the city and its densely populated central districts;

ii) The local road network, including several radial expressways, has been largely completed and is adequate for most normal workday traffic demands;

iii) The overall dependancy on private cars is gradually being eroded by restraining non-local through traffic in residential areas, thereby encouraging motorists to keep to the main road network or switch to public transport modes;

iv) Traffic restraint measures of this sort are allowing roadspace to be won back for pedestrians and residential parking or for the extension of the city's cycleway network;

v) The city centre has been protected from excessive terminating traffic by transferring the opportunities for long-term parking to the periphery.

The increasing public sensitivity to undesirable noise disturbance from road traffic and opposition to certain aspects of major construction works on

Table 16. MODAL SPLIT (MOTORISED TRIPS) IN THE CITY OF MUNICH, 1965-1980 AND FORECASTS

Percentage

	1965	1970	1974	1977	1980	Forecast for 1990-2000
Public transport	50	37	44	45	46	53
Private transport	50	63	56	55	54	47

Sources: MVV, 1983 City Development Plan.

environmental grounds have meant, however, that only gradual progress in completing the city's planned main roads is possible unless the City authorities are prepared to adopt much more costly engineering solutions. Noise abatement measures and the tunnelling of new road sections are being implemented but resource constraints make their general application impossible. Road development funds are therefore being allocated to those projects which can best improve traffic and environmental conditions.

Road Development Policy

A road network with its capacity geared to the absolute peak volumes, as generated by car commuters or holiday markers, can no longer be considered justifiable when the local public transport system provides reasonable alternative travel facilities, opportunities. The costs of catering to such heavy demands, in terms of space, financial resources and environmental damage of catering to such heavy demands now often make them physically and politically impossible. A good road network is required however for handling the essential private traffic and for keeping through traffic away from densely populated areas. The latter still affects extensive parts of the city.

The thinning out of the main road network, as provided for in the 1975 Transport Development Plan, was accepted and implemented through appropriate resolutions when construction plans were being modified. Thus, compared with the 60s fewer resources are now available for road works. In the case of main roads, the development of the middle ring-road has priority.

The development of the inner local network has to be based on the fact that traffic restraint measures are becoming possible in dwelling areas. Traffic forced out of such areas should be streamed by efficient peripheral roads if it cannot be diverted in ways more advantageous to the environment.

The service network is to be defined or expanded as sparingly as possible. Rebuilding operations on the service network contribute to traffic restraint, the improvement in living conditions and environmental protection. When a decrease in road dimensions is considered, this means that former traffic areas can be reduced in size, closed off or used for other purposes. However, implementation of these demands for environmental protection may well seriously affect the traffic flow in the short-term and in some cases also in the medium-term.

Noise Protection on Roads

Noise protection in connection with new roads involves a great deal in terms of planning, funding and legal implementation. Although road works sponsors show the greatest consideration for these demands, it sometimes takes decades to overcome the opposition from those directly involved. On the many roads with high housing density and heavy traffic, it will still take long-term and intensive financial efforts to achieve more than sporadic improvements. It would be most helpful if appropriate federal legislation could be introduced as soon as possible.

The Promotion of Pedestrian and Cycle Traffic

Ten years ago the bicycle was considered almost exclusively as a means of transport for leisure activities. Therefore no cycleways were planned in Munich's inner city. A change has occurred here, as the cycle is being increasingly used for the journey to rapid rail stations, shopping, or going to school and work. This change is taken into account in planning and construction operations. Basic arrangements must be made in urban development measures for central points as well as leisure facilities, local tourist areas and rapid rail stations to be reached more easily by bike or on foot. The actual precedence of the car over non-motor traffic should be limited case by case. More pedestrian areas and cycleways in town are being constantly provided. An important factor here is the elimination of network gaps and segments which are accident black spots. With this approach public and private expenditures on transport and the environmental pollution caused by it can also be reduced.

Unfortunately a considerable increase in accidents is also associated with the greater use of the bicycle. The accident rate has probably not risen to the same extent as bicycle utilisation, but precise studies have not yet been conducted on this subject.

The extensive pedestrian precincts in the old town are still greatly appreciated. Heated criticism of expected developments (for example too many musicians and so on) helps the municipal authorities to maintain the high standard of the precincts. All planning partners, especially the public's spokesmen in the outer districts of the inner city, have in the meantime learnt to accept the fact that expectations concerning these precincts must not be seen in isolation but only in conjunction with traffic restraint or preferably traffic improvement. Wishes for extension of the precincts in the peripheral districts of the inner city but also in outlying districts are therefore now seldom expressed.

Traffic Management Policy

An efficient main road network is essential for a balanced traffic restraint policy designed to improve living conditions in densely populated areas. Special importance has been attached to improving road safety when introducing such restraint measures. Another major concern has been to avoid the uncontrolled diversion of traffic to neighbouring areas or cause unreasonable obstacles to goods transport and delivery services. Attempts are therefore being made to achieve the desired general reduction and diversion of car traffic through a series of measures and incentives, with particular attention being paid to the complementary role of public transport. Extensive long-term traffic improvement programmes are thus being developed with the aim of using every possibility of rebuilding and developing roads and allocating scarce financial resources on an optimal basis. The renovation of individual streets in densely built-up inner city districts is bound to remain the exception, even in the future, because of the limited resources available and because of the more cautious attitude taken by local residents. More frequent use is being made of traffic restraint possibilities in newly developed areas. The conflicting views on the evaluation of differing street functions have been smoothed over. However, the discussion will go on — even in the more distant future — as to which of the many restraint measures might be the most appropriate in particular localities.

Parking Policy

Controls on parking in the city centre have proved effective. The number of long-term parking spaces open to the public has been frozen, and available parking and loading space is being limited to the essential needs of business and shopping traffic and of inner city dwellers. The task of reorganising parking arrangements, especially in inner city and peripheral districts, has acquired more importance. As far as possible, additional parking space is being created for residents in these densely built-up areas. The parking licence system has so far been tried out in only one outer district of the inner city. The findings of this experiment indicated that schemes of this sort are not a universal remedy for resident's parking problems. A very cautious approach is therefore being taken before transposing this technique to other districts of the city. Partly as a result of this experience, the necessity for limitations on kerbside parking is now being more closely reviewed and as far as possible reduced. The exemption from the obligation to provide parking space in new offices and apartment buildings in the city centre is still being pursued and the funds collected in this way are being used to finance a programme for the provision of underground parking garages in inner city districts. A big effort is also being made to direct more commuters to the local public transport and park-and-ride system. Finally, in compliance with the legal requirements, additional parking space is being provided for lorries to keep them away from residential areas.

Environmental Protection Initiatives

At the beginning of the last decade environmental awareness did exist in planning policy, although it found scarcely any expression and had lower priority than today. At present a certain indifference and at the same time a tendency to over-react is to be observed among the population, depending upon whether it is directly concerned or not. A newly created Department for Environmental Protection is trying with imagination and realism to give impetus to local initiatives in this field. In so doing it does not shy away from micro-measures to inform citizens and stimulate their readiness to co-operate. It is realised that the main emphasis in all efforts must lie in the improvement of vehicle technologies. All other action to improve environmental conditions through transport policy can never achieve anything but limited success. Nevertheless, the example set by Munich shows that initiatives paving the way for broader actions can be developed and implemented at the local level — where such phenomena have their worst effects — and usefully complement policies being pursued at the national and international level.

IX. CONCLUSIONS

During the period from 1966 to 1972 Munich, among cities of similar size, was regarded throughout the Federal Republic of Germany and in other countries as presenting a remarkable example of an enthusiastic urban planning policy and consistent but moderate application of experience acquired by other urban areas. In the years following the Olympic Games in 1972, when the breakthrough of new technologies was expected in the transport sector, and when general environmental awareness was beginning to develop into a political factor, Munich was above all concerned with pursuing the earlier momentum.

This consistency in the city's land-use and transport policy is partly explained by the local system of government. Both the Lord Mayor and the City Council's 80 members are voted into office simultaneously at public elections held every six years. Even when the Lord Mayor and Councillors change, the length and

Table 17. AVERAGE YEARLY CONCENTRATION LEVELS OF POLLUTANTS IN THE CITY OF MUNICH, 1974-1980

	SO (mg/^{3})	Partic. (mg/^{3})	CO (ppm)	HC (ppm)	NO (ppm)	NO (ppm)	NO$_2$ (ppm)
1974	0.034	0.23	1.9	–	0.08	–	–
1975	0.044	0.17	2.5	2.84	0.10	–	–
1976	0.034	0.14	2.1	1.92	0.13	–	–
1977	0.03	0.12	2.2	1.2	–	0.06	0.03
1978	0.04	0.10	2.1	1.2	–	0.07	0.02
1979	0.03	0.05	2.3	1.0	–	0.05	0.02
1980	0.03	0.05	2.2	0.9	–	0.03	0.02

Source: 1983 City development plan.

synchronisation of their periods of office makes for very effective government and has undoubtedly been an important stabilising factor in the planning and development of local transport over the past 20 years.

Although transport policy is only one facet of the actions being undertaken to improve the quality of the environment, it is easy to imagine how much more difficult the task would be if public transport in Munich had been left to decline in the face of the upsurge of private motorisation. The indications of a gradual improvement in air quality, for example, result at least in part from the City Government's decision to encourage citizens to continue using public transport rather than changing to private cars (Table 17).

The present phase of transport development in Munich is also being decisively influenced by the first signs of new conflicts of interest resulting from current economic and social problems. Economic growth, for example, is now associated with a flagging labour market and demands concerning the living environment conflict with a serious housing shortage. These challenges require rapid action. In these areas of tension, a new justification is given for the established practice of striking a balance between the high volume of investment in local public transport on the one hand and improvements in the possibilities for private car travel on the other.

The now obvious necessity of reducing environmental damage places new, heavy emphasis on this balance. The policy implemented for a decade has thereby been given a new, additional impetus and the first signs of special innovations in the environmental protection field can be perceived. In the next few years these innovations can only be developed into actions with more extensive effects if the environmental awareness of the political world and the population increases accordingly.

REFERENCES

1. G. Meighörner and H. Doleschal, "Munich", Case Study included in the Proceedings of the OECD Conference "Better Towns with Less Traffic", held in Paris, 14-16th April 1975; OECD, Paris, 1975.
2. D. Lippert, "Munich", Case Study included in the ECMT Round Table 47, "Scope for Railway Transport in Urban Areas, ECMT Paris, 1980.
3. H.J. Simkowitz, "Munich" Case Study included in the U.S. Department of Transportation report "Innovations in Urban Transportation in Europe and Their Transferability to the United States", Urban Mass Transit Administration Office of Service and Methods Demonstrations, Washington, D.C. 20590, February 1980.
4. Landeshauptstadt München, "Stadtentwicklungsplan 1983", 1983 City Development Plan, 2 Volumes, Referat für Stadtplanung und Bauordnung, Munich, July 1983.
5. Münchner Verkehrs – und Tarifverbund GmbH, "MVV On Show", MVV information brochure, Munich 1983.
6. Münchner Verkehrs – und Tarifverbund GmbH, "MVV Dates and Facts", MVV information brochure, Munich, Summer 1983.
7. Münchner Verkehrs – und Tarifverbund GmbH, "MVV Report '82", MVV annual company report, Munich 1983.

Chapter 7

NEW YORK*

I. INTRODUCTION AND CONTEXT

For several years, New York City faced, and has now emerged from, a financial crisis threatening the future of the City as a place in which to work and live. Following years of deferred maintenance and neglect brought on largely by the fiscal crisis, the region's subway, bus and commuter rail systems were on the verge of collapse. The origins of the public transportation crisis and the innovations developed to finance the rehabilitation of the transit system are the main focus of this study. New York City has embarked upon a five-year, 8.5 billion dollars capital programme that relies heavily on private sector financing techniques and funds. The goal of the programme is to improve the reliability and level of service of public transportation in the City and its region.

An accelerated programme for refurbishing the city's roads and a road capacity management programme are also under way. The benefits to be derived from transit rehabilitation are strongly linked to the success of these road improvements.

The New York Metropolitan Area (NYMA) does not meet national ambient air quality standards for carbon monoxide and ozone. The effect of transit rehabilitation and other transport on these pollutants are discussed.

New York City

New York is the most populous city in the United States. Its 7 million inhabitants account for 40 per cent of the population and income of New York State, and it is the focus of the nation's largest metropolitan area. It is a major centre for finance, banking, insurance, communications, publishing, fashion, design, retailing, advertising and tourism.

---
* Case study prepared by Messrs. M. Downey (United States), H. Simkowitz (United States), and R. Prudhomme (France).

New York City covers 1 044 square kilometres and includes five boroughs. The greatest concentration of activities and residential development is in the 57 square kilometres of the Borough of Manhattan, an island connected to the remainder of the region by bridges and tunnels.

The intense activity of Manhattan, and particularly of its central business district (CBD), would not be feasible without public mass transit. 85 per cent of the 1.5 million peak-hour commuters to the CBD use it; 5 per cent ride privately operated buses and the other 10 per cent travel in cars. 70 per cent of the public transit riders arrive on the subway and another 10 per cent arrive on commuter trains.

Public Transit

Metropolitan Transportation Authority

The Metropolitan Transportation Authority (MTA), an agency of New York State, was created in 1965 to develop and implement a unified mass transportation policy for New York City and seven adjacent counties. The MTA is a holding company for a number of subway, bus, and commuter rail undertakings as well as tunnels and bridges.

The MTA is responsible to all parts of its region, and its board consists of a chairman and 13 members. The Governor of the State appoints the chairman and five board members. The other members are nominated by the Mayor of New York City and the seven suburban countries. The members serve for a period of six years after confirmation by the New York State Senate.

New York City's transit system operates 24 hours a day, seven days a week. The MTA services carry 5.6 million passengers on an average weekday. This is one-third of the public transit riders in the United

States. In 1984 the MTA's passenger fare and toll revenues exceeded 2.2 billion dollars.

Subways

The subway network is 368 kilometres long, has 465 stations and is served by 6 125 cars. It employs 28 000 people and carries nearly one billion passenger per year.

Even though New York City is dependent for its existence on transit, subway ridership has declined steadily from a peak in 1947 due to economic and demographic changes. It is now at the lowest level since World War I (Table 1). While many riders are captives, others are "marginal" for non-commuting trips or have alternative means of transport. Many persons, especially those who travel off-peak, appear to have been lost to the subway system.

Table 1. TOTAL NYCTA REVENUE PASSENGERS
Thousands of US$

	Subway	Bus	Total
1950	1 680 844	653 564	2 334 408
1960	1 344 953	431 014	1 775 967
1970	1 307 387	823 815	2 131 202
1975	1 077 595	738 230	1 815 825
1980	1 037 602	621 395	1 658 997
1981	1 026 572	575 189	1 601 761
1982	990 849	504 435	1 495 284
1983	1 005 344	523 198	1 528 542
1984	1 000 640	501 408	1 502 120

Buses

The New York City Transit Authority (NYCTA) bus system includes 225 local and express routes running along 1 598 kilometres of streets. The system employs about 14 000 people and runs more than 3 800 buses. Most routes are designed to serve people travelling within a particular borough or to bring them to subways. The system carries approximately a half billion pasengers per year. Bus travel has been decreasing less fast than subway riding.

Bridges and Tunnels

The Triborough Bridge and Tunnel Authority (TBTA) operates all the intrastate toll bridges and tunnels within New York City, the Eastside Airlines Terminal, the New York Coliseum (a large exhibition hall), and two car parks. The TBTA has been authorised to issue bonds in the financial market at preferential rates of interest that are exempt from Federal income tax. TBTA bonds have been and will continue to be issued to finance a variety of subway, bus, and rail capital projects. The TBTA is authorised to expend up to 1.1 billion dollars of bond proceeds for transit projects.

The TBTA has the distinction of being the MTA's only consistent money-maker, collecting tolls from an average of more than 650 000 vehicles each weekday. The TBTA is financially independent of the MTA although it is required to transfer its operating surplus to the MTA to be used for operating commuter railroads, subways and buses. In the future, an increasing share of the surplus will go to supporting debt issued for public transit purposes.

Suburban Services

The MTA's commuter railways run over 2 472 kilometres of track and carry about 450 000 passengers on an average weekday. MTA also operates certain suburban buses.

Roads

There are more than 9 920 kilometres of streets and highways in New York City, occupying 23 per cent of the land area. Currently about 16 per cent of the people entering the Manhattan CBD between 7 am and 10 pm do so by car, taxi, or truck. Vehicle volumes to and from Manhattan have been climbing and reached a new peak in 1982. Between 1980 and 1982 traffic flow jumped by 118 000 river crossings daily (an increase of 8.1 per cent), despite a 25 per cent toll increase. The increase in vehicle usage has been attributed, in part, to the deterioration of mass transit.

New York's roads, like its transit, have been subject to deferred maintenance. According to a 1979 survey, some 5 280 kilometres of the City's streets required resurfacing or reconstruction. While the street repairing cycle has been reduced from 135 to 90 years over the past five years, most streets in New York City will require complete reconstruction in the next 50 years.

The New York City Department of Transportation (NYCDOT) estimates that it would take about 4.3 billion dollars to rebuild the City's streets. However, this expenditure is currently unachievable. In 1983 274 million dollars were spent, the highest amount in several years. Increases in Federal gasoline tax and other Federal assistance will result in 386 million dollars a year being available on average for the next four years.

Environmental Conditions

In 1970 the United States Congress established national ambient air quality standards for ozone, carbon monoxide, nitrogen oxide, sulfur ioxide, particules and lead (Table 2). The Clean Air Act required each state to develop a State Implementation Plan, a blueprint for its efforts to meet the national standards.

Concentrations of air pollutants have been decreasing in the New York Region over the past 15 years, and the New York Metropolitan Area now meets national ambient air quality standards for lead, nitrogen dioxide, sulfur dioxide and total suspended particulates. However, carbon monoxide and ozone concentrations exceed national ambient air quality standards.

The entire Metropolitan area is part of a larger ozone non-attainment area that includes northeastern New Jersey and southwestern Connecticut. The New-New York Jersey-Connecticut Air Quality Control Region has a long history of frequent and widespread violations of the standard.

The non attainment area for carbon monoxide is defined as all of New York City and much of the nearby surrounding area. The long term trend in eight-hour carbon monoxide concentrations has shown a decline. However, violations of carbon monoxide standards in the NYMA occur frequently in localised area (known as "hot spots") as a result of a buildup of emissions from congested traffic.

Table 2. US AMBIENT AIR QUALITY STANDARDS

Pollutant	Concentration	Average time
Ozone	0.12 ppm	1 hour
Carbon Monoxide	3.5 ppm	1 hour
	9 ppm	8 hour
Nitrogen Oxide	0.05 ppm	Annual
Sulfur Oxide	0.14 ppm	24 hour
	0.03 ppm	Annual
Particulates	260 ug/m^3	4 hours
	75 ug/m^3	Annual
Lead	1.5 ug/m^3	3 Months

Note: ppm = parts per million.

II. TRANSIT REHABILITATION PROGRAMME

Origins of the Financial Crisis

In the 1960s and, to an even greater extent following the fiscal crisis of the 1970s, New York City's transit policy was to reduce capital investment and maintenance spending on transit. Maintenance was deferred and maintenance work forces were reduced while equipement aged and service deteriorated. A decline in business activity in New York City has been attributed, in part, to the resulting decline in the level of transit service. While 128 of Fortune Magazine's for 500 companies had their headquarters in New York City in 1959, there were by 1981 just 72. Of the 50 largest retail companies, 14 were headquartered in New York City in 1959, but only six in 1981.

During this period, public officials generally invested in new and visible projects, rather than on the rehabilitation and replacement of existing assets. Pressure to keep fares at artificially low levels led to policies that siphoned off available capital funds to pay operating expenses. Starting in the early 1970s, annual operating losses were offset by surplus toll revenues and subsidies from the Federal government, New York State, and the city of New York. Since 1974, annual operating assistance payments appropriated by State and local governments for the MTA have increased from about 100 million dollars to an estimated 1.2 billion dollars for 1984. In 1984, fare box and toll income paid for more than half of all MTA operating costs, with the balance being made up by subsidies (Table 3). This is a high fare box contribution compared with many other US cities.

Changes in Policy

The result of these policies was to bring the transit system dangerously close to collapse. In 1980 the MTA reported that 14.4 billion dollars (1980 dollar) would be required over ten years to restore the public transportation system to a condition of "good repair".

Table 3. TRENDS IN NYCTA OPERATING COSTS NEW YORK CITY TRANSIT AUTHORITY
Dollars in millions

	Farebox revenue	Total operating revenues[1]	Operating expenses[2]	Operating deficit[3]	Total revenue passengers[4]	Fare recovery ratios[5]	Deficit per passenger[6] (cents)
1976	$ 678.7	$ 690.4	$1 114.3	$(423.9)	1 663.135	60.9%	25.5
1980[7]	$ 724.2	$ 742.5	$1324.1	$(581.6)	1 614.865	54.7%	36.0
1984[8]	$1 230.4	$1 261.3	$2 154.6	$(893.3)	1 487.979	57.1%	60.0

1. Includes farebox revenue and other operating revenue; excludes reimbursements for school children and elderly and handicapped citizens.
2. Excludes "specific expenses".
3. Operating revenues minus operating expenses. In 1984, deficit also includes $21.8 million of uncollectable accounts receivable.
4. Excludes add-a-ride passengers.
5. Farebox revenue divided by operating expenses.
6. Deficit divided by revenue passengers.
7. Fiscal period July - June.
8. Calendar year.

In 1981 the State declared a "transportation emergency" and passed the Transportation Systems Assistance and Financing Act of 1981.

The 1981 Act made possible the bulk of the capital funding which the MTA is expected to receive over the next 5 years. Previously, non-Federal capital funding for the MTA was provided by annual appropriations from State and City capital budgets, including proceeds of general obligation bonds. The new Act enables the MTA to obtain additional operating funds through tax revenues as well as additional capital funds through the sale of bonds. Such a combination of MTA planning and State fiscal backing is unprecedented.

Under the new arrangements the MTA is expected to receive 8.5 billion dollars over the five years 1982-1986 or over 1.7 billion dollars per year. This contrasts with the 200 million dollars granted annually prior to 1979 and the 601 million dollars annual average for 1979 - 1981. Whereas before 1982 the Federal contribution to capital funding had been 58 per cent, the Federal share is now only about 30 per cent.

Along with this new influx of funds, New York has reversed its policy of deferring maintenance in order to reduce operating expenses. The expansion of the subway system has been ended except for the completion of existing projects.

Priority is now placed on safe and reliable operation rather than artificially low fares. The focus is on the rehabilitation and replacement of subway cars and buses and of stations, workshops, yards, equipment, and signal and power systems.

Five Year Investment Programme

The Capital Programme of the MTA has the following goals:

— Re-establish, then maintain, the reliable operation of existing subways and buses;

— Ensure long term survival of the existing public transportation system and its safe, reliable operation at reasonable cost;

— Implement a Noise Abatement Programme and other improvement to the existing system;

— Continue "new routes" projects currently underway, but add no new ones.

The five year programme provides 5.8 billion dollars for the subways, 6 billion dollars for buses, and 2.1 billion dollars for commuter railway expenditures.

Purchases will include 1 375 new subway cars (1 405 million dollars) and 1 700 new buses (298 million dollars). The MTA will also spend 761 million dollars to rehabilitate existing cars and 687 million dollars to modernise subway workshops and storage yards.

By replacing old cars and buses with new ones, and by expanding inspection and maintenance, the NYCTA expects to improve the quality of its service to the public. These capital investments have been matched by a new organisational structure which stresses management accountability and strong supervision. New performance measures have been

Table 4. 1982 CAPITAL PROGRAMME PCAN PERFORMANCE AND TARGETS FOR MTA OPERATIONS

Measures	1981	1987
Subway		
Key location throughput (% of scheduled trips)	89	98
Terminal abandonments (trains/day)	190	30
Enroute abandonments (trains/day)	140	65
Mean distance between car failures (kilometres)	10 720	24 000
Annual delay caused by defective car doors	10 000	4 500
Spare factor (%)	20.8	17
Air conditioning		
Percent of fleet equipped	48	74
Percent of equipment working	88	95
Station modernisation		
Percent of station modernised since 1980	1.9	12.7
Bus		
Trips cancelled at origin (%)		
Manhattan	1.7	0.1
Other boroughs	1.2	0.1
Trips cancelled at route (%)		
Manhattan	1.9	0.9
Other boroughs	0.7	0.3

introduced to enable progress to be charted. Management goals are being established and refined on an annual basis (Table 4).

Noise Abatement

In 1982, the New York State Legislature enacted a Bill which established a rail transit noise code. In response, the NYCTA has developed a noise abatement programme and by 1st August 1983, 134 million dollars had been committed to it. Of this the Federal Government will contribute 51 million dollars. The programme has several components. New subway cars will be air conditioned and insulated to provide a quiet interior and have special features to reduce exterior noise. (Air conditioning permits operation with windows and doors closed and lowers interior noise by 8 decibels). Wheels with flat spots are 10 to 15 decibels noisier than smooth ones and also cause vibration which annoy people living nearby. Approximately 60 000 wheels are trued each year to remove flat spots. Grinding corrugated rail was begun in 1955 and can reduce noise by 10 decibels. Grinding reduces vibration transmitted to nearby buildings, too. Rail welding, track lubricators, rubber rail seats and acoustic barriers at stations are also in the programme.

III. FINANCING TRANSIT REHABILITATION

The MTA receives funding from the Federal Government, the State and the City (Table 5). For example, in 1984 Federal Government assitance accounted for 3 per cent of operating revenue while State, City and regional operating assistance covered approximately 32 per cent of it.

The introduction of a 1 cent per gallon gasoline tax for mass transit and new formulas for allocating transit aid approved by Congress in December 1982 are expected to provide further capital assistance. The MTA estimated that between 1983 and 1986 it will receive an additional 150 million dollars a year in capital funds plus a continuing 120 million dollars per year in revenue subsidy a result of his legislation. The increased capital funds will be used to accelerate the renovation of 50 subway stations and modernise an aged signal system.

Since capital investment of only about 500 million dollars a year was projected from existing Federal State and local sources (with a substantial amount dedicated to existing projects) other sources of money had to be found if modernisation was to be expanded.

The MTA has accordingly sought to develop new sources of funding in the private sector and using techniques typically found in that sector. These techniques include:

— Lease financing;
— Vendor financing;
— Long term bonds, notes and other credit mechanisms designed to gain access to capital markets;
— Joint venture.

Table 5. FINANCING OF PUBLIC TRANSPORT
Millions of dollars

	Road Operating	Road Capital	MTA total Operating (1984)	MTA total Capital (1982-1986)
Central Government	–	204	106	2 611
State	–	42	110	2 368
Local	44	141	378	629
Regional taxes	–	–	684	–
Transit bonds	–	–	–	1 897
Other bonds	–	–	–	1 244
Users	–	–	2227	–
Other sources	–	–	118	760
Total	44	386	3624	8 509

Lease Financing

In 1981 the US Congress enacted "Safe Harbor" leasing legislation. This enables firms without federal tax liabilities to sell equipment to corporations with taxable income. Such corporations are then able to depreciate the assets and reduce their Federal income tax liability. A provision for accelerated depreciation in the Act makes these arrangements even more attractive.

A special provision in the law enables public transit agencies to make use of this mechanism, and the MTA has actively sought to use it to finance its capital programme. By mid-1985, it had entered into transactions involving 1.1 million dollar worth of buses and rail cars, obtaining 242 million dollars in capital financing.

A typical transaction involves the "sale" of buses or rail cars to a corporation which makes a cash payment of 10 to 25 per cent of their value and furnishes a promissory note (a legal document acknowledging the debt) for the balance. The MTA then leases the equipment from the investor for payments exactly equal to the annual repayment of the principal and interest payments on the note.

In fact, except for the initial payments, no cash actually changes hands. The investor's return comes through his ability to enter the transaction on his Federal income tax return. In the early years he can deduct the depreciation on the vehicles as well as the "interest payments" on his note, thus significantly reducing his Federal tax liability. In later years, these deductions are reduced, and his income from lease payments becomes taxable. However, for an aggressive investor, the return on the use of funds from the early tax savings more than offsets future liabilities.

At the end of the lease period, the equipment is transferred back to the MTA.

During the five year period beginning 1982 the MTA expects to save over 500 million dollars through the sale and leaseback of 2.5 billion dollars in rail and bus equipment. In practice the MTA has been able to reduce the cost of buses and rail cars by more than 14 and 22 per cent respectively. Because of federal tax legislation executed in 1982, this source of financing will not be available after 1987.

Vendor Financing

Vendor financing necessitates the abandonment of competitive bidding as used by nearly all US public agencies. Purchases are made instead via negotiation with potential contractors, thus enabling the terms of financings to be included in the deal. Competing cars builders are asked to offer terms for loans, load guarantees or other credit arrangements. The MTA has made significant use of this mechanism, but it should be noted that it is not available if Federal funds are being used. Special State legislation was also needed.

The most advantageous financing terms have been offered by foreign manufacturers supported by their Export-Import banks. The MTA has, as of June 1983, secured credits which will exceed 650 million dollars, at a rate of 9.7 per cent.

Access to Capital Markets

The MTA is attaining capital funds using a variety of borrowing mechanisms. While this, too, mirrors private practice, the "earnings" being pledged to pay back the borrowings consist of fare box revenues and subsidies. Investors are given an incentive to buy such bonds by the exempting of interest payments on them from Federal, New York State and New York City income tax. Approximately 4.8 billion dollars of the MTA's total 8.5 billion dollars capital programme will be derived from such bonds. Some will be secured by the toll income of the Triborough Bridge and Tunnel Authority, some by future streams of capital assistance from the State of New York, and some by fare-box income from transit itself.

Almost 1.6 billion dollars in projects will be financed by the third method using what are called Transit Facilities Revenue Bonds. This is the first recent use of this mechanism in the United States, and the bonds are unique in that they contain no governmental guarantees. They are based solelyon the revenues of the transit authority itself. The MTA has fully documented this borrowing and secured a favourable investment-grade, rating. A similar credit based on commuter railway revenues has also been successfully established and will provide 350 million dollars in capital.

Joint Venture

Joint venture schemes are being developed by the MTA to obtain private investment at transit stations. Investment of this kind in station improvements is best brought about through public incentives of which four types exist and have been used to generate projects valued at 80 million dollars. They are:

- Comprehensive Public Development;
- Negotiated Amenity Package;
- Special Zoning Districts;
- General Zoning Provisions.

Comprehensive Public Development

This technique is appropriate when the public sector controls an area redevelopment plan. At Times Square, for instance, prospective developers will be required to contribute some 27 million dollars toward subway improvements. The MTA has programmed 12.5 million dollars for public works, giving a total investment of 40 million dollars. A comprehensive design is being developed that will integrate the new buildings with the subways lying immediately below them. One building actually sits on the middle of the subway complex, and its reconstruction will allow the station to be given a central focus in what is now the building basement.

At Columbus Circle, developers of a new building on the site of the former New York Coleseum will modernise two adjacent subway stations at an established cost of 25 million dollars in return for development bonuses. In addition the developers paid 455 million dollars for the land, with the proceeds also being used for transit rehabilitation.

Negotiated Amenity Package

Improved amenities can be negotiated when a developer needs something (such as a change in zoning) from a municipality and can be charged a price for it. One example is a proposed development of 5 000 apartments plus commercial facilities on a west side site that is zoned commercial. The developers have offered an "amenity package" to the City and the local community worth approximately 100 million dollars in return for the necessary changes in the zoning requirements. Parks, a waterfront explain, street work, and a major subway improvement are contained in the package.

The 72nd Street subway station, while not directly on the site, is expected to bear a significant portion of the travel generated by the proposed development. The developers have therefore agreed to contribute 30 million dollars toward reconstructing the station, although this amount will fund only half the needed work. One of the funding mechanisms being considered is a letter of credit from the developer guaranteeing payment of the 30 million dollars to the MTA.

Special Zoning Districts

Special zoning districts require building developers to provide free easements for the construction of subway station mezzanines, entrances and concourses and to provide cash contributions or actual works in return for floor area bonuses.

General Zoning Provisions

A general zoning for Midtown Manhattan enacted in 1982 provides for floor areas bonuses in return for major subway improvements. The zoning also requires sidewalk subway entrances to be relocated in new buildings.

The MTA has been developing a master plan for the use of the new Midtown bonus provisions. This will specify exactly what improvements should be included at each redevelopment site. The MTA should thus be able to renovate the whole downtown subway environment through a carefully staged series of public and private investments. The plan anticipates the generation of 15 to 20 million dollars private funding for station improvements. The MTA would like the bonus provisions for Midtown to be extended to other development areas in the city.

IV. OTHER ACTIONS

Other actions taken by the MTA include labor-management cooperation and automatic fare collection. New three year labour contracts offer management the ability to make changes that will enhance the efficiency of the system and the productivity of the work force in conjunction with the Authority's upgraded supervisory initiatives. Automatic fare collection employing some form of encoded fare cards and peak period pricing is planned. Monthly fare pass, which would be made available through employers or the person's financial institution, are also envisaged.

V. ROAD PROGRAMMES

Capacity Management

The New York City Department of Transportation (NYCDOT) has been working closely with the MTA to discourage auto use, particularly for commuting into Manhattan, and to improve the reliability of the bus system by providing preferential treatment. The NYCDOT has evolved a principle of Capacity Management (CM) which attempts to regulate both vehicles and space. The number of vehicles in an area, for example, can be controlled by adjusting entry and exit rates. Similarly, although the amount of space in a street is absolute, its ability to carry vehicles is variable and is affected by parking and turning restrictions, street direction, signal timing and lane width. CM shifts emphasis from keeping traffic moving to moving people and goods by assigning priority to modes of transport which make the most efficient use of the street. Thus in office and shopping districts, where walking is often the most efficient means of transport, CM may dictate the reassignment of street space to pedestrians.

CM underlies all NYCDOT street policy, from the determination of vehicle priorities to the special attention given to the repair and reconstruction of streets with bus lines. The application of CM has been most highly developed in the CBD where NYCDOT has joined the Triborough Bridge and Tunnel Authority and Port Authority in using lane reversals to control vehicle entry and departure from Midtown. Furthermore, computerisation of the CBD's signals in the near future will permit NYCDOT to meter traffic, slowing entry from the periphery and speeding departure from the congested area. The City also continues to study the feasibility of restricting single-occupant vehicles at the East River crossings during peak hours.

Operation Clear Lanes

Buses are recognised in New York City as making the most efficient use of street space for moving people. The NYCDOT has developed the "Operation Clear Lanes" programme to keep curb lanes free of parked and waiting cars in order to decrease bus travel time and improve reliability. It includes a no exception tow policy, new parking regulations, 100 dollars ceiling for parking tickets, tickets to taxis that do not pull up to the curb, and tickets and disqualification points on drivers licenses for blocking moving lanes. Since 1978, sixteen special busyways have been established, ranging from "Red Zone" curb lanes (kept clear of standing and all but turning vehicles) and exclusive bus lanes (both concurrent and contraflow) to crosstown transit corridors on 49th and 50th Streets and a transit/pedestrian mall in Downtown Brooklyn. Table 6 indicates the effect this programme has had on bus speeds.

Table 6. BUS COMMERCIAL SPEEDS (km/h)
Exclusive of stopping

	1972	1980	1983
Avenues	17	12.6	15.2
Streets	10.1	7.5	9.4

Bond Issue

In November 1983, the citizens of New York State passed a Transportation Infrastructure Renewal Bond Issue that will provide 303 million dollars to New York City. When matched with available Federal-aid and regular State appropriations, it will finance a 2.7 billion dollars, five-year programme for the improvement of the city's roads. A portion of the proceeds will be used for MTA projects.

VI. RESULTS

Financing

The MTA is generally on schedule in obtaining the funding for its capital programme. Through July 1, 1985 the MTA had received 67.9% of its planned funding for the Five Year Plan. In August 1985 the Authority sold an additional 150 000 million dollars in Commuter Facilities Revenue Bonds and in September, 125 million dollars in Transit Facility Revenue Bonds.

Through the middle of August, the MTA had committed 3.67 billion in capital projects for the New York City Transit Authority. This represents 58.3% of the TA Capital Plan. On the commuter railroad side, 1.35 billion dollars has been committed or spent, representing 63.1% of the commuter railroad capital plan expenditures.

Purchases of new rolling stock and buses are closest to full commitment with over 90% of the funds in these categories committed. Line structure and line equipment repairs, shop reconstruction, and depot modernisations have lagged due to design delays and longer than expected lags in beginning construction. Passenger station modernisation was delayed due in part to disagreements over handicap accessibility which have now been resolved.

Obstacles

One problem faced by the MTA was a US Commerce Department ruling that the Canadian government unfairly subsidised a deal for 825 new subway cars. Another negotiated contract with a Japanese firm came under Federal scrutiny. The MTA has since waived the Japanese offer of financing because low interest rates are now available.

While many subway cars are being manufactured by foreign firms, they contain a substantial proportion of American parts. The MTA has applied pressure on manufacturers to maximise the New York State content and to have the vehicles assembled in the State. As a result, 225 subway cars ordered from a French-Amercian consortium will contain 20 per cent New York parts and be assembled in a factory in New York City. In a recent negotiation, the MTA arranged for the purchase of 54 commuter rail cars under a contract in which the manufacturer voluntarily meets the "Buy America" standards.

Quality of Service

The new programme has already resulted in a turnaround in service quality (Table 7). Kilometres between equipment failure increased by 34 per cent for subway and by 34 to 300 per cent for buses between 1981 and 1984.

Table 7. KILOMETRES BETWEEN EQUIPMENT FAILURE

	Subway	Bus Manhattan	Bus Other boroughs
1978	21 552	1 565	640
1979	17 536	1 445	510
1980	13 136	1 149	504
1981	10 624	1 210	526
1982	11 458	1 738	741
1983	13 939	1 239	2 294
1984	14 207	1 620	2 204

VII. ENVIRONMENTAL IMPLICATIONS

Travel Conditions

The travel conditions experienced by New York City transit commuters are not pleasant. Subway stations are dark and dingy. Trains are noisy, dirty and overcrowded. Packed buses bounce along poorly maintained streets.

The Transit Rehabilitation Programme offers much needed improvements. The impact of the Noise Abatement Programme has already been discussed. Joint development and subway stations will help pay for some environmental improvements as well as provide better connections to street level activities. The actions being taken by the New York City DOT to rebuild the road infrastructure will complement the new bus purchases. Bus travel will be smoother, buses less prone to breakdown and service less overcrowded.

While these improvements to travel conditions are significant, the impact of the transit improvement programme on the region's environment is less clear. Table 8 summarises trends in air pollution emissions in the New York Metropolitan Area.

Transit Improvements and Air Quality

The 1970 Federal Clean Air Act required each state to develop a State Implementation Plan (SIP) to outline its efforts for meeting national air pollution standards. The authors of the New York State SIP accordingly consider the potential of transit improvements to contribute to cleaner air. This necessitated looking in turn at the role of transit in the region's travel, the effect of increases in transit use on emissions and the scope for increasing travel by bus and subway. In pursuing this set of interactions the SIP authors noted that the MTA's rail transit systems serve primarily Manhattan work travel, and that even a rail system as extensive as New York's is ill-suited to compete for the dispersed trips of non-central business district commuters. This is borne out by travel patterns in the New York Air quality region where the 10% of commuters to the Manhattan central business district who drive to work represent only about 11% of all car commuters in the region. Having established the role of transit in the region the authors of the plan estimated that every 1% increase in MTA transit ridership due to service improvements would result in a daily reduction in regionwide volatile organic compound (VOC) emissions of 0.18 tonne.

Estimating the increase in ridership likely to follow from MTA service improvements over the next five years proved more difficult but a 5% increase was considered optimistic. The authors of the SIP accordingly concluded that the potential transit improvements to improve air quality was small. "To the extent that transit improvements result in automobile drivers switching to transit, automobile emissions would, of course, be reduced with resulting air quality benefits. However, there are reasons to doubt that rehabilitating the transit system and making it more reliable will have any significant effect on automobile use".

The SIP also looked at the likely effects on air quality of such transportation system management measures as exclusive bus and carpool lanes, parking controls and pedestrians zones. Analysis was confined to control measures that could be quantified and the conclusion was that they would bring about less than a one percent reduction in VOC emissions. Furthermore, it was felt that even this small reduction could only be achieved by imposing measures so onerous as to be infeasible.

The transport measure identified as having the greatest potential to improve air quality was inspection for hydrocarbon and carbon monoxide tailpipe emissions and mandatory repairs begun in the New York Metropolitan Area on 1st January 1984. Under this programme failed vehicles must be repaired and pass a re-test in order to receive a registration renewal. These tests are expected to cut hydrocarbon and carbon monoxide levels significantly by 1987. Hydrocarbon emissions from vehicles subject to the programme are expected to be reduced by 38 and carbon monoxide emissions by 34 per cent. These reductions, when combined with the larger reduction expected from normal vehicle turnover, should result in an overall cut in hydrocarbon emissions of 65.5 and that in carbon monoxide emissions of 44.9 per cent in the Metropolitan Area. In 1980 highway vehicles contributed 47 per cent of the VOC emissions in the NYMA. By 1987 their contribution is expected to have fallen to only about 26 per cent.

As a result of vehicle exhaust controls, most hot spots are expected to reach attainment by 1987, and many sooner. Carbon monoxide emissions are expected to be cut nearly 50 per cent by 1987 and

Table 8. TRENDS IN AIR POLLUTANT EMISSIONS
IN THE NEW YORK METROPOLITAN AREA

10^3 metric tonnes

	Mobile sources		Stationary sources		Total	
	Mid 70s	Early 80s	Mid 70s	Early 80s	Mid 70s	Early 80s
NO	168	132	138	142	306	274
HC	156	71	—	—	156	71
CO	2 234	1 438	—	—	2 234	1 438
VOC[1]	170 771	71 251	161 264	138 312	332 035	209 563

1. Volatile organic compounds.

there will be few, if any, hot spots that cannot meet the standard by 1987.

Even though the authors of the SIP do not foresee that transit rehabilitation will have much effect on the region's environment, the 1979 fuel crisis demonstrated the potential role of transit to improve environmental conditions. During the peak of the crisis, weekday traffic volumes were down by 15 per cent and speed in midtown Manhattan had jumped by nearly 40 per cent. This combination of increased speed and decreased traffic volumes resulted in a dramatic decline in emissions. It is estimated that carbon monoxide and hydrocarbon emissions were down by nearly 40 per cent while nitrogen oxides emissions declined by approximately 20 per cent. During this same period subway and bus ridership in New York city increased by almost 10 per cent on weekdays and between 15 and 20 per cent at weekends.

VIII. EVALUATION

New York's Dependence on Transit

There is unanimous agreement that an efficient mass transit system is essential to the future of New York City. The street network is clogged with cars and the continuing boom in office building in Midtown Manhattan is adding to already intense land use. In response to the enormous financial burden that this entails, the MTA has developed an innovative funding programme that employs private sector techniques and relies heavily on the market for funds.

Pros and Cons of Private Financing

Private sector financing offers both advantages and disadvantage. On the positive side, it injects public authorities with market place disciplines and a concern for return on investment that is often lacking. In the case of the MTA, since it will need to retain access to capital markets, it will be forced to continue to pay close attention to financial and operational results, and to give great consideration to the services offered and the efficiency with which they are undertaken.

The private sector has proven to be the only source of funding sufficent to meet New York City's enormous financial requirements. Existing Federal assistance programmes could not have done it. Therefore, the MTA has used private sector funding to supplement, not to replace, government grants. The latter, with their associated "red tape" and Federal "Buy America" requirements are still used for projects where the impact of such constraints can be minimised.

In order to minimise costs, the MTA has sometimes had to ignore stated nation goals such as "Buy America". This may have had a negative impact on employment in the United States, but it has had positive implications for international trade. Furthermore New York has structured several of its large procurements to that the equipment will be partially manufactured or assembled in the State.

Lease financing has altogether different side effects. While the MTA derives benefit through reduced capital costs, the Federal Treasury loses tax revenues that would have been paid by the purchaser of the equipment. The MTA must, of course, share the benefits of each transaction with the purchaser of the equipment. In addition, a fee must be paid to the organisations involved in arranging the transactions.

A different concern is the high debt service expense that future MTA budgets will have to carry if interest rates rise. The MTA is exploring various devices to shorten the length of the debt commitments and stretch out the issuance period. Vendor financing and the use of bank-backed notes exemplify such devices.

The Consequences of a Fall in Revenue Subsidies

The private funding obtained by the MTA is dependent on certain Federal State and City contributions to transit operating costs. If these subsidies were to fall short of expectations, the MTA would be forced to make up the difference by raising fares. Thus the State's 1981 Transit Legislation authorised the MTA to issue bonds backed by the gross revenue of the transit system, and ultimately secured by fare revenues. Likewise the rate convanent of the 2 billion dollars of obligations, a key financing element for the Capital Programme, specifies that the MTA will, if necessary, raise either mass transit or commuter rail fares to cover operating costs and bond debt service for each of the systems. Elected officials would no doubt attempt to minimise the impact of any such fare increases on the public. However, a revenue feasibility study shows that in the worst event — the halting of unrestricted governmental transit subsidies — the city subway and bus system could still retain enough riders to meet costs, even if the fare was boosted from its present 0.75 dollar to 3.04 dollars in 1992 (the equivalent of a 1.41 dollar fare in 1982 dollars).

Beyond the Current Capital Programme

The current five-year capital programme will provide only half of the 14 billion dollars in funding required to complete the transit rehabilitation. Further funding will therefore be needed to improve safety, security, comfort and reliability to levels that promise to increase ridership.

New fare schedules based on trip distance and time of travel, are seen by the MTA as a key method of attracting new ridership, particularly during off-peak periods. Automatic fare collection, which could be used to tie fares closely to the marginal costs incurred by a particular trip, will be necessary to achieve this objective. Assuming that more revenue can be generated in this manner, then more resources will be available for improvements in service.

Agency Co-operation

The rehabilitation of the New York City bus service is greatly dependent on measures undertaken by the New York City Department of Transportation. Key measures are discouraging auto use through capacity limitations and strict enforcement of parking regulations. Such interagency co-operation is an essential part of the transit improvement programme.

Econometric analysis has shown that transit riders in New York have few alternative ways of travelling and are generally more sensitive to changes in service than in fares. Furthermore, ridership gains due to capital improvements will generate additional travel then will more than make up for losses due to increased fares. Similar studies are being prepared for the commuter railroad capital programme.

Environmental Impacts

Transit improvements are essential to maintaining current levels of bus and subway ridership and to holding down auto usage, but they are not expected to have significant influence on cleaner air. Mandatory inspection of tailpipe emissions and repairs is a more effective technique and can be expected to bring about significant reductions in hydrocarbon and carbon monoxide levels. Public transit can, however, be expected to reduce the extent to which streets are clogged with vehicles and so reduce the extent of local environmental degradation.

IX. CONCLUSIONS

New York City is the centre of the largest urban area in the United States and commuters to its central business district are heavily dependent on rail and road transit.

Subway travel has been in decline since 1947 and bus travel since 1970 although the latter has declined less severely than the former. Commuter rail ridership has increased over the last 10 years.

Concentrations of air pollutants have been decreasing in the New York Region for 15 years and national ambient air quality standards are now being achieved for lead, nitrogen dioxide, sulphur dioxide and suspended particulates. However carbon monoxide and ozone concentrations still exceed national ambient standards and carbon monoxide 'hot-spots' are frequent.

A transfer of funds from investment and maintenance spending to operating subsidies caused transit service to deteriorate throughout the 1960's and 1970's and led in 1981 to the declaration of a "transportation emergency". This led to decisions to undertake an 8.5 billion dollars capital improvement programme from 1982 to 1986.

Noise abatement is an important part of the renewal programme and will cover the modification of subway rolling stock, wheel truing, rail grinding and acoustic barriers at stations.

The enormity of the investment programme means that funding is being sought in the market place as well as from Government grants and subsidies. Amongst sources of private funds being tapped are sale-and-lease-back of rolling stock, vendor financing of new equipment, bonds backed by farebox revenue and subsidies, and the involvement of adjoining property owners in station modernisations.

The tieing of bond issues to fares and subsidies is expected to be a discipline on New York transit operating companies and will necessitate a raising of fares if revenue falls below expected levels. Assuming all subsidies were withdrawn the flat fare would rise from 75 cents of 1.41 dollar (1982 dollars) over the ten year period to 1992.

Financial collaboration with property owners over station improvements, including certain noise abatement measures, is dependent on zoning changes or other financial benefits being offered to the property interests.

The improvement of bus services is heavily dependent on the regulation of traffic and on bus lanes.

The transit investment plan is expected to bring about a major improvement in subway and bus travelling conditions in New York. Services will be more dependable, riding quieter and smoother, stations and stops newer and cleaner. It is not, however, expected to make a major contribution to clean air since 89 per cent of car commuters in the New York Region do not have Manhattan destinations and have journeys ill-suited to the existing radial rail system.

The transport measure identified as having the greatest potential to improve air quality is the inspection of vehicle exhausts for hydrocarbons and carbon monoxide emissions coupled with mandatory maintenance. Such a programme was got underway in the New York region in January 1984. These vehicle exhaust controls, coupled with normal fleet renewal, are forecast to cut HC emissions by nearly one half and to eliminate most 'hot-spots' by 1987.

The effects of the 1979 fuel crisis indicate that a reduction in weekday traffic volumes of 15 per cent brings about speed increases of nearly 40 per cent in Manhattan. Emissions of CO and HC would be cut by about 40 per cent. NO_x would be cut by 20 per cent and weekday transit riding would increase by nearly 10 per cent.

A combination of vehicle exhaust controls and traffic limitation measures would be the quickest and most effective way to reduce air pollution in the New York Region and in particular in Manhattan.

REFERENCES

1. Downey, Mortimer L., "Generating Private Sector Financing for Public Transportation", Metropolitan Transportation Authority, New York, 1982.
2. Selsam, Robert E., "Private Investment in Transit Stations", Metropolitan Transportation Authority, New York, 1982.
3. "Metropolitan Transportation Authority: Finances of Mass Transit Services in New York City", prepared by Bear, Stearns and Company, New York, 1982.
4. McNutt, Kenneth A., "Rapid Transit Noise Code Study Progress Report", New York City Transit Authority, June 1983.
5. Dooley, William J., "Update of the Noise Abatement Programme", New York City Transit Authority, August 1983.
6. "Performances Progress Report", Metropolitan Transportation Authority, New York, March 1983.
7. "Automatic Fare Collection Technology Study, Statement of Work", Metropolitan Transportation Authority, May 1983.
8. "New York State Air Quality Implementation Plan", Control of Carbon Monoxide and Hydrocarbons in the New York City Metropolitan Area, New York State Department of Environmental Conservation, Volume 1, revised June 1982 and revised September 1983.
9. Draft proposals for the New York State Implementation Plan Regarding Carbon Monoxide in New York City, submitted to the New York State Department of Environmental Conservation by the city of New York, September 1983.
10. "City Streets: A Report of Policies and Programmes", New York City Department of Transportation, July 1983.
11. "Operation Clear Lanes", New York City Department of Transportation, 1983.
12. "Fuel Crisis 1979 Impact on Traffic Flow", New York City Department of Transportation, October 1979.
13. "The 1980 Transit Strike: Transportation Impacts and Evaluations", New York City Department of Transportation, April 1980.
14. "Recent Trends in Traffic Volumes and Transit Ridership", New York City Department of Transportation, July 1983.

ANNEX
KEY FACTS AND FIGURES ON NEW YORK CITY AND ITS TRANSPORTATION SYSTEM

	Factor/area	Fact/figure	Comments/units
Population			
1983 (est.)	New York metropolitan area	13 206 600	within MTA Service boundaries
1983 (est.)	New York City (1 044 km^2)	7 073 500	within boundaries
1983 (est.)	Manhattan (57 km^2)	1 425 300	within boundaries
Employment			
1985 (est.)	New York City	3 708 800	all employment
1985 (est.)	Manhattan	2 509 000	all employment
1984 (est.)	CBD	2 030 000	all employment
Density – Population			
	New York City	6 775	persons/km^2
	Manhattan	25 005	persons/km^2
Travel pattern			
1980	*Journey to Work trips:*		US census data
	pedestrian, cycles, mopeds	161 907	(Manhattan bound)
	by car or taxi	210 136	
	by public transportation	1 101 151	
	Manhattan-based		
	Modal split: Work trips		
	Manhattan-based	N/A	public transport/cars
	CBD-based	90/10	"based" origin and/or destination in the area
Public transport system			
	Subways:		
	Total length of routes	368 km	
	Passengers per year	1 billion	
	Buses		
	No. of routes	244	
	Total length of routes	1 598 km	
	Travel speed	12.3 km/hr	Avg. of avenues and streets
1983	*Car transport system*		
	Car Ownership – Manhattan	161 494	Passenger cars only
	Private cars per 1 000 inhab.	111.3	
	Car Ownership – New York City	1 514 391	
	Private cars per 1 000 inhab.	214.1	Passenger cars only
	Road network		Streets and Highways
	Length of network	9 920 km	
1979	Freeway route kilometres	346.7	New York City only
	Safety		
	Ratio No. of people killed or injured in road accidents to the total population	.0129	New York City
1983	Travel speed Manhattan:		
	North-South Avenues	9.8 mph	Weekday 7am-7pm avg.
	East-West Streets	6.0 mph	Weekday 7am-7pm avg.

Chapter 8

OSAKA*

I. COMPREHENSIVE TRANSPORTATION POLICY

Context

Since 1975 the population in Osaka City decreased by 130 000 people (5 per cent), and it was at 2.65 million in 1980. However, from 1982 to 1983, it has increased, though only very slightly, by 3 200 people, and the central part of the city is also seeing some growth (Table 1).

Some 2.47 million people (as of 1981) work in Osaka City. Only 900 of these people are in primary industries, as compared to the 700 000 (28 per cent) in secondary industries and 1.78 million (72 per cent in tertiary industries, indicating significant growth in the latter two of these. Approximately 1.02 million (41 per cent) work in the city's central four wards, which cover no more than 9 per cent of the total city area (Table 2).

The number of people who commute from the outside of Osaka City to work or go to school every day was 1.25 million in 1980, an increase of 30 000 over 1975. On the other hand, the number of people who leave the city every day to work or go to school was 240 000, an increase of 20 000 people. As a result of this daily exchange of people, the population of Osaka City in the daytime increases by about 1.01 million people to 3.65 million, or 1.38 times the night-time population. (In 1975, this ratio was 1.36 times).

Particularly, this influx of people from neighbouring areas to the city's central four wards is 860 000, or 69 per cent of the total influx, increasing the population over that at night by 6.1 times. (In 1975, this ratio was 5.1 times). This daily flow of people concentrated in specific areas presents serious problems for the urban transportation.

These problems are as follows:

- The risk of activities in the city stagnating due to the decline in the quality of urban transportation, including overcrowding on the trains during rush hours and incessant backups and slowdowns in traffic flow;
- The undesirable effects of motor traffic on the living environment (such as vibration, noise and air pollution; traffic accidents; deterioration of conditions for pedestrians; the passage of motor vehicles into and through narrow streets in residential areas);
- The vicious circle created by the financial difficulties of operating public transportation systems and the deterioration of services caused by users' distrust and flight from public transportation.

Table 1. RESIDENT POPULATIONS AREA AND POPULATION DENSITY

	1975	1980	1982	1983
Osaka City				
Area (km^2)	208	211	210	210
Population (in 1 000 of people)	2 779	2 648	2 623	2 626
Population density (people/km^2)	13 353	12 554	12 367	12 383

* Case study prepared by Mr. K. Morita (Japan).

Table 2. WORKER POPULATIONS
In 1 000 of people

	1975	1981
Osaka City:		
Central 4 wards	979	1 023
Remaining part of Osaka City	1 394	1 450
Total	2 373	2 473
of which:		
Primary industries	0.9	0.9
Secondary industriies	786	697
Tertiary industries	1 585	1 776
Osaka metropolitan area		
(primary perimeter)	1 819	2 045
Osaka metropolitan area		
(secondary perimeter)	2 203	2 344
Total	6 395	6 862

Aims and Objectives of the Comprehensive Transportation Policy

Osaka has a variety of urban facilities at its disposal as a result of past investments. The stimulation of diverse cultural, economic and social activities could be achieved by better use of existing facilities. This would also reduce the need for additional investments and resources. In the past the transportation system has provided a strong support for convenient living and vigorous economic activities and has served greatly in the development of the city. If future urban management goals are to be achieved, the following aims should be given careful consideration:

— Providing better accessibility for everyone living and/or working in the city;
— Providing transportation services which will meet public requirements of convenience, reliability, frequency and comfort;
— Abating pollution associated with or generated by transportation and traffic;
— Utilising resources and facilities to the full.

It is very important to establish the basic concept that motor cars, buses and rail services should all be treated as elements of the overall transportation system and not as independent and unrelated modes. Moreover, a comprehensive urban transportation policy should be formulated in close coordination with land use planning.

The general objectives of Osaka's transportation policy are as follows:

— Dependence upon motor cars should be decreased;
— Public transportation systems should be drastically improved, both in terms of quality and quantity, to stimulate the change from motor cars to public transportation;
— The living environment, particularly around streets and roads, should be improved to cater to the needs of residents;
— A highly accessible transportation system should be created to foster and meet the demands of more vigorous and better quality urban activities.

More specific targets of the *comprehensive transportation policy* include the following:

— Betterment of public transportation facilities: The improvement and strengthening of public transport by incorporating the existing railways, buses and so forth into a new city-wide Ride-and-Ride system;
— Ways and means of restricting use of motor vehicles: The total volume of motor vehicle traffic within the city is to be limited to 22 million vehicle-kilometres/day and should be accommodated on the trunk roads where some environmental protection measures have been taken. In addition, freight movements should be improved by measures such as the creation of distribution depots, consolidated distribution and joint home delivery systems, together with parking controls;
— Environmental measures: Strong measures should be taken to eliminate air pollution, noise and vibration by enforcing zoning restrictions in residential areas so as to effectively protect the living environment from motor traffic.

II. TRANSPORTATION

New Aspects of the Movement of People

In the Kyoto-Osaka-Kobe metropolitan area (called the Keihanshin area) of which Osaka City is the centre, two person-trip surveys were conducted, one in 1970 and one in 1980 (Table 3). Significant changes from 1970 to 1980 revealed by these surveys are as follows:

Table 3. PERSON-TRIP SURVEY RESULTS
Trips related to Osaka City

A. Transition of trips by purpose and transportation mode

In 1 000 person-trips/day

Purpose		Railway	Bus	Car	Walk or 2-wheeled vehicle	Total of all modes
Commuting	1970	1 120	95	272	312	1 798
	1980	1 208	47	335	316	1 906
Business	1970	367	57	766	557	1 747
	1980	430	34	809	768	2 041
Personal business and recreation	1970	432	92	181	1 463	2 168
	1980	447	50	159	1 577	2 233
Total of all purpose	1970	3 708	475	1 721	4 743	10 648
	1980	3 960	260	1 865	4 888	10 973

B. Transition of modal split of transportation to and from the train station

In 1 000 person-trips/day

	Bus	Taxi	Car	2 wheeled vehicle	Walk	Total
1970	438	44	25	16	4 357	4 881
	(9.0 %)	(0.9 %)	(0.5 %)	(0.3 %)	(89.3 %)	(100 %)
1980	302	35	22	209	4 495	5 066
	(6.0 %)	(0.7 %)	(0.4 %)	(4.1 %)	(88.7 %)	(100 %)

C. Modal split of transportation at peak and off-peak hour, 1980

	Railway	Bus	Car	2 wheeled vehicle	Walk
Peak (8:00 - 9:00)	49.5 %	2.5 %	10.0 %	8.7 %	29.3 %
Off-peak (10:00 - 11:00)	26.5 %	2.5 %	19.6 %	18.7 %	32.7 %

— Despite the 330 000 decrease (11 per cent) in population from 2.98 million to 2.65 million during this decade in Osaka City, there was a slight increase in the total number of person-trips related to Osaka City of 320 000 trips/day, from 10.65 million trips/day to 10.97 million trips/day;
— The use of railways, motor vehicles, walking and two-wheeled vehicles as a means for making these trips increased, but the use of buses declined drastically by 45 per cent, from 480 million trips day to 260 000 trips/day;
— The use of two-wheeled vehicles increased 2.5 times, from 550 000 trips day to 1.4 million trips/day. Of these, the use of two-wheeled vehicles for transportation to the train station increased 13.1 times, from 16 000 trips day to 209 000 trips day. Moreover, excluding motor bikes, the use of bicycles alone increased amazing 14.5 times;
— A look at the percentages of the different means of transportation according to purpose tells us that walking to work decreased from 12 to 7 per cent, indicating that the percentages for the other means of transportation increased. When considered by region, the use of trains for commuting to the central eight wards in Osaka City increased, and the use of

two-wheeled vehicles and motor vehicles for commuting to the peripheral areas within the city increased. On the other hand, walking as a means to conduct business increased from 25 to 32 per cent, and the use of motor vehicles as a same means, though the number of motor vehicles did not change, dropped from 41 to 36 per cent;

The situation in 1980 had the following characteristics:

- The 10.97 million trips/day for Osaka City alone accounted for 25 per cent of the 43.5 million trips/day for the entire Keihanshin area;
- Of these, about 5 million trips/day, or about half, were for the central eight wards (approximately 18 per cent of the total city area);
- Of particular interest is that more than 1/3 of the trips made for business in the Keihanshin area were for Osaka City and that 2/3 of these were for the central area accounted for the largest share of total business traffic;
- A look at the use of different transportation in different areas shows that the ratio of the use of trains to the use of motor vehicles was 3/1 for the central eight wards and 4/3 for the peripheral areas within the city, indicating a much higher rate of utilisation of trains in the central eight wards and significantly higher use of motor vehicles in the peripheral areas;
- This same ratio for commuting was 9/1 for the central eight wards and 3/2 for the peripheral areas, indicating an even greater difference between the two areas.

Public Transportation

Railways

The total length of railways operated in Osaka City as of 1983 was 229 km. Of this length, 53 km was operated by Japanese National Railways (JNR), 78 km by private railway companies (suburban lines: 67 km; streetcar lines: 11 km) and 98 km under municipal management (subway: 91 km; automated guideway transit systems: 7 km) (Table 4).

In recent years all new lines have been opened under municipal management, and JNR and private lines have not constructed any new lines since 1970.

In order to supplement and strengthen the existing network of railways, subway lines are being extended both within and outside of the city. As of 1984, two new lines leaving the city and one intracity line, totalling 10 km, are under construction. These lines will either run in cooperation with or be connected

Table 4. RAILWAYS NETWORK
Except streetcar lines and automated guideway transit systems

	1979	1982
Number of lines		
Osaka metropolitan area	68	72
Osaka City	25	25
JNR	6	6
Private lines	13	13
Municipal subway	6	6
Length (km)		
Osaka metropolitan area	1 286	1 366
JNR	480	484
Private lines	725	781
Municipal subway	81	101
Osaka City	196	209
JNR	53	53
Private lines	67	67
Municipal subway	76	76
Number of stations		
Osaka City	160	172
JNR	35	35
Private lines	64	64
Municipal subway	61	73

with other lines, thereby expanding railway service considerably. Combining lines currently under construction with existing lines, the total operating length of subway lines will exceed 100 km, thus fulfilling, for the most part, the first stage objective of satisfying major transportation demands. In the second stage, remodeling of train stations to accommodate more cars/train, construction to promote development in the city and construction to eliminate areas presently lacking railway service will be undertaken (Table 5).

JNR and private lines on the other hand, do not build any new lines. They are constructing elevated crossings where existing railway cross roads, in order to 1. eliminate ground level crossings that greatly hinder automobile traffic and 2. introduce higher speed to these lines and thereby improve transport efficiency. From 1971 to 1982, 60 ground level railway crossings on 6 lines were eliminated. This construction of two-level crossings is meant to improve traffic flow on roads, but it is also used to renovate railway facilities. The costs of this construction, in the case of JNR lines, are born 90 per cent by the organisation managing the road and 10 per cent by JNR; and in the case of private lines, 93 per cent by the organisation managing the road and 7 per cent by the private line. As of 1983, 339 ground level railway crossings still remained in the city. In the future 117 of these will be eliminated by either elevating or putting underground 37 km on 12 lines on which trains operate at high speeds and which largely hinder the flow of traffic. Currently, JNR is working on 6 km (two lines) and private lines are working on 8 km (four lines). When this construction is completed, 50 crossings will be eliminated.

Railways are built or expanded based on a long-term plan approved by the central government, but since the first stage of the current plan is nearly complete, Osaka Prefecture and Osaka City submitted a proposal in 1982 for a future railway network in Osaka City and the surrounding area. This proposal calls for a total of 390 km on 24 lines to meet the increase in transport demands expected by the year 2 000. It is necessary that this long-term plan be authorised.

Table 5. TRANSITION OF PUBLIC TRANSPORT NETWORKS
Osaka City-managed

	1970	1980	1982
Subway			
Operating length (km)	64	86	89
Number of stations	52	70	73
Number of cars	629	792	798
Distance travelled (in 1 000 car-km/day)	148	179	195
Passengers (in 1 000 people day)	1 908	2 183	2 288
Minimum fare (yens)	30	100	120
New tram[1]			
Operating length (km)		7	7
Number of stations		8	8
Number of cars		52	52
Distance travelled (in 1 000 car-km/day)		8.5	8.5
Passengers (in 1 000 people day)		21	28
Minimum fare (yens)		120	120
Bus			
Distance length (km)	427	438	437
Number of stations	899	910	913
Number of cars	1 596	1 098	1 030
Distance travelled (in 1 000 car-km/day)	175	98	91
Passengers (in 1 000 people/day)	797	393	351
Minimum fare (yens)	30	130	140

1. 1980 data refer to 1981

Buses

Bus lines in Osaka City are almost all managed by the municipal agency. A look at the transition of these municipal bus lines over the past 10 years shows that the number of buses has decreased by nearly 40 per cent and the operating mileage by nearly 50 per cent. At nearly 60 per cent the decrease in the number of passengers is even greater. Particularly in the central eight wards, bus users dropped in half (Table 5).

However, in some peripheral areas within the city where there is no railway service, bus users hardly dropped at all, and buses have been playing a significant role as feeder transportation for railway. In contrast to this, bus users in areas with railway service has dropped largely due to the prevalent use of bicycles and other two-wheeled vehicles to get to train stations.

Ride-and-Ride Concept

The objective of the Ride-and-Ride concept is to carefully coordinate linear transportation facilities (such as railways, trunk bus lines and automated guideway transit systems) with areal transportation facilities (such as zone buses). This will give continuity to public means of door-to-door transportation and make them more convenient and easy to use.

The important point of this Ride-and-Ride concept is to make it possible to transfer between linear transportation networks and areal transportation networks quickly and easily. To accomplish this, we are constructing transfer terminals, locating bus stops near subway stations, and installing escalators and elevators in subway stations. The contruction of 25 transfer terminals in the city, of which 7 had already been completed by 1984, is planned. Of these, 6 connect railways with trunk bus lines and 1 connects trunk bus lines with zone buses. Furthermore, the locating of bus stops near subways is underway or completed for all of the subway stations for which it is physically possible. And, there are 82 subway stations (including stations for automated guideway transit systems) in the city; of these, a total of 85 escalators are installed in 59 stations, and 11 elevators are in 8 stations, leaving 22 stations without either an escalator or elevator (as of 1983).

Another important point of this Ride-and-Ride concept is to establish bus priority measures to stabilize zone buses of areal usage and secure punctual service.

Zone buses are currently under operation in 13 areas. The running intervals average 15 min. during peak hours and 30 min. during off-peak hours (as of 1981), which is rather low when compared to he respective 5 and 15 min running intervals that existed at the beginning of the plan. Bus priority measures include 1. three traffic regulations: exclusive busway, exclusive bus lane and bus priority lane, and 2. signal control that gives priority to buses. As of 1983, the former traffic regulation provided 9.5 km of exclusive busways, 74.0 km of exclusive bus lanes, and 9.3 km of bus priority lanes, for a total of 92.8 km and a 33.6 km increase over 1975. The latter signal control was introduced in 1977 but had only been extended to 25.0 km as of 1983.

To promote the utilisation of buses, a bus location system has been introduced and the buses themselves have been improved. This bus location system uses an indication system that informs the passenger waiting at the bus stop when a bus is approaching and its destination, and was designed to relieve passengers to be who is queuing up for bus from some irritation. It was first introduced in 1981 and is set on 11.6 km on two bus routes. The improvement includes the introduction of buses with low floors and wide doors to facilitate easy boarding and debarkation by passengers and air conditioning to make the passengers' room more comfortable (average temperatures in Osaka City range from 29°C in the summer to 4°C in the winter). The number of buses with air conditioning is 649 (67 per ent of the total buses) in 1984.

Another important point is the improvement of the operating system in order to make transfers easier. In other words, to offer discounted fares when transferring between trains and buses. This system is now being used for connections between the subway and municipal buses, some connections between the subway and private lines, and some connections between JNR lines and private lines. In Japan, it is generally the case that JNR, private and municipal transportation facilities, fares are independent of each other, which proves to be very inconvenient for passengers transferring between these different facilities. To solve this problem, for example, a common fare system making it easier to transfer between the different transportation facilities should be studied and then introduced. With regard to municipal transportation facilities, a system that made transfers between trunk line buses and zone buses free has been in use since 1974. A discount system, which reduces the combined fare by Yens 60 (14 to 23 per cent) for transfers between the subway and buses, was introduced in 1979. Commuter passes that permit unlimited use of a specified period for the subway and buses have been sold since 1978, but they were used little (approximately 13 000 in 1979) when the first went on sale because of their high cost. However, by lowering their cost 22 per cent 10 months after they went on sale, varying the expiration from one month to one, three and six months in 1982, and carrying out a sales promotion campaign, their use has increased greatly (88 000 in 1983, up 6.8 times).

Bicycles

Recently in Osaka City, the use of bicycles as a convenient means of transportation has been on the increase. The number of bicycles owned in the city has reached 1 300 000, or 1.4/household (as of 1980). Bicycles, of course, are used to carry out daily errands, but their use as a means of transportation to train stations has increased significantly recently. The distances bicycles used to get to train stations are in the range of from 400 to 1 500 meters; 61 per cent of these are ridden for more than 800 meters, which is generally considered the walking distance limit. These figures indicate that bicycles are replacing the bus as a means of access to train stations. This trend is supported by the particularly high use of bicycles in areas where train stations are far apart and bus service is not so frequent.

As of 1982, the number of bicycles at the 144 stations in the city reached 127 000. To accommodate these bicycles, Osaka City is seeking to establish bicycle parking facilities that can handle a total of 78 000 bicycles at 115 stations in the city. Facilities that can handle a total of 51 000 bicycles at 90 stations were already completed by 1982.

Automated Guideway Transit Systems (New Tram)

Automated guideway transit systems serve as a link between buses and trains. Construction of the New Tram automated guideway transit system in Osaka City, was commenced in 1978 to provide a link between new residential areas on reclaimed land and the subways. It began operation in 1981 (total length: 6.6 km) (Table 5).

The New Tram is composed of an elevated track on which a train riding on rubber tires and powered by electricity travels. The cost of construction is much less (about 30 per cent) than that required for subways. And the New Tram is safe, reliable and pleasant to ride on. Noise is also greatly reduced, and it does not pollute the air. Furthermore, the train itself, ticket gates, remote monitoring via TV camera, and opening/closing of the doors on the train and corresponding doors on the platform (track and platform are separated by a glass screen) are all automatically controlled by computer, thereby reducing labour and energy consumption. Unmanned operation of the trains is possible, but currently they are manned to assure correct operation of the relatively new system.

One train can be composed of as many as 6 coaches, but currently only four coaches are used. One coach is designed to carry 75 people, so one train can carry 300 people. The current transport capacity of the system is a maximum 10 000 people hour. As of 1982, the system handled 28 000 people a day, but future demand is expected to rise to 72 000 people a day.

Taxis

Taxis are often used on the following occasions:

- To go to places not readily accessible by trains or buses or at times trains and buses are not operating;
- To get somwhere in a hurry;
- When carrying luggage or parcels;
- When physically handicapped people find it difficult to ride trains or buses.

Of these reasons, the use of taxis to and from train stations and hospitals and in business districts is most prevalent.

According to a trip survey conducted in 1980, the taxi trips in Osaka City are 120 000 trips day, and 96 000 of these, or 80 per cent, are for the central eight wards of the city where most pivotal business is concentrated. Of all trips made using transportation facilities, the share of trips made in taxis is 1.6 per cent for the entire city and 2.5 per cent for the central eight wards. With respect to trips made only for business with the aid of taxis, the number of trips for the entire city is 48 000 (3.3 per cent share) and 42 000 for the central eight wards (4.6 per cent share) (see also Table 6).

Table 6. OSAKA CITY-MANAGED PUBLIC TRANSFORTATION FACILITIES

In millions of yen

	1975	1980	1982
	A. Operating income[1]		
Subway			
Income (A)	55 977	99 586	118 916
Expenditures (B)	62 630	101 561	122 523
(A) - (B)	6 653	1 975	3 607
Accumulated debt	65 608	48 151	50 570
Income (C)	—	49	1 250
New Tram			
Expenditure (D)	—	56	4 943
(C) - (D)	—	7	3 693
Accumulated debt	—	7	7 742
Bus			
Income (E)	18 314	24 676	26 680
Expenditure (F)	29 770	26 201	24 164
(E) - (F)	11 456	1 525	2 516
Accumulated debt	48 216	51 222	44 109
	B. Percentage of Operating Costs Covered by Fare Income Osaka city-managed public public transportation facilities)		
Subway	64.3	71.3	75.5
New Tram	—	87.5	16.7
Bus	37.3	51.2	60.6

1. Building costs and interest is not included in expenditures.

Motor Vehicle Traffic

Road Networks and Motor Vehicle Traffic

The total length of roads in Osaka City is approximately 3 850 km. Of this length, trunk roads wider than 20 m account for 306 km and urban expressways account for 61 km (as of 1980). A look at road construction over the past 10 years shows that this total length has only increased by 7 per cent, but urban expressways have increased by 150 per cent. In 1980, area of the roads was 36.3 km^2, and the road rate was 17.2 per cent. (From 1980 to 1982, 13 km of new urban expressways were constructed bringing the total to 74 km).

The total volume of motor vehicle traffic has levelled off recently, but the share of urban expressways has risen from 12 per cent of the total volume in 1970 to 26 per cent in 1980.

As of 1983, the length of trunk roads under construction was 51 km and urban expressways was 16 km; an additional 112 km of trunk roads is planned (Tables 7 and 8).

Freight Movements

Since 1975, no surveys concerning the movement of freight have been taken in the Keihanshin area, and so up-to-date details cannot be given here. (A major survey is planned for 1985).

According to the 1975 survey, the volume of freight movement in this area was 2.32 million tons/day, and the number of freight movements was 1.17 million cases/day, with Osaka City alone accounting for 28 per cent (653 000 tons day and 43 per cent (505 000 cases day), respectively.

In Osaka City, 94 per cent of the movement of goods is performed by motor vehicles, of which 92 per cent are trucks. Half of the freight movements in the city, or 251 000 cases day, are related to the central area, and the majority of the freight comprises industrial products such as machinery. Privately owned and operated trucks account for 91 per cent of these trips. However, the average load on these vehicles is small, only 0.4 tons, compared with the average 3 tons carried on trucks owned by freight agencies. Most of the private operators use vans and small trucks and run rather short distances with light loads.

Though consolidated distribution systems reduce the number of trips required to move freight and thereby decrease the amount of traffic on the roads, there are many instances where they are not used, even though the companies sincerely want to consolidate freight shipments, due to the desire to protect business secrets in business fields with keen competition. However, in Osaka 3 consolidated distribution

Table 7. NUMBER OF REGISTERED VEHICLES AND TRAFFIC VOLUME
Osaka City

	1970	1975	1980	1983
Total number	532 653	653 189	703 328	750 722
Passengers cars	165 940	257 187	307 525	342 818
Trucks	206 792	243 672	251 015	236 938
Buses	5 055	5 607	3 958	3 749
Others	154 866	146 723	140 830	167 217
Number of vehicles 1 000 personnes	179	234	264	286
Volume of motor vehicle traffic[1] in 1000 veh.-km/day (%)	15 800	18 000	17 700	–
On urban express ways	1 900	3 600	4 600	–
On trunk roads	13 100	13 500	12 300	–
On others[2]	800	900	800	–

1. 1975 data refer to 1976.
2. Volume on others was calculated assuming it accounted for 6.5% of volume on trunk roads.

systems have been introduced since 1980, bringing the total number to nine. Of these three, two were started by textile wholesalers and one by trucking companies.

A recent significant development in the freight shipping business is the rapid growth of home delivery services, which were first started in 1974. These home delivery services are characterised by 1. soliciting of freight deliveries by phone or an easy-to-use freight collection system that works through rice dealers, liquor stores, and convenience stores, 2. large terminals and radio dispatched delivery trucks that deliver anywhere in the country in 24 to 48 hours, 3. easy-to-understand charges for each categorized items based on regions. Freight handled by home delivery services is generally less than 20 to 30 kg, and the sum of the three dimensions is less than 1.2 or 1.5 m. The growth in home delivery services has been particularly remarkable since 1977, with the volume of parcels handled around the entire country exceeding that handled by JNR in 1980 and that handled by the post office in 1982. The number of parcels handled in 1982 was 170 million, which accounts for an average of 2.5 parcels household delivered for the entire year.

Parking Measures

In terms of both frequency and duration, traffic backups in the city are on the rise, and the crowding of roads is continuing. The crowding on narrow roads in the centre of the city where motor vehicles are concentrated is truly serious. The cause for this crowding in the centre of the city is a result of 1. the remaining of old divisions in the city that do not provide sufficient parking space for buildings and 2. while many wholesalers and retailers who have frequent and large-scale commodity movements are concentrated, they do not have sufficient facilities for the sale and disposal of goods and so must often park on the street to load and unload their goods.

Parking on the street is a problem that requires immediate solution in order to eliminate this crowding in the centre of the city. However, to make new parking lots requires that buildings be reconstructed which is an extremely difficult problem in space, finance and hours. Therefore, we are only left with the use of existing parking lots. As a result of a survey of existing toll parking lots, it was discovered that there

Table 8. DISTRIBUTION OF BURDEN OF OPERATING COSTS RELATED TO ROADS
Osaka City
In millions of yen

	1975	1980	1981
National	16 755	28 210	24 368
Osaka Prefecture	–	–	–
Osaka City	28 692	38 086	37 480
Others	858	3 186	1 190
Total	46 305	69 481	63 039

Note: Excluding urban expressways.

were a number of open spaces during the day. Therefore, if a system could be made that would tell drivers where parking lots are available and how to get to those lots, then vacant parking spaces during the day could be used effectively to relieve some of the downtown crowding.

Traffic Control System

A traffic control system has been instituted and is in use in the Osaka area as a means to effectively control motor vehicle traffic on existing limited road networks. This traffic control system 1. has vehicle sensors placed in locations where traffic backups are expected to occur in order to 1 collect information on the number of vehicles on the road and crowding conditions, 2. collects that information in the traffic control centre and analyses it with the aid of computer, 3. automatically controls traffic signals and variable road signs, and 4. thereby attempts to prevent or alleviate traffic backups. The same analysis and control system is used for backups on urban expressways and occasionally restricts the number of vehicles that enter the urban expressway.

As of 1983, a traffic control system covered the trunk roads in almost all areas in Osaka Prefecture (324.5 km) and, particularly, covered the area (136.5 km^2) in the city where traffic backups are frequent. The traffic control centres comprise one main centre in Osaka City and three sub-centres.

The terminal equipment used by this system includes 2 299 traffic signals, 4 460 vehicle sensors, 108 bus sensors '55 centrally controlled variable traffic signs, and 62 traffic monitoring television cameras.

New International Airport

Currently in the Keihanshin area there is only one international airport. This airport is small with an area of only 317 hectares, one runway 3 000 m long and one runway 1 828 m long. Besides this, because it is surrounded by a densely populated area, taking off and landing at night and early morning (9 p.m. to 7 a.m.) is prohibited to reduce the undesirable effects of noise made by airplanes. As a result, there are only 130 000 departures and arrivals a year, which is not enough to satisfy the air traffic demands of a metropolitan airport.

To alleviate this problem, a project to build a new international airport has been started. Considering such environmental factors as the propagation of noise based on the direction of the wind, the constructin of a new airport on a new man-made island 5 km off shore in Osaka Bay and 35 km south of the city centre is planned. A major feature of this new airport will be around-the-clock operation, which will make it possible to also accommodate rapid growth in air cargo. The current plan calls for the airport to be completed in some stages. In the first stage, a 3 300 m runway will be built on about 500 hectares. This will permit 160 000 departures and arrivals a year. Finally, on a land area of about 1 200 hectares, two main runways 4 000 m long and one auxiliary runway 3 400 m long will be built, thus making it possible to handle 260 000 departures and arrivals a year.

Opening of the airport in the first stage is expected in 1992. This requires the reclamation of an island, construction of the runway, the terminal and other related facilities and the access road from the city centre to the new airport, and the development of the area around the airport in a very short period of time. The cost of just the first stage alone, excluding the access road and developent of the area around the airport, is expected to be Yens 820 billion. The new airport comes under the class 1 airports that require the government to construct and maintain them; however, this airport will be constructed and maintained by a new special company that the central government will establish by the special law and provide more than 50 per cent of its capital.

Table 9. TRAFFIC BACKUPS OSAKA CITY

	1970	1975	1976	1977	1978	1979	1980
Frequency of backups (cases)	6 771	15 943	17 857	20 737	28 374	32 380	33 154
Time of backups (hours)	8 863	24 944	31 524	41 980	46 119	53 788	53 718

Note: A line of vehicles longer than 500 m and waiting more than 30 min is considered a backup.

III. ENVIRONMENT

Pollution

Noise

The motor vehicle noise along trunk roads is generally high ranging between 60 and 75 phons and averaging 69 phons in Osaka City (as of 1982). These levels have pretty much remained the same since 1971 (Figure 1).

As a countermeasure against the source of this noise, gradually stricter regulations directed at reducing the noise generated by motor vehicles have been successively passed since 1971. The first stage of these regulations has ended, and from 1982 to 1984, the second stage of regulations, which excludes large trucks and small compact cars, has been established.

As a countermeasure for motor vehicle noise along trunk roads, the law was promulgated (1980). This law requires the appropriate and efficient use of adjacent lands in order to reduce the undesirable effects of motor vehicle noise. And studies concerning the introduction of buildings for use as buffers along these roads are in progress.

Urban expressways in the city are almost elevated and many motor vehicles pass on these expressways. As countermeasures for motor vehicle noise along these expressways, the following are taken: 1. plastic panels are attached to the top of walls on the side of the urban expressway, 2. trees and shrubs are planted on a 10 to 20 m wide area along both sides of some routes as a buffer, 3. funds are also provided to help people living along noisy routes insulate their homes and install air conditioning equipment or ventilation fans to prevent excessive noise from entering their homes. Noise prevention panels are attached to all expressways passing through residential areas, and panels that are becoming old are being replaced by panels that are taller (2 to 3 m). Also, 4 km of green buffer zones along the urban expressway are planned, and 2 025 homes have been insulated against noise in some way (as of 1983).

Furthermore, together with the elevation of railway lines to eliminate railway crossings, roads more than 6 m wide are also being constructed along the sides of these elevated tracks. As of 1983, the construction of five side roads such as these, totalling 16 km along three railway lines, were planned; 8.9 km of the 9.8 km planned along three routes have been completed.

Air

Environmental standards with respect to carbon monoxide (CO) and sulfur dioxide (SO_2) are achieved through countermeasures directed at the source.

Based on both measurements by general environmental monitoring stations located on the roofs of schools and by motor vehicle exhaust gas monitoring stations located at the sides of roads, the average

Figure 1. **NOISE ALONG TRUNK ROADS**
(within Osaka City)

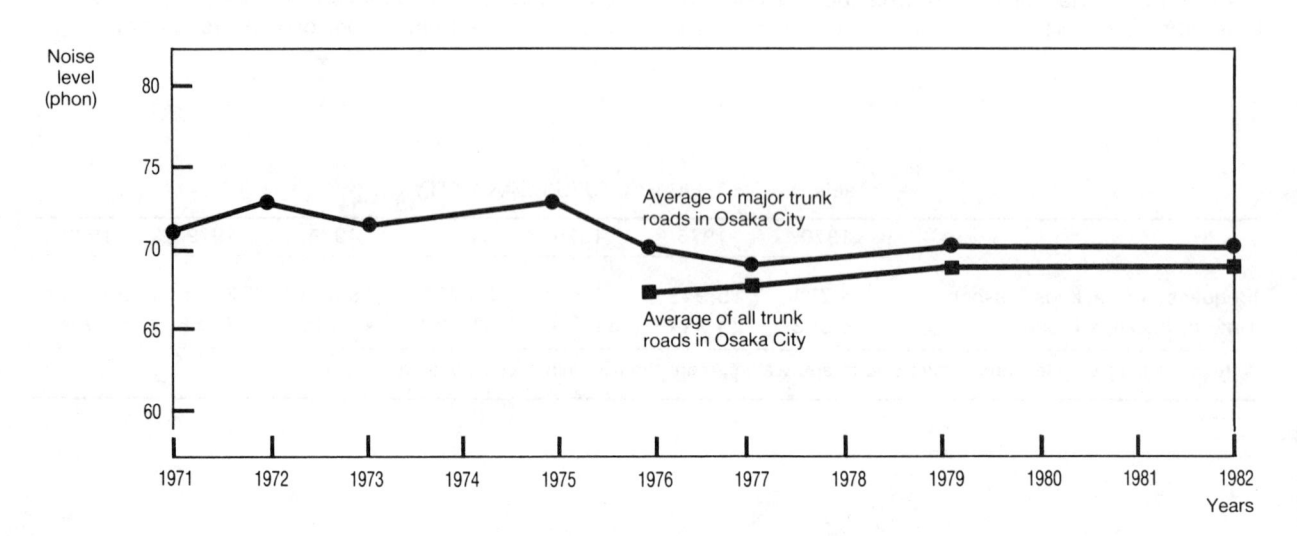

Figure 2. **CONCENTRATIONS OF ATMOSPHERIC POLLUTANTS**
(within Osaka City)

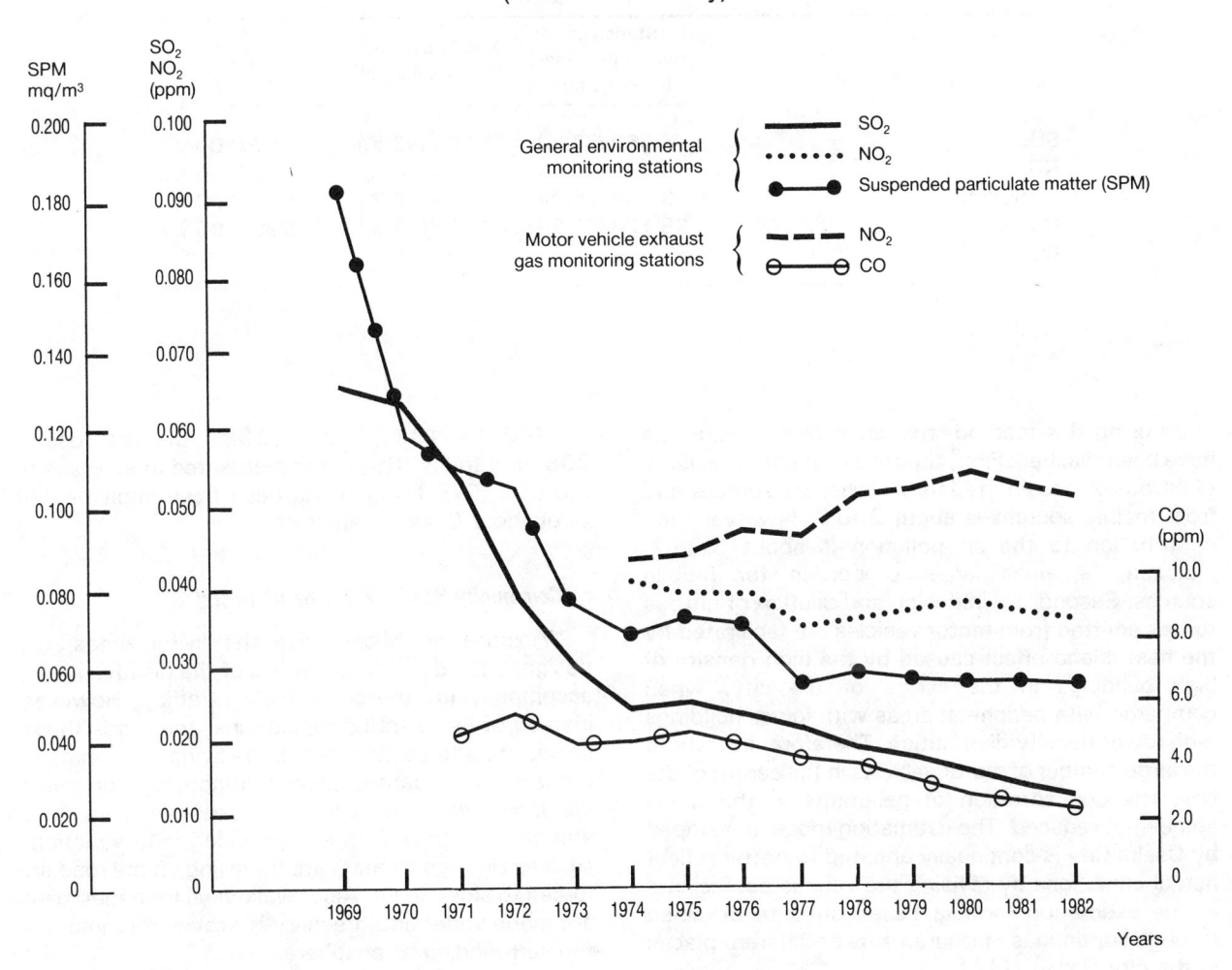

yearly levels of nitrogen dioxide (NO₂) have dropped since 1981. However, when compared with the environmental standard, only of 7 general environmental monitoring stations out of 12 reported levels within the standard, while all 11 motor vehicles exhaust gas monitoring stations reported levels exceeding the standard (1982) (Figure 2).

Factories are being directed to undertake the following two measures as countermeasures against stationary sources of nitrous oxides (NO$_x$):

- Convert fuel from oil to natural gas;
- Attach denitration devices to the exhaust ports for burnt gas.

Moreover, the countermeasures against mobile sources are divided into stages depending on the type of motor vehicle. Although they are being changed to strictly regulate motor vehicle exhaust, as the durability of motor vehicles has been greatly improved and extended it will be sometime before all motor vehicles will conform to these regulations. (Motor vehicles produced before application of these regulations are exempted).

Osaka City is faced with the localisation of meteorological conditions and the complexity of having various sources of nitrogen oxides widely spread throughout the city. In an attempt to get a quantitative understanding of air pollution caused by nitrous oxides emitted from both motor vehicles and stationary sources (such as factories), and then establish well-planned countermeasures against nitrous oxides, the city tried to develop a simulated nitrous oxides estimation technique and, under the cooperation of specialists, has finally developed a practical estimation technique.

Table 10. DISCHARGE OF POLLUTANTS IN THE ATMOSPHERE
Osaka City

	Stationary sources (factories, buildings etc)	Mobile sources (motor vehicles)	Total
SO_x	4 360 (57 %)	3 290 (43 %)	7 650 (100 %)
NO_x	10 370 (40 %)	15 670 (60 %)	26 040 (100 %)
Particulates	3 340 (76 %)	1 070 (24 %)	4 410 (100 %)
HC	39 120 (83 %)	8 170 (17 %)	47 290 (100 %)
CO		55 190	

By using this method, two important conclusions have been reached. First, the ratio of the total volume of nitrous oxides emitted from stationary sources and from mobile sources is about 2 to 3; however their contribution to the air pollution is about 3 to 7, indicating a much larger proportion for mobile sources. Second, the diffusion and dilution of nitrous oxides emitted from motor vehicles are facilitated by the heat island effect caused by the high density of high buildings in the centre of the city, when compared with peripheral areas with lower buildings with lower density distribution. Therefore, in spite of the large number of motor vehicles in the centre of the city, the concentration of pullutants in the air is somewhat reduced. The estimation model developed by Osaka City is continually updated to better reflect actual conditions by revising the various coefficients in the estimation formula based on data obtained through continuous measurements at different places in the city (Table 10).

Measures for the Living Environment

Traffic Regulations in the "Living" Zone

Traffic regulations in the "living" zones are being promoted by the Osaka Prefectural Public Safety Council (Osaka Police Headquarters) with the objective of ridding residential areas of the disadvantages of motor vehicles traffic by restricting the passage of motor vehicles through residential roads to make them safer for pedestrians and bicyclists. These living zones are set within D.I.D. areas in about 1 km² units surrounded by trunk roads. In these zones, traffic regulations such as one-way streets, restriction of large trucks, reductions in speed, no parking, temporary stopping at intersections, and designation of pedestrian roads are incorporated depending on the needs of the zone.

In 1978, there were 188 of these living zones in the city, but with the addition of 12 new zones in 1979, 6 in 1980, 1 in 1981, and 1 in 1982, the total rose to 208, and today they are implemented in all areas of the city. (579 living zones have been implemented throughout Osaka Prefecture).

Community Roads (Yuzuriha no Michi)

"Yuzuriha no Michi", like the living zones, are designed to rid residential areas of the disadvantages accompanying motor vehicle traffic. However, instead of using traffic regulations, the roads themselves are altered to improve the living environment. ("Yuzuru" in Japanese means compromise or give a way). For example, one-way-roads are kept to a standard width of 3 m and provided with variations such as zig zags so that cars travelling on the road are forced to slow down. Also, walkways for pedestrians are made wider and greenery is provided to improve the surrounding atmosphere.

Approximately 15 km on 40 roads are planned for the five year period from 1981, when construction of "Yuzuriha no Michi" first began, until 1985. As of 1983, 6 km on 20 roads has already been completed.

Auto Free Zones

A typical example of restricting motor vehicle traffic in residential spaces is the auto free zone in Nanko where the new Nanko Port Town, planned for 40,000 inhabitants, is being built. From the first stage of planning, it was decided that this town would be an auto free zone. Endorsed by the Osaka Prefectural Public Safety Council (Osaka Police Headquarters), all motor vehicles, including motor bikes, are restricted from passing though during the day. Even emergency vehicles, which are of course excepted from this restriction, must travel no faster than 20 km/hour.

This zone:

1. Is the result of the planned development of the Nanko Port Town;

2. Is surrounded by earthen banks to prevent noise from the outside of the town and to restrict access by cars;
3. Has all the facilities necessary to daily life within walking distance; and
4. Is linked to the subway by an automated guideway transit system (New Tram) that passes through the centre of the town.

Pedestrian Space

Besides taking the above measures to restrict the motor vehicle traffic, efforts are being made to provide pedestrians with safe, pleasant places to walk.

As of 1983, 750 km of roads had been provided with sidewalks to improve the appearance of the city and make the sidewalks easier and more enjoyable to walk on. They have been paved with coloured, water-permeable blocks and decorated with short and tall trees and shrubs. Particularly at intersections with oblique corners in the centre of the city, traditional gardening techniques are being used to make miniature gardens that transform street corners into pretty street parks. A total of approximately 115 km, including that to be landscaped, is being planned for roads for exclusive use by pedestrians, and of this 115 km, construction on 25 km on 11 roads is underway; as for 1983, 20.5 km have been completed.

The historical promenades (walkways connecting historical sites) are planned for a total length of 150 km. In 1983, 46.4 km on three courses were under construction, of which 23.4 km were completed.

Bicycle Roads

The use of bicycles became popular rapidly in recent years, and to accommodate this rise in bicycle traffic, bicycle lanes on roads are being established and roads exclusively for bicycle use are being constructed. A total of 450 km of bicycle lanes is planned, and as of 1982, 364 km (81 per cent) were completed. In addition, 80 km of roads for exclusive bicycle use are planned, of which 28 km (35 per cent) were completed in 1982.

Beautification of Bridges

For centuries Osaka City has been referred to as the "City on the Water" because many rivers and canals flow through the city. The bridges that cross over these waterways are not few. However, many of these bridges are old and narrow, and are not pleasant to use. Osaka City is attempting to improve the appearance of these bridges by repaving them, and modernising the design of the balustrade and lighting for the enjoyment of the pedestrian. As of 1983, six bridges were completed and four were under remodeling.

Traffic Accidents

Traffic accidents in Osaka Prefecture have been on the rise since 1976, until which time they had tended to decrease. The tendency of the number of traffic deaths to decrease also ended in 1979 and has been increasing ever since (Table 11).

A significant feature of this rise in traffic accidents is the rapid rise in the number of accidents imputed to minibikes and motorcycles. The total number of accidents from 1979 to 1983 rose 1.31 times, while the number of accidents imputed to minibikes and motorcycles rose 1.85 times. Moreover, the total number of deaths rose 1.28 times, while the number of deaths imputed to minibikes and motorcycles rose 1.99 times. The percentages of accidents and deaths that minibikes and motorcycles account for in the total figures for these two statistics are 76 and 79 per cent, respectively. Implementing traffic safety measures to reduce these figures is a task that requires top priority.

Table 11. TRAFFIC ACCIDENTS

	1960	1965	1970	1975	1980	1981	1982	1983
Osaka Prefecture								
Accidents (case)	–	29 604	52 968	31 630	35 734	37 747	41 862	44 421
Deaths	935	699	848	433	364	383	383	412
Injured persons	27 933	34 937	74 649	40 989	44 113	46 569	51 523	54 354
Osaka City								
Accidents (cases)	–	16 748	25 242	12 726	13 322	14 262	15 192	15 719
Deaths	467	292	274	130	104	128	131	117
Injured persons	18 325	19 786	34 642	16 029	16 163	17 476	18 448	19 220

IV. EVALUATION

Railways

The development of the railway system in Osaka is progressing satisfactorily, and this is helping to promote the utilisation of mass transits instead of motor vehicles. According to the results of a person-trip survey, the utilisation of motor vehicles in the entire Keinhashin area has grown faster (1.43 times) than the utilisation of railways (1.12 times), while in transportation related to Osaka City the utilisation of both railways and motor vehicles has grown about the same (1.07-1.08). Particularly with regard to the central area of the city, the rise in the utilisation of railways for commuting and business (1.10 times for commuting, 1.23 times for business) is quite a bit greater than that for motor vehicles (0.95 times for commuting, 0.96 times for business). However, in peripheral areas of the city, the rise in the utilisation of motor vehicles for commuting (1.25 times) is quite large. To further promote the utilisation of mass transits instead of motor vehicles, efforts must be made to substantially improve railway service.

Buses

The municipal bus system is currently in a critical situation. With a total debt of approximately Yens 44.1 billion (as of 1982), operation of the bus lines is becoming increasingly difficult. To cope with this situation, the management has attempted to adjust assets by cutting the number of drivers (3,174 in 1970 to 1,929 in 1983) and reducing the number of buses while raising fares (though consumer prices have risen only 2.3 times over the past 10 years, bus fares have risen 4.3 times). As a result of this effort to bring financial balance into management, income from fares was able to cover 61 per cent of the operating costs in 1982 as compared with 37 per cent in 1975. However, such measures reduce the frequency of bus operation and adversely effect service, which in turn, deters people from riding buses. In order to save municipal buses, firm management that will increase productivity is required together with the establishment of a financial assistance programme that will improve and strengthen public transportation services to bring city functions in line with the demands and needs of the times (Table 6).

One scheme to increase the utilisation of buses that is already in effect is the bus location system, and it is showing results. Separate surveys of 1 800 passengers taken on a weekday and a holiday showed that 91 per cent of them thought the system was useful and that 65 per cent of them watched the bus indicator while waiting for the bus. A comparison of the number of passengers before and after the system was installed shows that passengers increased by 11 per cent after bus indicator installation.

The distances of the roads where bus priority measures are being applied is increasing all the time. However, bus priority lanes, for example, are employed on only 21 per cent of the entire length of bus routes. Expansion in this area should be required.

Ride-and-Ride Concept

To further the Ride-and-Ride concept, the installation of facilities (transfer terminals, location of bus stops near subway stations, installation of esclators) that make transferring between means of transportation easier is progressing all the time, and the system is gradually becoming more and more functional. However, in contrast to initial expectations of the plan, we have not been able to use zone buses as we would have liked to in our approach to linear transportation using railways and trunk line buses due to the low level of zone bus service resulting from management difficulties. Moreover, bicycles are replacing buses as a means of areal transportation, and a Ride-and-Ride system in which transfers are made between bicycles and trains is becoming more common.

Due to this Ride-and-Ride plan, facilities that make transfers easier are becoming more common, simultaneously becoming more convenient for the people who use them. However, though these facilities are employed for municipal bus lines, they have not been extended to private bus lines. Furthermore, though there is a national subsidy programme, construction of new terminals by bus lines already far in debt is difficult, and expansion of this subsidy programme is required.

Fare systems that make it easier to transfer between different means of transportation and transportation facilities under different managements are progressing favourably between the subways and municipal buses. Seasonal passes for all municipal bus and subway lines, in particular, have come into common use due to discounted prices, variable use periods and an active sales campaign. Moreover, a discount fare system for transfer, though the discount is small, is being introduced for transferring among JNR, private lines and subways.

Bicycles

In contrast to the sudden decrease in bus utilisation, the rapid increase in the use of bicycles, particularly to and from train stations, is especially remarkable. Reasons for this include 1. the normal distance bicycles are rode (400 to 1 500 m) conforms with the average distance between train stations (1 km in the city centre, 1.5 km in the peripheral areas) and 2. decreased bus speeds and services have caused people to switch to bicycles. Whatever the case, bicycles conserve resources and do not pollute, and they are starting to play a significant role in the Ride-and-Ride system. Osaka City has been working to develop bicycle parking facilities around train stations, and a large number of bicycles can now be accommodated. However, facilities are still insufficient, and many people still park their bicycles on sidewalks or roads.

Furthermore, to protect bicyclists, bicycle lanes are being secured on roads to separate bicycle and motor vehicle traffic. This programme is progressing satisfactorily, but these lanes can only be secured on roads wider than 25 m. On narrower roads we have to alter sidewalks to eliminate kerbs and ridges and bicycles and pedestrians must use the same space. It protects the bicyclists from motor vehicle traffic, but the bicyclists then endanger the pedestrians, which presents a problem we have not yet to answer.

In addition to being used for commuting to work or school, bicycles also serve as a means of healthful exercise and leisure. Bicycle roads are being built along river banks and other such places for people with these objectives. Many bicycle roads have already been completed, and are being well received.

Taxis

Taxis serve to fill the gaps left in the transportation networks of trains and buses. They are also commonly used by people unfamiliar with the area, people in a hurry and people with physical handicaps.

Road Networks and Motor Vehicle Traffic

The total volume motor vehicle traffic in the city has tended to level off. This is due to measures which attempt to reduce utilisation of motor vehicles by improving railway facilities, promoting the Ride-and-Ride concept, and improving the service of public transportation facilities. These measures have made it possible to restrict motor vehicle traffic without the use of direct controls to limit the amount of motor vehicles entering the city.

Urban expressways in Osaka are elevated toll roads, and compose a road network separate from that of trunk roads. Utilisation of these urban expressways is gradually increasing, and they are relieving some of the strain from trunk roads. As a result, some of the trunk roads are being made smaller and the sidewalks wider, greenery is being planted, block pavement is being used, and the corners of intersections are being landscaped, which is all being well received by citizens. This has also made it possible to increase the number of bicycle lanes.

There are currently nine consolidated distribution systems to move freight within the city. In addition to these are there new, fast growing distribution systems, referred to as home delivery service, that perform a function different from that of conventional distribution systems. These home delivery services help to reduce the number of trips by grouping vehicle trips related to freight movements. Furthermore, by taking advantage of the recent remarkable developments in electronic data processing and communication systems, it is becoming possible to achieve certain objectives without moving freight within the city, thereby improving distribution efficiency.

Crowded traffic conditions are a continuing problem, and are adversely effecting the environment. The frequency and duration of traffic backups in Osaka City are increasing. In order to prevent and eliminate traffic backups, efforts are being directed toward the construction and improvement of road networks, and perfection of the traffic control system. Traffic backups on the urban expressways are most common on the central loop, and in order to prevent these backups, we are planning to build new off-ramps on expressways before its connection with the central loop. In this way, we can divert the traffic on these expressways, which does not necessarily use the central loop, before they reach the central loop, and thereby reduce some of the traffic on the central loop. The traffic control system on the trunk roads is gradually being perfected, and as a result, it is contributing greatly to the improvement of traffic flow. Travelling speeds increased by 20.7 per cent, the frequency of stops due to backups decreased by 30.2 per cent, and travelling time was shortened by 16.3 per cent (as a mean value from 1972 to 1982).

Pollutions

The noise problem along trunk roads is as serious as ever. However, we have entered the second stage in which stricter regulations on the level of noise produced by motor vehicles are being enacted as countermeasures against the sources of noise. Furthermore, a system is being developed that would establish buildings along roads to act as buffers. So we do have some hope for the future. The countermeasures against noise be taken along urban expressways are yielding positive results.

One problem that remains with respect to air pollution is the measures needed to reduce nitrous oxides. In an attempt to regulate the overall discharge by stationary sources and motor vehicles are being implemented. As a result, the average yearly levels of nitrous oxides have tended to come down over the past few years, indicating some progress in the reduction of air pollution.

Measures for the Living Environment

The implementation of traffic regulations in "living" zones regulations, too, is showing some positive results. A comparison of the three months before and after they were enacted in 65 districts in Osaka Prefecture in 1982 shows that cases of traffic accident decreased by 9 per cent, deaths by 11 per cent, and injuries by 10 per cent. Damaged property, too, decreased by 5 per cent.

"Community Roads (Yuzuriha no Michi)" make residential areas safe and pleasant, and are helping to expand comfortable living spaces. These changes are welcome by people living along the roadways.

The car free zone being implemented in Nanko Port Town has eliminated traffic accidents and street parking, which is a common nuisance in other areas. However, parking areas on the periphery outside the town were designed on the assumption that only 20 per cent of the households would own a car. The fact is that 40 per cent own cars, and parking capacity is not sufficient.

The improvement of pedestrian space and beautification of bridges is progressing well and is providing a safe, pleasant place for pedestrians to stroll. The "street park" at the corners of intersections are particularly well loved, because they play a role of oasis among the concrete, glass and steel buildings in the city.

Relative safety has improved since the implementation of living zone traffic regulations in all of the districts in Osaka City. However, traffic accidents in the whole city area are on the increase, and there has been a sharp increase recently in the number of accidents imputed to minibikes and motorcycles. It will be necessary in the future to educate the people who ride minibikes and motorcycles, particularly housewives who often ride minibikes, while taking measures to separate automobile and two-wheeled vehicle traffic.

V. CONCLUSION

In order to solve the problems concerning traffic and pollution, Osaka has implemented many measures that have taken advantage of and added improvements to legacies from the past and created something new. For example, by 1935 more than 90 per cent of today's railway system for the metropolitan area was completed, and this is serving a vital role as the skeleton of the metropolis infrastructure. Since then, as the metropolis has grown, new lines have been constructed to handle the increased demands for transportation, but emphasis has been placed on strengthening the transport capacity of existing lines (installing two-track lines, increasing the number of coaches on a train, building new stations). And now, efficiency has been increased by such things as automated ticket gates. And to make riding the trains a little more pleasant, trains are seasonally heated or cooled, and there is even an undertaking to install televisions on some coaches to provide diversion during long commuting rides. It is through these kind of measures that we hope to make public transportation more attractive to passengers. In other words, following original and additional construction of railway lines, we are now at a stage where we must improve their quality. The same can be told for roads. After first increasing roads thus improving road ratio, we have been constructing trunk roads, converting railway crossings into two-level intersections and introducing urban expressways to cope with ever increasing motor vehicle traffic. We are now adding such amenities as landscaping corners of intersections, providing sidewalks and widening present sidewalks. It is the objective of the comprehensive transportation policy in Osaka City to restrict the increase of motor vehicle traffic and switch some of the burden to public transportation, thereby stimulating the potential of the city. This policy is gradually succeeding. And, as a result, though there still remain some problems, not only pollution but also living environment is improving. Housing policy, too, which is closely related to the comprehensive transportation policy, is effectively working to get people back to Osaka City by developing well-organised housing projects with parks, schools, roads and other facilities at the sites of evacuated factories near the centre of the city. The key point to all of these measures is continuous improvement. By continually improving onwards, the city will be able to be a good place to reside, work and enjoy.

Chapter 9

PARIS*

I. TRANSPORT IN THE ILE-DE-FRANCE REGION: PAST DEVELOPMENTS AND PRESENT SITUATION

Travel within the Region

In 1982, the total population of the Ile-de-France region was approximately 10.07 million, including a labour force of 4.8 million, spread over an area of 12 000 km² — a population density of 840 inhabitants per km². Over 18.5 million motorised journeys[1] are made each day by this population, 22 per cent of them, i.e. over 4 million, between 5 and 7 p.m. Paris is the focal point for these journeys, approximately 6.6 million of them having their starting point and/or destination there. This is because of the population/jobs imbalance between Paris (2.18 million inhabitants, 1.82 million jobs) and the suburbs (7.9 million inhabitants, 2.76 million jobs). Commuter travel (home/work journeys) accounts for 36 per cent of the total, and 38 per cent of commuter journeys are in the peak period. The private car is the main transport mode (56 per cent of all journeys, against 29 per cent for public transport and 15 per cent for two-wheelers and school or works buses). During the peak period, the car share falls to 54 per cent, while that of public transport rises to 34 per cent of the total and to over 80 per cent on radial Paris-suburb links.

The average duration of commuter journeys is about 34 minutes, the longest (50 minutes) being Paris-suburb journeys and the shortest suburb-suburb (the overall average is 26 minutes — by car 21 minutes and by public transport 44 minutes).

The very substantial economic activity of the region involves goods transport, a great deal of it by road (over 70 per cent of the 230 million tonnes a year traded in the Paris region).

Transport Networks

Dedicated Public Transport

There are two types of public transport in the region, rail networks (SNCF suburban railways, RER, metro) and bus services (RATP urban and suburban routes, APTR suburban routes[2] (Table 1).

The standard-gauge rail network, essentially a radical system, is for the most part operated by the SNCF; but the RATP, which already operated the Sceaux line, took over first the Boissy-Saint-Léger line — the first section of the Réseau Express Régional (RER) — in 1966, and then the Saint-Germain line in 1972.

The SNCF suburban network (36 lines) comprises 927 km of track, 875 of which are electrified (99 per cent of the traffic). The two RATP RER lines (A and B) at present in service total 103 kms[3]. The network is relatively dense — virtually no point in the conurbation is more than 3 km from a suburban line. In 1982 the SNCF carried 444 million passengers on 5 200 trains a day, while the RATP carried 246 million.

The narrower-gauge metro is operated entirely by the RATP. It is located mainly within the city of Paris (158 of its total 192 km). The network is very dense (one is rarely more than 600 metres from a metro line), especially in the central area, and carries 1 130 million passengers a year on its 11 000 trains a day.

Public Surface Transport

The bus routes operated by the RATP cover two zones, roughly separated by the external boulevards. The urban network which serves the city of Paris comprises 55 routes which complement the metro network either by covering links the metro does not cover or by duplicating overloaded metro lines (lines 1

---

* Case study prepared by Messrs. J. R. Fradin (France), B. Pearce (United Kingdom) and Ph. Bovy (Switzerland).

Table 1. MAIN CHARACTERISTICS OF PASSENGER TRANSPORT IN THE ILE-DE-FRANCE REGION, 1982

Transport Mode	Network	SNCF Rail	RATP RER	RATP Metro	RATP Bus Urban	RATP Bus Suburban	APTR Bus	Taxis	Total	Route Car[1]	Route Two-wheelers
Number of lines	(A)	36	2	15	55	141	465	–	–	–	–
Number of stops	(B)	326	65	360	1 663	4 098	7 000	1 000[2]	14 497	–	–
Total kilometres	(C)	927	103	192	509	1 928	7 041	–	–	45 500	45 000
Passengers per year (millions)	(D)	444	246	1 130	314	410	153	130	2 842	4 500	500
Vehicle stock	(E)	3 001	620	3 548	1 375	2 488	1 569	15 985	–	3 610 000	2 600 000
Km x places offered (millions)	(F)	48 600	15 822	30 784	3 043	6 957	4 644	2 300	110 200	125 000	2 300
Vehicles x km (millions)	(G)	240[3]	57	191	43	100	58	670	1 359	28 000	2 000
Passengers x km (millions)	(H)	7 670	2 703	5 546	758	1 333	950	625	19 585	35 000	2 800

1. This table does not include figures relating to goods transport (utility vehicles), which involves about 4.5 million vehicle/km a year.
2. Number of taxi ranks in Paris and the three départements of the inner suburbs.
3. This figure is estimated, as the SNCF publishes only the total number of train/km on its entire network.

and 4 in particular). The urban routes generally start from one of the "Portes" (city gates) or from a metro terminus just outside and run through the area bounded by the Paris main line stations to terminate at one of them. Totalling 500 km, urban bus routes carried 314 million passengers in 1982. The suburban bus service comprise 146 routes totalling 1 648 km, which carry passengers from areas poorly served by the SNCF or RER to the metro termini or act as feeders for the standard-gauge radial lines (suburban SNCF, RER). 410 million passengers used these services in 1982.

The APTR network, which comprises 63 private bus companies, carries 153 million passengers a year, using 1 500 vehicles on 478 routes totalling over 7 041 kms. The APTR also operates transport services, for the RATP and under its control, in the new towns.

There are, too, "special" bus transport services: school buses (142 000 children carried each day in the Ile-de-France region, or over 25 million journeys a year) and works buses (about 35 million journeys a year) run by firms to carry their employees.

Taxis

Paris taxis are operated either by owner-drivers (8 750) or by taxi companies (7 500 drivers). A total of 15 985 taxis (of which 14 300 in Paris and the three départements of the inner suburbs) carry about 130 million passengers a year. There are about 1 000 taxi ranks in Paris and the inner suburbs.

The Regional Road Network

The road network includes 45 500 km of roads and 578 km[4] of motorways and expressways. Like the rail network, it is still mainly radial, axial routes often being inconvenient and of very variable capacity. The network has to accommodate some 4.2 million vehicles (of which 3.6 million cars and vans), comprising the vehicle stock of the Ile-de-France region. This explains the large volume of road traffic (over 10.4 million journeys a day, about two-thirds of the total vehicle stock being used on week-days, with expressways and motorways carrying about one-third of the total traffic expressed in vehicle-kilometres), and the relatively slow traffic speed (18 km/h in Paris, 20 km/h on the axial networks). This overloading of capacity also affects parking, particularly in Paris, where the excess of demand over supply results in an unduly large amount of illegal parking. The roads are also used by two-wheelers (1.8 million journeys a day, or almost 10 per cent of the total.

Table 2 indicates that in 1982:

— Average rate of use of public transport capacity was only 18 per cent (of seats available);
— This rate is substantially lower than that for private cars (1.25 passengers per 4.5 seats, i.e. about 28 per cent).

There is a very widespread idea that the car is badly utilised; while this is true, the same applies to public transport.

Table 2. BREAKDOWN OF PUBLIC PASSENGER TRAFFIC IN THE ILE-DE-FRANCE REGION AND MAIN FEATURES OF NETWORK UTILISATION (1982)

Transport mode	SNCF Railway	RATP RER	RATP Metro	RATP Bus Urban	RATP Bus Suburban	APTR Bus	Taxis Cars	
Modal Split								
Trips per year (%)	15.6	8.7	39.8	11.0	14.4	5.3	4.6	100 %
Passenger-km (%)	39.1	13.8	28.3	3.9	6.8	4.8	3.3	100 %
Ratios[1]								
Average places offered per vehicles (F)/(G)	203	279	161	71	70	80	3.4	(83)
Average places occupied per vehicle (H)/(G)	31.9	48	29	18	13	16	0.9	(15)
Average capacity utilisation (H)/(F)	0.16	0.17	0.18	0.25	0.19	0.20	0.27	0.18
Average length of journey (H)/(D)	17.3	11.0	4.9	2.4	3.2	6.3	4.8	6.9

1. The characters in parenteses refer to Table 1.

Table 3. SUMMARY ESTIMATE OF THE OVERALL BREAKDOWN OF JOURNEYS
ACCORDING TO TYPE OF INFRASTRUCTURE

	Average number of O-D trips per day at present	Main type of infrastructure used for journeys each day[1]	
		SPECIALISED (Dedicated infrastructures: rail network and motorways)	NON-SPECIALISED (Ordinary roads)
Journeys			
By car	10 400 000 56 %	1 450 000 14 %[2]	8 950 000 86 %
By public transport	5 400 000 29 %	3 560 000 66 %[3]	1 840 000 34 %
Bus, works bus	2 700 000 15 %	135 000 5 %[4]	2 565 000 95 %
Totals	18 500 000 100 %	5 145 000 28 %	13 355 000[5] 72 %

1. These figures are indicative only, the breakdown of multimodal and multinetwork journeys being very complex and difficult to assess.
2. The share of regional journeys using for the most part the motorway network proportion calculated on the basis of vehicle-kilometres adjusted by the average kilometrage for each type of journey).
3. The traffic share of the rail network — SNCF, RER and Metro — calculated on the basis of journeys mainly using the network concerned.
4. Estimate.
5. 13 million motorised journeys each day, to which should be added between 5 and 10 million O-D journeys on foot and probably over 40 million terminal trips and transfers between mechanised modes made on foot.

Both systems have substantial "latent reserves". At present, however, the biggest difference between the two modes is that only the latent capacity of public transport is filled (or even over-filled) in peak periods, while the capacity utilisation of cars remains practically constant throughout the day.

Table 3 shows estimates of the percentage of traffic using two types of infrastructure:

— Dedicated infrastructure (including motorways);
— Ordinary roads.

The fact that according to this estimate 72 per cent of daily journeys[5] use the ordinary road network underlines the importance of having a policy geared to road traffic management.

Brief Survey of Transport Policy in the Ile-de-France Region

The organisation of the passenger transport system in the Ile-de-France has two major characteristics:

— The number of bodies dealing with transport problems, the complexity of the relations between them and the interpenetration of their services;

— The predominant role of the State as regards both decision-making and financing, counterbalanced by the increasing influence of the regional authorities, responsible since July 1976 for defining and implementing a regional transport policy and strengthened in this by the decentralisation policy progressively implemented since 1981.

This state of affairs largely explains why it is virtually impossible, at any given moment, to define the content or even the broad lines of a transport policy. An attempt can be made to determine them after the event by comparing the measures taken or in hand with the proposals in the only document to date covering the whole field of transport, i.e. the overall plan drawn up by the Préfecture of the region in 1972[5]. This plan, which was part of the master plan for the development of the region (Schéma Directeur d'Aménagement et d'Urbanisme de la Région d'Ile-de-France — SDAU), proposed a phase series of measures for immediate, medium-term (1975) and long-term (1985) implementation.

Since the plan was drawn up, a number of factors have either modified the priority ranking of the objectives recommended by the plan or introduced considerations which were not taken into account at the time (e.g. energy). As regards Paris, moreover, as a result of the change in the city's status and the establishment of the traffic plan, the provisions of the overall plan in this sector have become more precise.

Accordingly, from the short and medium term standpoint, it has been found in recent years that:

As Regards Traffic Management and Parking:

- The principles of an order of priority and complementarity in road use have been applied on a very piecemeal basis, particularly in the suburbs, where there has been no improvement in the operating conditions for bus services (very inadequate kilometrage of reserved bus lanes);
- Still very inadequate parking control and not enough pavements returned to pedestrians, despite the recent establishment of a number of pedestrian precincts and the introduction according to plan of park-and-ride facilities in the suburbs according to plan. Some 75 200 parking places were available at 233 stations at the end of 1982 and there will be about 100 000 places in 1989.

As Regards Modernisation and Infrastructure Extension:

- There has been substantial modernisation of the metro and of the bus fleet, but also difficulty in reorganising the road network, especially in the suburbs;
- Very marked slowing of the motorway programme (it had been planned to double the network), particularly as regards the A86 and A87 bypasses, mainly because local populations are increasingly opposing the construction of major infrastructures;
- Normal progress of the railway network extension programme, especially the prolongation of metro lines into the near suburbs;
- Higher priority given to completion of the standard-gauge RATP-SNCF link (Luxembourg-Châtelet, Gare de Lyon underground) and the SNCF network connections (Invalides-Orsay, Massy-Pont de Rungis), scheduled for completion under the IXth Plan.

As Regards Fares and Financing:

- The objectives set out in the overall plan (moderate fare increases, a fairer distribution of costs) have been more than attained as a result of the introduction of the transport levy and the Orange Card and partial reimbursement of employees' season tickets by employers.

As Regards Institutions:

- A progressive approach has been adopted towards the problems of reorganising the regional transport system. Insofar as there has been a change in the responsibilities assigned to the Ile-de-France region in the field of transport and road traffic (July 1976), Paris has been given a new status under the legislation on decentralisation (1982), and by a bill concerning the reform of Paris transport, debated in Parliament at the end of 1983.

At the same time as changes were being made in the content of the programmes planned in 1972, the policies adopted for the medium and longer terms were reshaped and four aspects of them are likely to have a significant influence on future decisions:

- The need to ensure that major infrastructures are more effectively integrated into the environment, in view of the strong reactions when major projects are being implemented, and to improve environmental conditions (in particular noise levels, user safety, obstruction of pavements, etc.);
- The inevitable slowing of big road construction programmes owing to rising costs, the switch of priorities towards public transport, the redefinition of priorities at national level and the increasing consideration given to environmental issues; this should lead to the adoption of a stringent road management policy and recourse to additional sources of finance to extend the road network at a reasonable pace and give public transport a more important role in the suburbs;
- The steady and appreciable increase in the public transport operating deficit, which is likely to lead to efforts to find additional sources of finance if the government were to reduce the substantial share of the operating costs it bears at present;
- The continuing priority given to public transport investment, which should lead to continued extension of the rail networks, completion of the interconnection of these networks and the creation of lighter infrastructures reserved for public transport (buses, trams, etc.), mainly in densely populated suburbs.

The Institutional Framework

One of the paradoxes of the transport situation in the Ile-de-France is the contrast between the global nature of the problem and the fragmentation, both institutional and financial, of the relevant procedures and structures. The regional reform of May 1976 and the subsequent Act of March 1982 (transferring responsibility for the regional executive from the Prefect to the elected Chairman of the Regional Council) should have helped clarify the way in which transport is organised (the region having been assigned responsibility under the 1976 Act for "defining regional traffic and passenger transport policy after consultation with the General Councils and ensuring its implementation"). But the fact that existing positions and interests could not be reconciled has led to the prolongation from year to year of a complex system which is not even in accordance with the law!

There are therefore two main actors in the system: one – *the Region* – which in theory holds all powers over transport and traffic but is far from having all the resources needed to exercise those powers, and the other – *the State* – which has de facto control over the operation of the system through numerous channels, although its responsibilities are not always clearly defined.

The organisation of transport in Paris therefore hinges on these two centres of authority:

First, the central government, whose involvement is explained partly by historical reasons and partly by Paris' role as capital of the country. It exercises its control by:

- Co-financing investment;
- Fixing fares;
- Fixing the départements' share of the compensatory levy;
- Controlling the national transport undertakings (SNCF, RATP);
- Limiting the possibilities of loan finance (through the market or from the Economic and Social Development Fund – FDES[2]) for the undertakings and local authorities.

Secondly, there are:

a) The Ile-de-France region, theoretically in charge of regional passenger transport policy since May 1976, intervenes solely when it co-finances investment with the central government. This situation involves a double paradox, because on the one hand the region cannot in practice exercise the prerogatives assigned to it by law, and on the other it can take no part in the management of infrastructures, though this is what the rapid improvement of transport conditions depends on. What is more, any decisions the region may take regarding investment affect operating expenditure and, in particular, the deficit which the region does not help to cover. This can lead to fundamental inconsistencies as regards the internal workings of the transport system, policy orientations, priorities, etc.

b) The départements, which have to contribute to the operating costs. They also give their views on regional transport policy to the Ile-de-France authorities and are responsible for organising school transport. In the communes, the mayor has certain prerogatives with regard to local traffic and public transport services.

c) The transport undertakings involved include the SNCF and RATP – national undertakings with a regional function which carry out their activities within the Greater Paris transport region (and, in the case of the RATP, within a more restricted area) – and private carriers, who are represented by an association, the APTR. The difference in the nature of these two types of undertaking and the different way in which they are treated by the authorities mean that the quality of service offered to users within and outside the Paris transport region varies very considerably.

d) Lastly, to co-ordinate relations between all these bodies, there is a public board – Syndicat des Transports Parisiens (STP) – comprising representatives of the central government (who constitute the majority) and the départements of the Ile-de-France region, chaired by the "Commissaire de la République". This board is responsible for the general organisation of public transport services within the "Région des Transports Parisiens". Set up in 1954, it is responsible for "determining the routes on which services are to be provided, designating the operators, laying down the technical requirements and general conditions for the operation of services and fixing fares". Its main resource is the income from the transport levy, which is used:

- First, to offset in full the concessionary fares granted to employees;
- Then for specific investment in public transport (but in fact to cover some depreciation costs so as to reduce the share of operating costs borne by central and local government).

Financing Public Transport

Total expenditure on the Paris transport system is very considerable. In 1982, investment and operating expenditure together amounted to FF 75 billion onroads[7] and FF 20 billion on public transport. Less than 10 per cent of this total was spent on investment!

As regards public transport alone, investment amounted to about one-quarter of the overall turnover of FF 20 billion.

Financing Investment

Investment by public transport undertakings (FF 4 205 billion in 1982) can be split into two components:

- Renovation expenditure (replacement of rolling stock and fixed installations) is covered out of own funds, by loans granted by the *FDES*[2] and by increasing recourse to the money market. Such expenditure has grown rapidly in recent years;

- Extension of the rail networks (which now accounts for a little over 20 per cent of total investment as against 45 per cent a few years ago) is financed on the basis of 30 per cent by central government subsidy, 30 per cent by regional subsidy and the remaining 40 per cent by low-interest loans guaranteed by the Region[8].

Within the overall financing of these operations:

- The own-funds share has increased slightly in recent years but still remains low;
- Financing out of budget revenue — central government and regional subsidies — has decreased sharply (especially in the case of the SNCF) because investment in extensions and new infrastructures has slowed;
- Loan finance, particularly through the money market (where interest rates have risen at the same time as repayment periods have considerably shortened) has increased very sharply (especially in the case of the SNCF which does not benefit from FDES loans for its suburban network).

Given all these developments, the situation is now such that almost all investment is decided without any reference to the Syndicat des Transports Parisiens or to the Region, as it is the nationalised undertakings themselves which obtain their funds from the FDES according to specific procedures covered neither by legislation nor by administrative regulations!

Under the present system, moreover, procedures are made still more complex by the practice of co-financing, which increases the lead time between decision and implementation and gives rise to problems of co-ordination. In principle, these problems should be resolved by the conclusion of contracts between the central government and the Region in the context of the Plan.

Financing Operating Costs

Paris transport operating expenditure approached FF 16 billion in 1982 (15.3 billion of which was for the RATP and SNCF alone):

- Labour costs amount to almost 60 per cent of the total and have risen rapidly over the past two years, partly because numbers of staff have been increased;
- Financial costs now amount to over 10 per cent of the total and are also increasing steadily (having virtually doubled over the past four years).

While costs are rising sharply (having doubled in six years), revenues have followed somewhat contrasting trends:

- Direct revenue from users, which had dropped substantially between 1972 and 1975 as the result of fares being held down, now remains fairly steady at roughly one-third of total expenditure, but over the past two years has increased at a slower pace than expenditure;
- The transport levy, which covers a little under one-quarter of operating expenditures, is increasing far less quickly than those expenditures. In particular, the share of the transport levy allocated to depreciation fell in current francs in 1982, primarily because this levy on firms produced less in view of the present economic difficulties;
- The compensatory payment granted on the basis of 70 per cent by the central government[9] and 30 per cent by local government makes up the difference between expenditure and direct revenue and has increased considerably, particularly over the past two years.

Apart from the size of the sums involved, operating costs show a high rate of increase year by year; more particularly, there is considerable uncertainty as to the size of the shortfall between costs, which are difficult to control entirely, and revenue, which is tending to diminish steadily. Clearly the Region alone — or even in conjunction with lower levels of local government — cannot cover the total financial cost of operating public transport in the Paris area. Association with the central government is therefore inevitable. What is more, it is essential to have resources that can be rapidly adjusted to contend with the uncertainty specific to the financing of operating costs. In addition, the rapid increase in the cost of debt servicing, which weighs on the operating costs of transport undertakings, calls for consideration of the best way to finance infrastructure investment (entirely through subsidies?) and other investment expenditure (readier access to low-interest loans?).

Lastly, since transport undertakings' operating costs are increasing faster than inflation, it is clear that considerable efforts to improve internal as well as external productivity are called for.

II. ACTION TO PROMOTE PUBLIC TRANSPORT, AND NOISE ABATEMENT MEASURES

A Financial Measure: the "Transport Levy"

The transport levy is a major innovation for the purpose of financing public passenger transport in the Paris region.

The levy was instituted in great haste in 1971. The idea took shape at the beginning of the year, was implemented by the Act of 16th July 1971 which came into force on 1st September 1971[10] and contributed to the financing of public transport that autumn. It takes the form of an addition to the social security contributions paid by enterprises with ten or more employees, and involves about 55 000 firms employing 3.7 million people (the wages on which it is calculated being subject to a ceiling). It is now (since 1st July 1978) at the rate of 2 per cent in Paris and the three neighbouring départements and 1.2 per cent for the rest of the Paris transport area. The revenue from the levy is paid to the Syndicat des Transports Parisiens (STP) and allocated solely to public transport. More precisely, it is used to help offset the revenue lost thrugh granting concessionary fares (Orange Card, weekly season tickets, etc.), to finance transport undertakings' investment and to cover depreciation (Table 4).

This innovation is important in several respects:

— It brings in substantial sums, which vary with wage levels. This revenue grew, with the geographical extension of the area in which the levy was applied and with the increase in the rate, to reach a total of more than FF 4 billion in 1982, i.e. about 20 per cent of the overall financial cost of public transport in the Paris region;
— It produces revenue which is controlled by the authority responsible for the organisation of pulbic passenger transport in the Paris region. Thus it strengthens both the financial weight and the role of the regional authority — the one most appropriate for co-ordinating urban transport policies — and also accentuates the division between public and private transport;
— It makes employers pay part of the cost of commuter travel[11], in the same way as employees' health insurance or pension contributions. The introduction of the transport levy was, in fact, welcomed by both unions and employers. The phased nature of its introduction (in terms of geographical area and contribution rates) and its selectivity (exemptions for small firms and new towns, reimbursement for firms providing transport for certain employees) help explain the attitude of firms, which has to be seen in the light of the buoyant economic situation of the time;
— It helps spread the financial burden over three groups — users (who pay the fares), taxpayers (who pay taxes to the central government and to regional and local authorities), and employers. However, the reduction of the burden on the national taxpayer is not as great as it may appear at first sight, since company profits and the corresponding tax revenue are reduced by the levy. A more detailed analysis would also have to take account of the effects of the transport levy on the Value Added Tax received by the central government on the transactions involved.

Lastly, this innovation is easy to manage and does not involve high administrative costs. Moreover, although its incentive effect is small, it may help encourage firms to set up in the new towns or in small towns where commuter travel is relatively cheap.

Over the past few years, however, the advantages of the transport levy have tended to diminish, primarily owing to:

— The economic crisis which, through its impact on firms, reduces the base amount on which the levy is calculated and hence the revenue produced;
— The fact that this resource has since 1973 been increasingly used to cover depreciation, thus enabling the central and local governments to reduce their share of operating costs but, at the same time, reducing a source of investment finance (as from 1979).

The part played by the transport levy in financing Paris transport has therefore become progressively smaller. One of the purposes for which the levy is used is to help offset the cost of providing cheaper "social" fares. Only about 50 per cent of the levy was used in this way in 1975, but the proportion has increased considerably and now represents over three-quarters of the total revenue raised (whereas elsewhere in the country this "social" cost takes only 20 to 40 per cent of the total sums raised, the rest going to specific investment projects). In these circumstances, not only has it no longer been possible since 1979 to allocate the transport levy to financing investment but, given the way things are now going, the revenue will soon be used entirely to make up the income foregone by granting concessionary fares. This would considerably increase the compensatory payment, so placing a much heavier burden on central and local government.

Table 4. ANNUAL REVENUE FROM THE TRANSPORT LEVY AND PURPOSES FOR WHICH IT WAS ALLOCATED

Millions of Francs

	1971	1972	1973	1974	1975	1976	1977	1978	1979	1980	1981	1982
Carryover	–	55.7	284.5	207.9	232.8	163.6	151.8	19.1	20.2	–0.7	5.8	57.3
Net revenue	244.8	838.8	900.4	1 080.4	1 438.4	1 882.1	2 121.8	2 447.7	2 812.1	3 134.5	3 743.7	4 160.7
Total	244.8	894.5	1 184.9	1 288.3	1 671.2	2 045.7	2 273.3	2 466.8	2 832.3	3 133.8	3 749.5	4 218.0
Compensation for loss of fares revenue (season tickets and Orange Card)	189.1	575	612	666.5	807.5	1 203.3	1 546.2	1 793.9	2 103.5	2 471.5	2 786.7	3 284.5
of which: RATP	83.6	273.1	239.6	310	378.2	548.4	704.1	834.2	953.5	1 113	1 244.6	1 486
SNCF	105.5	301.9	313.6	341.3	364.6	561.1	684.1	803.6	940.5	1 103.5	1 242.1	1 437
APTR	–	–	8.8	15.2	64.7	93.8	158	156.1	209.5	255	300	361.5
Specific investment (depreciation and investment)	–	35	365	389	700.1	690	708	652.7	728.9	665.3	727.6	663.2
of which: RATP (depreciation)	–	–	30	190	270	325	380	355.7	470	468	505.7	462
SNCF (depreciation)	20	135	–	55	40,6	27	–	–	–	220.6	193	–
RATP (infrastructures)	–	–	170	88	130	145	108	169	238	197	–	–
SNCF (infrastructures)	–	15	30	102	233.1	180	173	120	–	–	–	–
APTR (infrastructures and rolling stock)	–	–	–	9	12	40.6	20	8	20.9	0.3	1.3	1.6
Sundry	–	–	–	–	–	–	–	–	–	–	–	7
Agreed contributions	–	–	–	–	–	–	–	–	0.6	0.4	0.6	0.8
Total	189.1	610	977	1 055.5	1 507.8	1 896.9	2 254.2	2 446.6	2 833	3 127.3	3 514.3	3 948.5
Solde	55.7	284.5	207.9	232.8	163.6	151.8	19.1	20.2	–0.7	–3.4	55	+269.5

The fact is that the scope for increasing the revenue produced by the transport levy is limited and the measures that might be possible are likely to be difficult to implement. These would be:

- An increase in the contribution rate, which is not very high as compared with the costs generated by commuter traffic on the transport network; given the prevailing economic climate, however, this measure would be difficult to implement;
- A broader basic assessment: the retail trade (supermarkets, hypermarkets, department stores, etc.) and firms with less than ten employees are at present exempt. It may also be possible to tax certain categories which can be singled out as deriving particular advantages from the transport system, but here again the economic climate does not seem very propitious for such a step;
- Removal of the ceiling on contributions. The only reason for the ceiling is to bring the levy into line with other types of employers' social security contributions, and there are no grounds for it in terms of transport itself.

All these solutions would be equally difficult to implement, particularly as the Paris transport system is at present in a period of transition.

A Fare Measure: the Orange Card[12]

Objectives and Planning of the Orange Card System

A major change in the fares structure of Paris-area public transport was brought about on 1st July 1975 with the introduction of a *non-transferable monthly (or yearly) season ticket*, a concept new to France. The Orange Card:

- Can be used for an unlimited number of journeys;
- Can be used on different modes (bus, metro, suburban train) and networks (RATP, SNCF, APTR);
- Provides for a concentric zonal fares structure;
- Has a social function, the aim being to bring down travel costs, in particular for radial commuting.

The aims of this change in the fare structure were:

- "For journeys considered as equivalent, to establish the same price for all travellers by setting fares according to the service rendered, irrespective of the transport mode or series of modes used";
- "To reduce the burden on the commuters with the longest journeys and have a fare structure that encourages inter-suburban travel";
- "To introduce a simple ticket system providing for the successive use of different transport modes".

The aims referring to uniform fare structure and simplified ticket system, were consistent with the planned interconnection of the SNCF and RATP networks and greater integration of the Paris regional public transport system. the aim referring to reduced burden fell within the context of social policy for public transport and was consistent with the radial development of the Paris conurbation. With respect to the third, it was at first intended to apply the system to all ticketing, but because of the technical problems associated with checking the validity period of tickets sold individually it was finally decided that the system should apply to season tickets only, at least for the time being.

The authorities determined the above characteristics for the Orange Card after carrying out a great many studies and analysing a broad range of proposals, which even included a one-zone ticket financed by employers and local authorities at a break-even level. The solution finally adopted has five concentric zones centered on Paris, and the ticket is valid for a month or a year. The price depends on the number of contiguous concentric zones in which travel is allowed. The price has been increased regularly each year since this system was introduced in July 1975 (Table 5).

Impact of the Orange Card on User Behaviour

The launching of this new type of ticket was preceded by a big advertising campaign (posters, folders, press and radio advertising). A second campaign was run three months after its introduction, when the summer holidays were over.

Reception by the public was excellent and the impact of the Orange Card was far greater than generally expected, since the number of tickets sold six months after its introduction exceeded the forecasts by about 50 per cent[13]. Today the number of tickets sold is still increasing, the average being 1.4 million a month and the maximum number 1.7 million (Table 6).

The Orange Card is particularly advantageous for users commuting between the suburbs and Paris. User surveys to determine the breakdown of sales also show that these positive effects relate mainly to the labour force, medium-income groups and service sector employees.

Five months after its introduction, an opinion poll among inhabitants of the Paris region showed that the Orange Card was very well known (by 9 people out of 10) and had an extremely positive image (9 out of 10 thought that "the 'carte orange' is a good idea", 8 out of 10 "feel freer", and 1 user in 2 had "the

Table 5. PRICE OF THE MONTHLY ORANGE CARD

Number of adjacent zones	Price of a second-class monthly ticket in current francs[1]								
	75/76	76/77	77/78	78/79	79/80	80/81	81/82	82/83	83/84
1-2	40	45	50	57	70	85	100	110	122
1-3	60	67	75	86	105	128	145	155	165
1-4	80	90	100	115	140	170	190	205	220
1-5	100	112	125	144	175	213	230	250	260

1. The price of a first-class ticket it 1Z1/2 times to twice that of a second-class ticket.
Source: DREIF (1981, 1982 and 1983), *Les Transports de Voyageurs en Ile-de-France*.

Table 6. TOTAL SALES OF THE ORANGE CARD: MONTHLY AND ANNUAL TICKETS

Millions of tickets sold per month

	1975	1976	1977	1978	1979	1980	1981	1982
January	—	907	1 180	1 339	1 419	1 442	1 453	1 544
February	—	930	1 178	1 338	1 400	1 436	1 463	1 543
March	—	968	1 209	1 338	1 427	1 463	1 445	1 528
April	—	972	1 071	1 265	1 353	1 312	1 339	1 381
May	—	995	1 193	1 286	1 387	1 374	1 414	1 522
June	—	953	1 120	1 261	1 334	1 373	1 352	1 443
July	190	732	870	950	1 013	1 026	1 027	1 112
August	206	472	573	606	621	634	746	775
September	429	810	968	1 029	1 067	1 070	1 086	1 301
October	705	1 070	1 257	1 330	1 363	1 356	1 430	1 599
November	835	1 148	1 299	1 380	1 404	1 448	1 518	1 699
December	880	1 117	1 277	1 370	1 405	1 484	1 478	1 728
Monthly average	541	923	1 100	1 208	1 266	1 281	1 313	1 431
Percentage increase		+ 70.6	+ 19.2	+ 9.8	+ 4.8	+ 1.2	+ 2.5	+ 9.0

Source: DREIF (1980, 1981, 1982), *Les Transports de Voyageurs en Ile-de-France*.

impression of travelling gratis for the whole month"). This shows the importance of the type of ticket for the way in which fares are perceived and highlights the first major effect of the introduction of the Orange Card, i.e. its positive impact on the way the inhabitants of the Ile-de-France region view public transport and their experience of it.

A second major effect of the Orange Card has been a considerable increase in the number of people using Paris buses — up 36 per cent from 745 000 to 1 010 000 passengers a day. The buses have benefitted by:

— A switch by passengers from metro to bus, partly due to the fact that before the introduction of the Orange Card the cost of bus travel had been considered too high compared with the metro, and perhaps also because of an improvement in the image of bus transport, which had previously been poor. This switch accounts for 38 per cent of the increase;

— A switch from walking, because the marginal journey is now free. This accounts for 30 per cent of the increase;

— A switch from the car, accounting for 14 per cent of the increase;

— An increase in personal mobility, with people making journeys they would not previously have made, particularly in off-peak hours. This accounts for 13 per cent of the increase.

The Orange Card has had a far less marked effect on numbers of passengers using the metro (up 1 per cent), the RER (up 5 per cent), suburban buses (up 5 per cent) and the SNCF (up 1 per cent).

The RATP was able to increase its vehicle fleet to meet this increased demand by delaying the withdrawal of the oldest buses and at the same time rapidly acquiring new ones. It was also possible to increase the number of drivers through a stepped-up training programme.

A third major effect of the Orange Card has been a switch from car travel to public transport, especially to the bus and metro within Paris, an effect that is thoroughly consistent with the Paris Region policy of encouraging this type of switch almost exclusively by means of measures to promote public transport.

The scale of the switch can be estimated on the basis of the available statistics, though these have their limits. It would seem reasonable to put forward a figure of about 70 000 daily journeys transferred from the car to public transport in Paris (half switched to buses and half to the metro). This represents a reduction in car journeys (Paris-Paris, Paris-suburb, suburb-suburb) of 2.8 per cent over the whole day, 4 per cent in the evening peak period (5 to 7 p.m.) and 1 per cent throughout the evening peak. This is a significant effect that should not be underestimated.

The reasons for the switch are no doubt to be found as much in the convenience of the new ticket and the improved image of public transport as in strict cost and time comparisons between car and public transport travel, especially as the publicity concerning the Orange Card had shied away from such slogans as "Travel free for a month on public transport for the price of a tank of petrol".

Recent Changes and Future Prospects

There have been two changes in the Orange Card system since 1982:

a) Partial reimbursement by the employer;
b) The introduction of a weekly ticket.

In August 1982, the French government decided to introduce a new scheme — partial reimbursement of the Orange Card by the employer — to replace an existing aid system for the labour force whereby employers paid a fixed sum towards the commuting expenses of all employees, whatever the mode used [14]. As from 1st October 1982, employers in the Ile-de-France had to pay 40 per cent of the cost of public transport commuting, and this rate was increased from 40 to 50 per cent as from 1st October 1983. Employees can thus claim reimbursement of half the purchase price of their Orange Card (monthly or yearly), the twelve-journey weekly workman's ticket or, since its introduction on 1st November 1982, the weekly Orange Card. This new ticket costs a little over one-quarter of the price of the monthly Orange Card and gives its holder the right to unlimited travel on the different public transport networks for one week.

The aims of this change were mainly to make employers in the region aware of the social cost of their choice of location and to further promote public transport. Under the old system, the employer paid a fixed sum (FF 23 monthly) to each employee, regardless of the transport mode used for commuting. Under the new scheme, only employees using public transport may have their transport expenditure subsidised by the employer. This also explains why the overall cost to the employer of this new system (when reimbursement was 40 per cent) was equivalent to the overall cost of the old system. Since the employer's share was increased from 40 to 50 per cent, however, certain firms have had to bear an additional financial burden.

The initial findings indicate that this reimbursement system had generated a limited amount of new traffic (new travel, modal switches) and substantial transfers from one type of ticket to another, especially from workmen's weekly tickets — which allowed only a limited number of journeys over a fixed route — to the Orange Card.

There are likely to be other changes in the fare system in the near future. Under the RATP plan for the period 1983-87, the zonal system will be simplified and introduced on a general basis. This is likely to involve several changes:

— A permanent season ticket, which will be introduced in the first year of the plan to replace the present yearly Orange Card;
— A daily ticket which will be introduced by 1985;
— A reduction from 5 to 4 zones (by amalgamating zones 2 and 3), which is bound up with extension of the flat-rate fare to suburban metro lines;
— Zonal fares for all tickets;
— New methods of payment.

All these fare measures, together with the incentive effect of the employer paying 50 per cent of the price of commuter travel, should increase traffic by an estimated 2 to 3 per cent (according to mode) over the period of the plan.

The partial restructuring of public transport fares and the Orange Card season ticket fare are therefore moves towards a "social" public transport policy and fare simplification to prepare the way for full integration of the Paris public transport system.

The interest of these fare measures lies not so much in their innovatory character as in their successful impact on the public, notably as regards:

- The improved image of public transport;
- The considerable increase in the numbers of passengers using Paris buses;
- The incentive to switch from the car to public transport.

A Promotional Measure: Improving the Image of Public Transport

Public Perception of Public Transport in Paris

In 1973, the RATP became concerned about its public image. Opinion polls showed that the RATP had no distinct corporate image and revealed the ways in which Parisians saw and judged metro and bus services. The public viewed the metro as an efficient and reliable mode, but the underground maze of tunnels and stations was seen as an unnatural, oppressive environment, disconnected from normal life above ground. The bus was considered to be a more agreable form of transport, open to everyday life in the city, but at the mercy of traffic conditions and hence inefficient and unreliable. Its network was also regarded as very difficult to understand. Many non-users were not consciously aware of either system. For them the metro was "invisible", while the bus was merely "part of the general surroundings". Neither mode entered into their travel decisions. The RATP realised that if it wanted to avoid a decline in the number of its passengers and to attract new customers, it had to improve the image of public transport. Over the past ten years its policy has been to influence the way in which its services are perceived by the public, through action in the following four areas:

1. Information and provision of interest and entertainment;
2. Advertising and communication;
3. Security; and
4. Public perception of costs and fares.

Information and Entertainment Policy

The RATP created a new "Transport Promotion Service" to spearhead an ambitious policy of public information to project a positive corporate identity and take potential customers by the hand. Information campaigns launched in the Paris media focussed on the opening of new bus services and major construction projects, such as the extension of metro lines and the renovation of stations. These campaigns sought to convey the idea that the RATP is a dynamic enterprise, actively engaged in developing and modernising its networks. A telephone information service was set up to provide users, in an efficient and courteous manner, with details of all the RATP's bus, metro and rapid rail services.

The public's reservations about the bus and metro systems were attacked by specific action designed to improve the image of each mode. Publicity campaigns stressed the improvements in the speed and reliability of bus services through the creation of pilot lines and reserved bus lanes. Information on routes, timetables and connecting services was provided in a simple, attractive form easy for passengers to use. Maps and service information were displayed at bus stops, in the buses themselves and in metro stations, and distributed to every household in Paris to drive the message home. Promotional offers encouraged people to use the bus for leisure and shopping trips during off-peak hours and to make Parisians more aware of the bus.

Various action was undertaken to overcome the public's unfavourable idea of the metro. The information available to passengers was improved by providing better maps and brochures and setting up a new reception service; the task of the extra staff assigned to the latter was to personalise the information service and provide users with a reassuring human presence. The installation of better lighting, colourful design and allowing shops to open in metro stations was also a psychological move to reassure travellers. The RATP also endeavoured to take people's minds off their daily cares by livening up the metro.

The aim of the RATP's new "*entertainment policy*" was to bring the city into the metro and to project the image of the metro as a lively and interesting place open to every citizen. The first event, "A Week of Music in the Metro", was organised at a major interchange station, Charles de Gaulle-Etoile, in 1977. It was widely covered in the local press and well received by the public. Subsequent events also met with success, and the RATP decided to make them a permanent feature of its information policy.

Since then shows and exhibitions have been organised in the metro with increasing frequency, so that today something new is presented almost every month. For example, performances have been given by famous orchestras and by dance, theatre, puppet and circus troupes. Exhibitions of classical art have been followed by others on amateur photography, the cinema, cartoonists and comic books. Topics covered have also included sporting events, and there have been meetings with organisations concerned with public health and safety and with micro-computer experts. The most recent initiative is a non-stop video-show in trains on the busiest metro line. The RATP considers that its entertainment policy serves to reduce anxiety and stress in the metro and to project a better image of the system.

Advertising and Communication Policy

In a first phase, the RATP decided to enhance the image of public transport and not to launch a direct attack on its principal competitor on the regional transport market, the private car. Rather than stressing the problems of driving in Paris, it presented public transport as the household's "second car". This campaign was well received by the public, but failed to attract motorists to public transport. An assessment of its impact revealed that while non-users' image of the bus and metro systems had improved, their image of public transport users had not; the non-users saw the users as socially inferior. The image of the car as a liberating mode and status symbol also remained strong. In a second phase, the RATP decided to change the image of public transport users and develop its own up-market style to compete with the car manufacturers.

A series of advertising campaigns was launched to project the metro as a place where happy, pleasant and interesting people were to be encountered. Public transport users were portrayed as young, athletic and fashionably dressed. The whole campaign was built around the ordinary magnetic ticket (yellow with a brown stripe down the middle), a symbol presented for the first time in October 1981 in a television spot and in cinema commercials. The "Ticket chic, Ticket choc" idea quickly became the RATP's hallmark.

Although it is hard to gauge with any exactitude the effectiveness of its advertising and communication policy, the RATP monitors public opinion by regular surveys and is convinced that it now sees it and its transport services in a more favourable light. Other indicators also suggest that there has been a slight increase in the number of people using public transport or declaring their intention to do so — a rise in interest that cannot be explained by other action (such as the Orange Card).

Intensive local promotional actions has also proved successful. A campaign based on the slogan "Paris is your neighbourhood" was run in Boulogne to coincide with the opening of an extended metro line. It brought higher increases in the numbers of passengers using the new section of line and the feeder buses than had been anticipated. Such action is quite costly, however, and of limited geographical impact. The RATP therefore plans to repeat it only where the potential markets for new clients are biggest.

The RATP's annual "communication" budget totals FF16 million, or 0.2 per cent of its operating budget, a figure that compares favourably with the 0.85 per cent average spent by French firms on advertising.

Security Policy

The main public transport security problem in Paris is theft, especially by pickpockets and bag and jewellery snatchers. There is no major crime and violent incidents are rare. In recent years, however, opinion surveys have revealed a growing feeling of insecurity among public transport users, particularly in the metro. The great majority of passengers say that they are afraid of being attacked; 55 per cent say that they know someone who has been, and 15 per cent claim that they themselves have been the victims of an aggression. That last figure would imply 300 000 attacks a year, or 200 times the number actually recorded in 1982. This difference between the statistics and the public's perception of the risks of aggression is related to the metro's image as a menacing underground world. Every incident in the metro takes on greater importance than similar incidents in the street. The public associates aggression not only with violence and theft, but also with inconsiderate behaviour by other travellers (e.g. jostling during rush hours, people who talk too loud or smoke, or the presence of a tramp or a drunk in their carriage). The extreme sensitiveness of metro users has also been heightened by exaggerated press coverage of a few serious incidents.

The RATP is trying to dispel the public's feeling of insecurity by reducing the opportunities for aggression and crime. Passenger comfort has been improved (by better quality service, cleanliness, renovating stations, allowing more shops to open in the metro, and improving passenger information and reception services). Crime prevention has been stepped up by raising the number of policemen assigned to patrolling the metro, by training and assigning more reception staff, by improving radio-telephone communication links between train drivers, platform staff and emergency services, and by extending electronic surveillance.

A better relationship with the media has also been established by creating a control group to advise the RATP on matters of security, by designating one station (Bastille) as a field laboratory for new security and passenger information techniques, and by organising regular seminars and conferences on general urban issues. The main aim is to show the public that the RATP is doing its best to improve security in the metro, while pointing out that insecurity is a general urban problem calling for joint action by all the authorities and individuals concerned.

Policy Regarding Public Perception of Fares

For almost a decade the RATP has sought, by simplifying its system of fare payment, to influence the way in which public transport costs are perceived. The Orange Card, for example, with its system of concentric fare zones, was designed to be economical to operate and easy to understand and use. Although its price has been put up each year since it was first introduced, the number of users and the revenue from it have increased. This inelasticity of demand in response to price increases is largely explained by the fact that most purchasers perceive the card as offering extremely good value for money:

a) for the price of a meal in a restaurant, it entitles them to unlimited travel on the RATP's bus, metro and rapid rail networks within the specified zones;
b) all 'extra' journeys are seen as being free;
c) it is simple to use, reduces the frequency of fare payment, and offers convenient access to and transfers between modes; and
d) it is personalised.

The RATP has tried to enhance these features of convenience and personalisation aspects by issuing a new, annual, version of the Orange Card, introduced on 1st May 1984. This "Integral Card" is the same shape and size as a credit card and fulfils a similar function. It can be paid for outright or by instalments, monthly deductions from the purchaser's bank account. This means that passengers have less need to carry cash or a cheque book, and gives fare paying a modern look, on a par with electronic money.

No matter how modern the fare payment and control system, some users will always try to avoid paying fares. Fraud prevention is costly, and in certain cases it may be more economical to tolerate a certain loss of revenue. But the dishonesty of certain passengers is strongly resented by others, who consider that they personally are obliged to pay for the cheats. In Paris it is easier to cheat in the metro that on buses, and in recent years this practice has become more noticeable. In 1982, the RATP was accused of laxism and of wasting public funds. The RATP admitted the seriousness of the problem, and invited discussion of ways of combatting fraud. It then launched an information campaign denouncing fare evasion in order to prepare the way for tighter control measures. New anti-fraud barriers were installed, surveillance by closed-circuit television was extended, and the number of random inspections on platforms and trains was increased. This action reassured users by showing that the RATP was concerned by the problem and was ready to take steps to overcomne it. Fraud control in the metro remains a difficult task, however, and new countermeasures are constantly required to keep fare evasion at an "acceptable" level.

Constructional Measures: Traffic Noise Abatement along Motorways and Urban Expressways

Exposure to Road Traffic Noise

Traffic noise on roads and railways causes widespread nuisance in the Ile-de-France region. According to the results of various surveys carried out between 1976 and 1980 [15, 16], well over a quarter of the urban area's 8.5 million inhabitants are exposed to what are now generally considered to be unacceptable noise levels, with daily L_{eq} (8 a.m.-8 p.m.) reaching 65 dB(A). Conditions are particularly bad within the city of Paris, where it is estimated that almost half of the 2.3 million inhabitants are exposed to such noise levels even in the late evening (Table 7). Moreover, in a high proportion of streets noise levels reach 80 dB(A) for at least five hours each day.

Some of the locations most exposed to traffic noise are to be found in the residential zones bordering motorways and urban expressways running from the

Table 7. EXPOSURE TO TRAFFIC NOISE
Estimates for different urban areas, 1980

	Population				
Urban Area/Region	Millions 1975	Percentage (%) exposed to various external noise levels [dB(A)][1]			
		55	60	65	70
Paris					
City (intra muros)	2.3	84	64	48	20
Suburbs	6.2	76	47	20	2
Conurbation	8.5	78	52	28	7
Other conurbations:					
200 000 to 2 million inhabitants	10.3	68	51	23	8
20 000 to 200 000 inhabitants	11.6	70	50	30	9

1. Data refer to *evening* noise levels, Leq (9 p.m.-10 p.m), measured in front of the most exposed façades of buildings.
Source: IRT-CERNE.

city outskirts to the regional boundary. There are some 578 km of motorways and urban expressways in the Ile-de-France region[17]. Although they represent only 1 per cent of the total network, they carry almost a third of all the traffic in the region. Similarly, the "boulevard périphérique"[18] and the "voie sur berge rive droite"[19] represent only 4 per cent of the city's roadspace, but carry 40 per cent of internal traffic. Not only do these routes carry intense flows of local and through traffic day and night, but the share of lorry traffic can rise to as much as 50 per cent of the total during the night hours, so that the people living along them cannot sleep. It is estimated that about 400 000 people live alongside the region's urban motorways and expressways, almost half of them in critical zones where the daily L_{eq} (8 a.m.-10 p.m.) is at or above 70 dB(A) and sometimes 80 dB(A).

Over the past decade public dissatisfaction with this situation has built up, from isolated complaints by a few people affected by particular noise problems to more general manifestations of discontent. Complaints made to the Nuisance Bureau of the Paris Police show that noise is the predominant environmental nuisance: 70 per cent of complaints in 1971 and just over 90 per cent in 1981[20]. The likelihood of even higher noise levels in Paris and the surrounding region is strong, unless radical action is taken. Since 1978 noise abatement has been a clearly stated national policy objective, and the Ministries responsible for transport, housing, industry, planning and the environment have stepped up the scale and diversity of their interventions in recent years. Regional and local authorities are also co-operating in the formulation and implementation of schemes to reduce the noise nuisance caused by road traffic. It has also been decided to make noise protection obligatory for new buildings to be constructed near an existing thoroughfare.

Traffic Noise Abatement Programmes

Two main approaches have been adopted:

— Preventing the creation of new noise black spots by incorporating protection measures in the design of new road infrastructure; and
— Undertaking a remedial noise protection policy for housing where:

a) Noise level at the façade exceeds 70dB(A); and
b) The dwellings were built before the road. (both criteria must apply).

a) Noise Abatement along New Roads

Since 1978 control of noise nuisance generated by traffic has been largely integrated into the highway design and planning process when new roads are being built. The Ministry of Transport's Directorate for Roads and Road Traffic has fixed the noise ceilings to be complied with by the constructors of new sections of the national road network, which come under its responsibility[21]. The creation of new roads must not generate L_{eq} (8 a.m.-10 p.m.) of more than 65 ± 5 dB(A) at the façade of existing buildings. In quiet residential areas the aim is not to exceed 60 dB(A), although in certain marginal cases a higher level of up to 70 dB(A) may be accepted. After only five years of application, the levels aimed at are essentially in the lower part of the range [60 to 65 dB(A)]. The techniques adopted seek, above all, to reduce noise emission by means of civil engineering works along the roadway (earth mounds, screens and in some cases, partial or total coverage). Where these measures are inappropriate, inadequate or too costly, the phonic insulation of building façades is improved (mainly by double glazing of windows). In certain cases anti-noise screens (2.5 to 3 metres high) are used to protect outdoor areas of special social or community value (e.g. sports facilities, playgrounds). On average, the incorporation of noise protection measures at the highway design stage adds about 10 per cent to the road construction cost per kilometre, but these additional costs vary, from as little as 5 per cent to over 40 per cent depending on the complexity of the case. Given the clear slowdown in the construction of new roads in France in recent years, traffic noise abatement policies are likely to concentrate on reducing noise nuisance along the existing network.

b) Noise Abatement along Existing Roads

Following a census of noise black spots and critical zones along the region's motorways, urban expressways and Paris' inner ring road, estimates were made of the likely costs of the protection needed to reduce external noise levels during the day to 65-70 dB(A)[22, 23]. In 1978 the Ministry of Transport undertook a remedial noise abatement programme along the motorways, with financial aid from the Ile-de-France region. The aim of this exercise is to protect the 90 000 persons living in the zones concerned. As with new roads, the technical solutions adopted focus mainly on the reduction of noise during its transmission by means of civil engineering works and structures along the roads, double-glazing windows on exposed building façades only where these external measures are inappropriate or inadequate. The remedial programme to abate noise along motorways and urban expressways, which was implemented first, is now half-way to completion.

In view of the heavy financial burden involved, it was decided to stagger implementation of the remedial action programme over several years and to try to cut down expenditure. Thus it was decided to confine the measures financed under the programme to critical zones and to dwellings that had already

Table 8. ESTIMATED COSTS AND BUDGET FOR NOISE PROTECTION MEASURES ALONG MAJOR ROADS IN THE ILE-DE-FRANCE REGION

Millions of Francs

Road category	Estimated cost	Credits allocated up to 1983
Radial motorways and expressways	870	380
Paris inner ring road	300	50
Total	1 170	430

existed before the roads concerned had been constructed (the principles of "priority" and "anteriority"). Furthermore, as road coverage is many times more expensive than building screens or insulating façades, preference was given to the latter, less onerous solution. The total costs of the exercise were estimated as shown in Table 8.

In 1982 a remedial noise abatement programme was begun around the Paris inner ring road as part of a joint action by the State, the Region, the City of Paris and the départements. This programme is scheduled to last six years, from 1983 to 1988, and so far FF 300 million have been allocated as shown in Table 9.

The decision whether or not to undertake the remaining road coverage (partial and total) along the boulevard périphérique is being left to a later date.

The cost of financing remedial noise protection measures is shared in the same way as that of noise protection incorporated in new road building[24] (Table 10).

The arrangement for financing the Paris inner ring road is exceptional because of its status as a municipal road (not a national road). Several Ministries agreed to contribute to the financing of the remedial noise protection measures along this ring road (Table 11).

Table 9. TYPE AND COST OF PROTECTION MEASURES ALONG THE PARIS INNER RING ROAD

Technique	Protection measure	Budget (millions of Francs)
Less costly	Anti-noise walls and screens (17 200 m)	100
	Double-glazing of windows (roughly 35 000)	100
More costly	Total and partial coverage of roadway (1 400 and 1 800 m respectively)	100

Table 10. THE FINANCING OF NOISE PROTECTION MEASURES ALONG MAJOR ROADS

Source of funds	Share of total cost by road category (%)	
	Motorways	Ring Roads
State	85	30
Region	15	70
	100	100

Table 11. THE FINANCING OF NOISE PROTECTION MEASURES ALONG THE PARIS INNER RING ROAD

Source of funds	Share of total cost (%)
State[1]	25)
Region	35) Suburban side
Department of Paris	40) Paris side
	100

1. The Ministries of Transport, of the Interior and of the Environment.

III. EVALUATION OF ACTION UNDERTAKEN AND CONCLUSIONS

Management of the Networks

The Urgency of Concerns Relating to Surface Networks

Following a period of spectacular expansion of its transport infrastructures (radial motorways and inner ring road, then the RER and suburban extensions of the underground), the Ile-de-France Region has substantially reduced the pace of such major development works:

— The lower-than-foreseen growth of the Parisian conurbation has led to the reduction of motorway programmes;
— The higher construction costs for urban infrastructures are less and less compatible with investment budgets, which are suffering from the backlash of the economic crisis that is currently limiting the financial resources available;
— Concern to rebalance respective shares in the financing in the Ile-de-France Region will mean less State aid for investment;
— Within the Region, the desired shift towards providing better facilities in the suburbs will involve the use of less costly development better adapted to lower population densities and more diffuse travel patterns.

Coupled with the changes that have taken place in recent years in the ranking of values relating to urban life (life style, environment), these factors make it even more necessary to adopt a policy whose thrust is towards better management of existing transport infrastructures, and in particular towards more efficient and better co-ordinated utilisation of road space. Until now, concern with this question had brought only isolated action in the areas where needs were most pressing.

The global transport plan proposed in 1972 by the Prefect of the Region already stressed the need for better management of the supply of existing transport. Its proposals on the utilisation of road space brought only the partial installation of a "controlled regional network" over which the Prefect of the Region had regulatory and policy control — a first step in a truly regional traffic plan which still remains to be devised.

The implementation since 1981 of a decentralisation policy that confers additional responsibilities for road traffic on local authorities has led to the withdrawal by the State of subsidies allocated specifically to the "controlled regional network" and to local traffic plans. In this new context, therefore, new mechanisms taking traffic management problems into account still have to be set up. The formulation of "urban travel plans" proposed in new legislation on transport (la Loi d'orientation des transports), but which in the specific case of the Ile-de-France region have not yet been finalised, could make a positive contribution, although this would require the regional authorities to succeed in arousing a genuine "regional awareness" of traffic problems.

So, although the fact was hidden behind achievements in the construction of major motorways and the improvement of rail infrastructures, road management in the Paris region today can now be seen to be lagging behind that of other European cities, in particular in the United Kingdom, Germany and Scandinavia.

Promotion of Public Transport

The promotion of public transport is one of the main planks of Paris transport policy, and has already been immensely successful. As well as having carried out a major programme of expansion and interconnection of the urban and regional rail network, the Ile-de-France Region has promoted public transport is typified by means of three spectacular moves which have been attempted practically nowhere else in the world:

— The transport levy (a financial initiative);
— The Orange Card (a fare-related initiative);
— The enhancement of the image of public transport (a psychological initiative).

This action and its effects have been described in Part II of this chapter, and its success is evident. Moreover, the first and second initiatives, coupled with the remarkable efforts of the RATP as regards information and renewal of stock, have contributed greatly to improving the image of public transport, and especially surface transport.

But much still remains to be done to increase the supply of transport in the suburbs, to simplify the use of existing lines and to create the image of a truly coherent public transport network covering the entire Ile-de-France Region.

Traffic Management and Selective Limitation

During a first phase, concern to increase the capacity of the road network left little room for concern with its management. In a second phase, the implementation of decentralisation and the consequent changes in modes of financial intervention by the State (replacement of specific subsidies by block grants) has led to the abolition of the few formerly existing instruments for the control of road use (controlled regional network, traffic plans).

One characteristic of the road network is *congestion*, due to the fact that demand for car journeys greatly exceeds the road space supply available. This leads to acute problems in certain spots (especially in neighbourhoods where congestion affects the activities and life of the inhabitants, at the Portes de Paris and on certain uncompleted bypasses in the inner suburbs) and at certain times of day (especially since peak hours have lengthened).

These problems of congestion and quality-of-life objectives lead to proposals that one major aspect of traffic management policy should be *limitation of the use of private cars when and where necessary*. This might mean:

a) Measures to improve local traffic conditions by encouraging the creation of pedestrian zones, bypasses to protect residential neighbourhoods from through traffic and the concentration of traffic on main roads; all these measures would have to be incorporated in future traffic plans;
b) Reducing the traffic flows entering Paris, especially at peak hours (generally by giving preferential treatment on the radial routes to public transport and high-occupancy vehicles, reducing the total capacity of the entry routes into Paris for general traffic and imposing parking charges);
c) Cutting down the increase in vehicle/km for private cars in the Ile-de-France region through measures to encourage better use of private cars and other alternative modes (public transport, two-wheelers, etc.).

Parking Policy

Driving in Paris is difficult, and parking there is even more difficult. In the absence of a coherent parking policy, there is uncontrolled growth of *illegal parking* (more than 60 000 spaces by mid-day). In addition, illegal parking has often been tolerated and sometimes even encouraged (certain pavements being "taken over" by the removal of no-parking signs).

Studies show that enforcement of parking restrictions is an effective instrument for limiting automobile traffic. If the present illegal parking could be cut by three-quarters, this would reduce the number of Paris-related car journeys by about 15 per cent and the number of vehicle/km travelled daily in the Ile-de-France region by 5 per cent, bringing appreciable gains (time, convenience, etc.) to road users and passengers on surface public transport.

This suggests that network management policy should include parking-related measures aimed at increasing the supply of spaces for residents (mainly off the streets)[25] and reducing the demand for commuter parking.

Demand could be reduced for example, by limiting the numbers of vehicles entering Paris, tightening the enforcement of regulations on main arteries and extending parking charges, as well as by providing park-and-ride facilities near stations, especially in the outer suburbs.

Use of Economic Instruments

Economic as well as physical and regulatory instruments can play an important role in network management.

Fares

The Ile-de-France Region is a case in point. The Orange Card, by simplifying the structure of fares even more than by changing their level, enhanced the image of public transport and made it much more convenient for users. It altered the structure of demand for transport, in particular by increasing the number of passengers on the bus network and bringing a transfer from the private car to public transport. Thus the intrinsic importance of fare structures is one of the main lessons to be learned from this experience.

This calls for three comments concerning the future:

a) The success of this simplification indicates that it should be extended to other fares (tickets bought singly, weekly cards, etc.) and that the Orange Card itself should be further simplified (by reducing the number of zones);
b) The price of the Orange Card could probably be increased (it is low compared with fares in other European cities); this would substantially ease the financial management of Paris public transport;
c) At the same time, in order to maintain some balance between transport modes, measures (fare-related or otherwise) should be taken to limit the use of private cars as and when necessary with a view, in particular, to bringing down the external social costs entailed by this type of transport.

The imposition of *parking charges* is a second economic instrument whose use, long deferred, is now being rapidly extended to dissuade all-day parking by commuters rather than parking by residents and is bringing in financial resources to enforce compliance with parking regulations. The extension of such measures in other European cities shows that this is an instrument which is most effective when used in association with other disincentives to home-to-work travelling by private car.

Generalisation of such measures, however, comes up against two major obstacles: the cost of their implementation and the fact that the heterogeneity of road users increases proportionately to distance from the central zone.

At present, *private motorists pay no charges for road use* in the Ile-de-France region. Though the introduction of tolls on the radial motorways came up against stiff opposition from elected representatives and users, and had to be provisionally abandoned, this remains a measure which would bring new resources and lower demand on the radial motorways, once a satisfactory public transport service is available as an alternative to the private car.

Other instruments, too, merit detailed study. For instance, use of motorways, driving in Paris and use of "park-and-ride" facilities could be linked to the purchase of an "Orange Card for motorists". Such instruments, by partially covering the external social costs caused by automobile traffic (accidents, noise, pollution, etc.) would bring in financial resources, induce users to take these external social costs into account and, by persuading them to switch to other modes, would establish conditions of competitiveness between modes, thus reducing public transport deficits.

Financing Public Transport

The economic crisis at present affecting the country has singularly increased the difficulties endemic to the system of financing public transport. Concern to maintain fares within socially acceptable limits, the fall in the yield from the employers' contributions that are meant to cover part of the cost of their employees' daily travel on the networks, the growing debts incurred by public transport enterprises in modernising and extending their networks, leave the central government and the "départements" to carry an increasingly heavy financial burden.

In the present difficult times, rather than calling for even greater efforts from those who now finance the system, it would perhaps be preferable to extend the group of "financiers" to include others who benefit from the existence of a highly efficient public transport system. Developers, property owners, shopkeepers and even motorists, all of whom gain substantial advantages in one way or another from the presence of such a system, might be called upon to contribute to it.

The way in which public transport is financed can also contribute to better utilisation of its networks. For instance, the replacement of the fixed sum of FF23 formerly paid by employers to all their workers by a roughly equivalent sum to reimburse first 40 and then 50 per cent of the cost of the Orange Card to employees using public transport channelled an existing financial contribution towards the promotion of public transport. So this measure introduced by central government, as well as installing a ticketing system more convenient for users, has brought a small but significant increase in the clientèle using public transport.

Tools for the Management of Regional Transport

a) Introduction of Transport Accounting for the Ile-de-France Region as a whole

The financial resources allocated by the Paris and Ile-de-France transport system to investment and operating costs are considerable. Although each of the enterprises involved is, of course, thoroughly conversant with its own financial situation, there is as yet no official overall accounting system for urban and regional transport as a whole. A single financial statement drawn up annually under the responsibility of the Region and setting out all the usual information[26] (on the debit side, expenditures on investment and operating costs; on the credit side, resources listed according to their origin: contributions by the various users, transport levy, State, regional and local authority aid, etc.) would provide an essential management tool. So far as possible, statement of expenditures should be sufficiently detailed to allow the impact of specific action to be assessed (e.g. the proportion of the police budget allocated to the enforcement of parking regulations).

b) Establishment of an Overall Plan to be Reviewed Regularly

The RATP establishes a five-year plan[27] that is revised annually. The SNCF does the same. A regional development and infrastructure work plan is prepared every five years, and various other services or government agencies all draw up plans, each with a different horizon.

It would certainly be better for all concerned and make for greater clarity and better understanding, if all these transport-related projects were combined in a single comprehensive five-year plan for the region as a whole and perhaps reviewed annually, as the RATP plan does.

Transport and the Quality of Life

Transport and the Urban Environment

Transport should enrich rather than usurp non-transport-related activities in public places such as streets and squares. The installation of pedestrian precincts on a permanent basis is a first, widely welcomed response[28] to the problems of revitalizing inner cities and fostering social, cultural, and commercial activities in conditions offering security and an attractive environment. The first Parisian pedestrian areas (around the Beaubourg centre and in Les Halles) have been designed with this in view and should lead to early implementation of some proposals that have been roughly outlined, for instance in the Paris traffic plan.

Subsequently, the realisation of such plans could provide the opportunity, in certain neighbourhoods, to adopt the "traffic cells" that have proved to be of equal worth[29] and, for a moderate cost, eliminate through traffic so that the space in the streets of the "cells" thus created can be given over to local traffic or provide places for local children to play and residents to meet, thus improving security and enhancing the environment.

Finally, the fact that Paris and the urban centres of the region are exceptionally densely populated and busy and the street is an important factor in urban life means that special attention must be paid to street furniture and the use of squares and pavements (where, for instance, cars should not be allowed to park) and to keeping traffic in green belts to a minimum.

Transport and Nuisance

a) Noise

The campaign to abate traffic noise should not only be waged as at national level but also stepped up in the Paris Region [30]. In already built-up areas the measures currently being taken to protect people living along expressways by putting up anti-noise screens and making compensatory payments to householders for the phonic insulation of dwellings are extremely expensive. Care also has to be taken to avoid noise emission on new infrastructures, and this again entails substantial extra cost. This involves complementary action of the following kind (some of which has already begun):

- Use of quieter surfacing materials on major roads (asphalt rather than cobblestones);
- Stricter checks on the noise emission levels of moving vehicles (anti-noise police brigade);
- Further selective action to rationalise the use of vehicles where and when necessary;
- Pedestrian precincts;
- Traffic cells;
- Regulating heavy lorry traffic;
- Gradual improvement of the vehicle stock and particularly of buses.

b) Air pollution

Air pollution (increasing for nitrogen oxides and high for carbon monoxide) needs to be fought not only at national but also at local level. The social costs of pollution, like those of noise, are particularly heavy in built-up urban areas such as Paris, where much of the population is exposed to high concentrations of pollution. Thus regulations on the emission levels of new vehicles have to be backed up by effective checks while vehicles are actually being used and by reducing traffic where pollution levels are particularly high.

When new road or rail infrastructures are being planned, it is important to study the alternative solutions in the light of the legislation on protection of the environment, to consider in detail how those infrastructures can be integrated and to assess impact on the environment so that the best possible choice of lie and action can be made.

Transport and Safety

Local measures can complement national ones. The safety of pedestrians, cyclists and motor-cyclists must be a priority, especially in Paris and the inner suburbs. For example, systematic improvement of signs on pedestrian crossings and the application of special regulations around pedestrian precincts, stations and other places where pedestrians congregate can contribute to safety. The separation of pedestrians and two-wheel traffic from automobile traffic should be encouraged. Effective implementation of traffic regulations and speed limits in built-up areas should help bring a gradual change not only in the behaviour of all road-users but also in that of decision-makers, so that human beings really do take priority over vehicles.

Transport and Energy

Giving priority to public transport and travel on foot rather than to the use of private cars can contribute both to the reduction of total energy consumption and to the use of electricity from various sources instead of the consumption of petroleum products. Action to promote more rational use of private cars has the same thrust. What must be done is to maintain or improve the mobility of the population while at the same time reducing traffic and energy consumption.

ABBREVIATIONS

APTR:	Association Privée des Transports Routiers
IAURIF:	Institut d'Aménagement et d'Urbanisme de la Région Ile-de-France
RATP:	Réseau Autonome des Transports Parisiens
RER:	Réseau Express Régional
SDAU:	Schéma Directeur d'Aménagement et d'Urbanisme
SNCF:	Société Nationale des Chemins de Fer Français
STP:	Syndicat des Transports Parisiens
DREIF:	Direction Régionale de l'Equipement de l'Ile-de-France
IRT:	Institut de Recherche des Transports
CERNE:	Centre d'Evaluation et de Recherche des Nuisances et de l'Energie
FDES:	Fonds de Développement Economique et Social

NOTES AND REFERENCES

1. A motorised journey is a complete journey from starting point to destination by a person within the Ile-de-France region using one or more modes of transport other than walking.
2. See Annex for abbreviations.
3. Three RER lines are at present operated:
   – two (A and B) by the RATP;
   – one (Line C) by the SNCF, its 131 km being included in the total SNCF suburban network (927 km).
4. Including the ring road (boulevard périphérique) and the riverbank expressways in Paris.
5. And 53 per cent of annual passenger-km.
6. Subsequently reproduced to a large extent in 1973 in the report on urban transport in the Paris region by the then State Secretary to the Minister of Transport, M. Billecocq.
7. This figure includes an estimate of the depreciation, maintenance and running costs for cars and vans borne by private vehicle users.
8. In the case of reserved bus lanes, the breakdown is different: 50 per cent central government, 50 per cent Region.
9. It should be noted that the central government contribution is really less than 70 per cent, in view of the taxes and charges collected. In fact, contributions by the different parties in 1982 amounted to: central government, 18 per cent; local government, 12 per cent; employers, 25 per cent; users, 38 per cent; miscellaneous, 7 per cent.
10. The transport levy first applied only in Paris and the neighbouring départements, but was extended on 1st August 1975 to the whole of the Paris public transport area, although not to firms in the new towns. As from 1st January 1974 it was applied in other French towns of over 300 000 inhabitants, and from 1st January 1975 to towns of over 100 000 inhabitants.
11. The employer in fact already granted a transport allowance, but the amount was small (FF 23 a month at 30th January 1970), and the employee did not regard the payment as being specifically for transport. After the 1982 Act obliging employers to reimburse 40 per cent of the cost of season tickets (Orange card, weekly tickets, etc.) used by their employees for commuter travel, this transport allowance – which cost employers about FF 950 million a year – was discontinued in almost all cases.
12. See references 17, 27, 31 and 32.
13. Preliminary studies estimated that 600 000 people were likely to prefer the Orange Card to other types of ticket on cost grounds. In fact, since the introduction of the Orange Card the weekly workman's ticket has been little used, except on the SNCF networks.
14. In practice, payment of the "transport allowance" is still possible, but no longer compulsory, so that few employers continue to pay it.
15. M. Maurin et al., IRI-CERNE, Lyons (1981), "Enquête nationale sur l'exposition des Français aux nuisances de transports".
16. J. Lambert, IRI-CERNE, Lyons (1979) "Recherche d'éléments de décision pour l'action de l'Etat en matière de réduction de la gêne due au bruit de la circulation aux abords des VRU".
17. DREIF, "Les transports de voyageurs en Ile-de-France, 1982" Paris, 1983.
18. The inner ring road that encircles the densely built-up city of Paris.
19. The right-bank expressway which follows the north bank of the River Seine from west to east through the heart of the city.
20. Association pour l'information municipale "La lutte contre le bruit", in Ville de Paris, No. 44, June 1983, Hôtel de Ville, 1983.
21. Ministère de l'Environnement (1982), "Le bruit", Mission Bruit, Neuilly.
22. DREIF, "Programmation pluriannuelle des protections phoniques en région d'Ile-de'France: Autoroutes et voies rapides", Paris, 1981.
23. DREIF, "Boulevard périphérique de Paris, protections phoniques – étude générale" Paris, 1982.
24. To discourage property owners from making unnecessary demands or proposing over-costly estimates for the acoustic insulation of building façades, they are asked to contribute about 5 to 10 per cent of the costs of double-glazing, etc.
25. Resident parking represents 75 per cent of parking demand in Paris.
26. On the same lines as the document published by the Service régional de l'équipement de l'Ile-de-France, "Un tableau

général du financement des transports terrestres dans la région parisienne en 1974".
27. RATP, "Plan d'entreprise 1983-1987", Paris, 1982.
28. OECD, "Streets for People", Paris, 1974.
29. OECD, "Urban Transport and Environment", Paris, 1979.
30. OECD, "Noise Abatement Policies", Paris, 1980.
31. L. Kelner, "Impact de la carte orange", Transport, Environnement et Circulation, juillet-août 1976, Paris, 1976.
32. OECD, "Evaluation of traffic policies for the improvement of the urban environment", Paris, 1976.

Chapter 10

SINGAPORE*

I. SUMMARY

Singapore is a city-state with a population of 2.4 million inhabitants and it is the second busiest port in the world. In 1975 the Government introduced the "Area Licensing Scheme" (ALS), a package of transport measures aimed at arresting the growth of car-ownership and use and at improving public transport and the environment.

A restricted zone of 620 hectares roughly corresponding to the central business district (CBD), was declared and cars and taxis obliged to buy special licenses costing S$3[1] daily and S$60 monthly to enter it between 7:30 and 9:30 a.m. The cost of the license was subsequently increased to S$5 and S$100 for private cars (the rates were doubled for company owned cars and cut to S$2 and S$40 for taxis). Also the period of restriction was extended to 10:15 a.m. Over the same period the number of buses in scheduled service was increased by one-third to 3 203. Other features of the transport policy package were free passage into the restricted zone for car-pools of four or more riders, increased parking charges in the CBD, park-and-ride services (which proved unsuccessful); strict enforcement at 28 points of entry to the restricted zone and progressively more onerous taxes on the import, purchase and registration of cars. (Figure 1).

Concurrently with the transport package the Singapore Government mounted a major programme to promote the development of offices, hotels and shopping centres in the city centre. Employment accordingly rose from 200 000 in 1975 to 268 000 in 1983.

Before supplementary licensing was introduced, about 74 000 vehicles of all kinds entered the restricted zone in the morning peak period. In 1983, notwithstanding substantial increases in employment and incomes, which might have been expected to increase car-ownership with corresponding inbound flows forecast to be well in excess of 100 000 vehicles, the flow actually counted comprised less than 58 000 vehicles. The morning traffic restraints brought about a switch in commuting from cars to car-pools and buses, but in the absence of comparable restrictions during the homeward rush, traffic conditions continued to be congested.

The area licensing scheme, coupled with area traffic control, in the CBD and the construction in highways by-passing the CBD made it possible to achieve free flowing morning rush-hour conditions at comparatively low cost. It is estimated that additional investment in new roads of the order of S$1.5 billion would have been necessary to cope with increased demand without the 1975 transport restraint package.

Taxes aimed at limiting the growth of car-ownership have meanwhile resulted in the Singapore car fleet growing from 136 000 in 1973 to 184 000 in 1982 when forecasts based on income/car-ownership relationships indicated a fleet of 270 000. Estimates based on comparisons with ownership rates in other South-East Asian countries suggest that the Singapore car fleet could be even larger and as high as 340 000 vehicles.

It is extremely difficult to single out from the recession and other economic changes the effects of the ALS on land values and business in central Singapore. Analysis of rent changes and interviews with businessmen revealed no identifiable effects of the ALS on rents, retail sales or hotel-room occupancy rates. However, labour availability appeared to have been improved by the development of bus and car-pooling services. ALS was thought to have had reduced the availability of taxis.

---

* Case study carried out jointly by the World Bank and OECD, and realized by Messrs. Behbehani, Pendakur and Armstrong-Wright (World Bank) with the contribution of the Ministry of Public Works of Singapore.

Figure 1. **RESTRICTED ZONE**

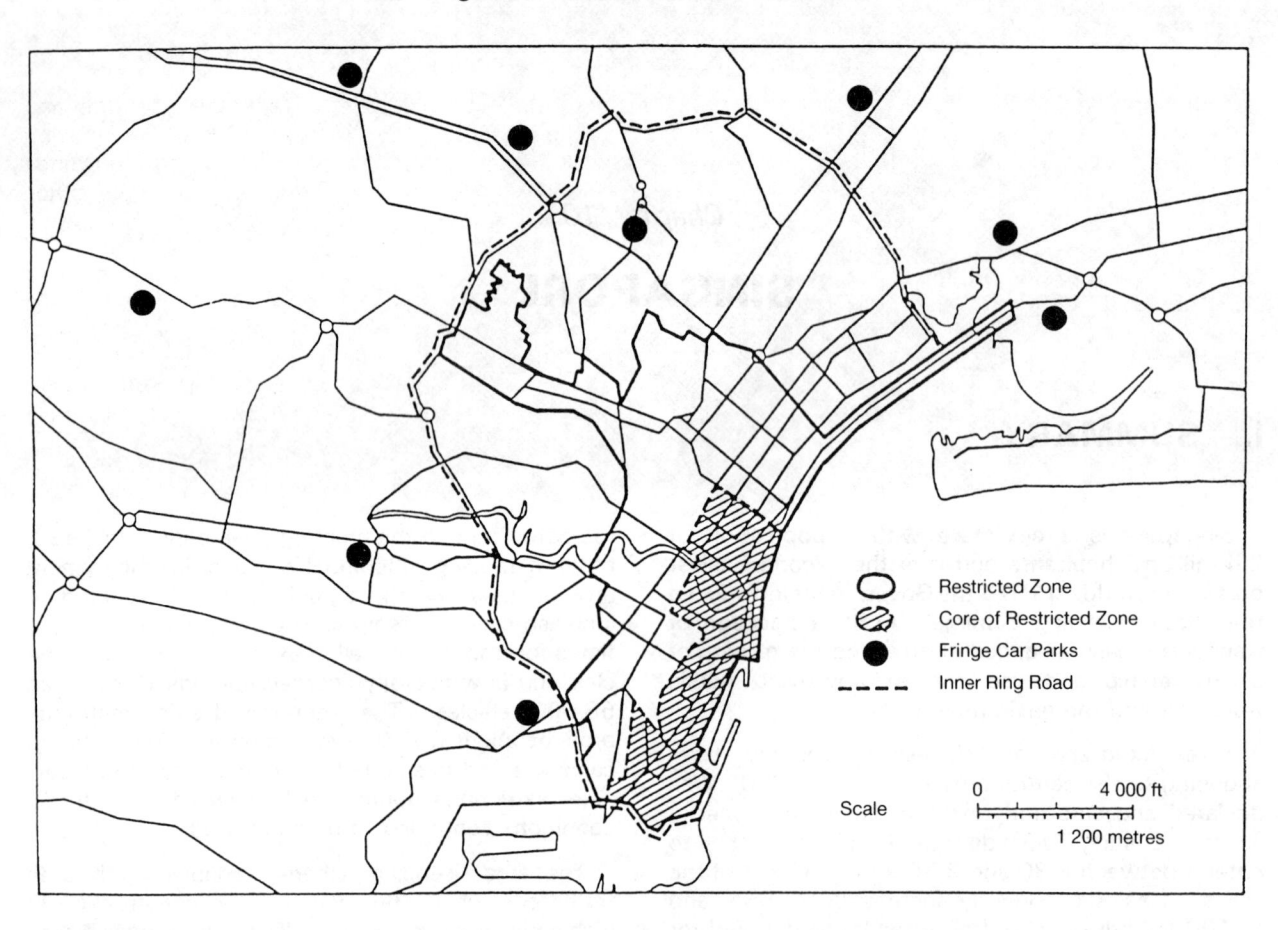

- ⬭ Restricted Zone
- ▨ Core of Restricted Zone
- ● Fringe Car Parks
- ---- Inner Ring Road

Scale: 0 — 4 000 ft / 1 200 metres

Measurement of the environmental effects of the ALS is confined to central area air quality monitoring. This shows that CO concentrations have increased, reflecting increased, day-long, traffic volumes. Reductions in total acidity, smoke levels and NO_x have, however, been recorded. These favourable changes are mostly attributed to better control of industrial pollutants, improved vehicle inspection and maintenance, and tax incentives to replace old by new cars.

II. INTRODUCTION

The Singapore Area Licensing Scheme (ALS) was the subject of a major before and after study undertaken by the World Bank. The findings have been widely reported[2] and their results have received wide publicity.

Similar traffic management schemes have been proposed in other ASEAN (Association of South East Asian Nations) cities, but have not been implemented. The Singapore ALS has now been in operation for nine years (1975-84). A study of its lasting effects is therefore timely and useful. However, the very substantial influence of other factors, such as the oil crisis, the world recession and rapid development in Singapore at barious times, have made it difficult to isolate and identify the effects of the ALS on the urban economy. As a result, this study is confined to a description of the package of measures that make up the Singapore Area Licensing Scheme, a review of any changes since 1975 and the impact on traffic movement and patterns, with a brief examination of the effect of the ALS on land use and values, business, employment and the environment.

III. THE DEVELOPMENT OF SINGAPORE

General Description

The Republic of Singapore has a population of 2.44 million (1981). The urban area of Singapore is located along the southern coast of the main island of the Republic and has a population estimated at 2.25 million.

Singapore has a strategic position in South-East Asia. Its port is the second busiest (i.e. tonnage) in the world and an active industrial policy has resulted in rapid growth, leading to a per capita GNP of 5 240 US dollars in 1981.

The Urban Redevelopment Authority

An official Urban Redevelopment Authority (URA) was set up in Singapore in April 1974 to carry out the comprehensive redevelopment of the central area. The URA lays down development guidelines, including parking requirement and has powers of expropriation which it uses for land assembly. Sites, once assembled and cleared, are auctioned off. The URA is responsible for defining and implementing the overall parking policy in Singapore. It also constructs, operates and maintains public car parks and controls street parking. The URA is the only permitted agency to collect parking charges on streets. Rent controls, tenancy and ownership laws make it very difficult for private developers to assemble land. The URA is therefore in a powerful position to control the timing and pattern of urban development.

Development Plans

Singapore experienced a very high and consistent rate of economic growth between 1975 and 1983 and during this period the pattern of urban development was heavily influenced by government policies. A Master Plan was prepared and considerable emphasis put on building New Towns and on developing the city as a major international centre for commerce, financial services and tourism. The government, through the URA, was the catalyst of these developments.

Urban Development, 1975-1983

During the period 1975 to 1983 Singapore's economic growth was sustained by government efforts to maintain construction at a high level. The result was remarkable rates of growth in privately built office buildings, retail shopping centres and hotels in addition to new towns built by the Housing and Develoment Board. Taking into account buildings completed, approved or under construction, office space will have more than quadrupled in the period 1975-1986.

Since 1975 there has also been a surge in the construction of new shopping centres, most of them in or adjacent to the central area. In 1975 the total retail space available was 0.25 million m^2. It grew to 1.0 million m^2 in 1983 and was 1.4 million m^2 in 1986.

With respect to hotel rooms, there were 8 200 hotel rooms in 1975, 11 687 in 1977 and 17 175 in 1983. Based upon hotels under construction, it is estimated that the number of bedrooms will considerably increase in the next few years. Average occupancy rates have been consistently high since 1975 rising from 83 per cent in 1977 to 86 per cent in 1981. Over a slightly longer period (1975 to 1983) the number of tourists visiting Singapore increased from 1.2 to 3.0 million p.a.

Central Area Growth

The Government's objective for the central area has been to renew and redevelop it into a strong central business district (CBD) analogous to those found in western countries. The Urban Redevelopment Authority (URA) has accordingly provided incentives for residents to move to the suburban new towns and attracted commercial enterprises to the central area. The official master plan calls for continued and increased residential population in the central area but the results have been otherwise.

Table 1 shows central area growth in relation to Singapore as a whole. In the period 1975 to 1983 the resident population fell from 220 000 to 153 000, a reduction of 30 per cent, while the total population of Singapore increased by about 10 per cent. As a result only 6 per cent of Singapore's population now live in the central area. This is a direct result of land assembly and urban renewal policies and for the residents who have moved it means that trip lengths and travel modes have changed.

Retail and office space meanwhile more than tripled while hotel rooms grew by two-thirds. Employment in the central area accordingly rose from 200 000 in 1975 to 268 000 in 1983 and additional people are now travelling to the central area during peak hours.

Table 1. CENTRAL AREA CHANGE 1975-1983[1]

Measures of change	1975[2]	1978[2]	1983[2]
Residential Population	220 000 (2.3 million)	178 000 (2.4 million)	153 000 (2.5 million)
Employment	200 000 (0.83 million)	224 000 (0.94 million)	270 000 (1.1 million)
Retail space (million m²)	0.17 (0.25)	0.2 (0.4)	0.6 (1.0)
Office space (million m²)	0.5 (0.7)	1.1 (1.4)	1.7 (2.2)
Hotel rooms	6 100 (8 200)	7 410 (12 562)	10 800 (17 175)

1. Data in this table is drawn from documents prepared by the Ministry of National Development and reconciled for presentation.
2. Figures in brackets () are for Singapore total.

Urban Transport

Two transport studies were carried out in Singapore during the period 1967 to 1974. The first, land use and transportation study, had the objective of preparing long range development plans. The second involved a detailed examination of public transport requirements and was aimed at selecting a mass transit system. Both studies independently reached the conclusion that restraints on car-ownership and usage would be necessary before 1992. This implied a need for radical changes in government policies and in public attitudes towards car-ownership and usage and a re-orientation of both towards greater use of public transport.

As in most other large cities, urban growth has been associated in Singapore with worsening traffic problems. Although the road network has been continually expanded and improved car-ownership has grown too.

From 1962 to 1973 the average annual growth rate of the car fleet was 8.8 per cent. More recently, strong government efforts to limit the growth of car-ownership have stabilized the fleet at around 184 000 cars or about one car per 14 persons. No comparable stabilization is apparent amongst motorcycles or goods vehicles (Table 2).

Table 2. MOTOR VEHICLES IN SINGAPORE BEFORE AND AFTER INTRODUCTION OF AREA LICENSING AND OTHER RESTRAINTS

Type of vehicle	End 1974	End 1978	End 1982
Motor cars	143 767	136 654	184 150
Motor cycles	84 849	98 248	136 899
Goods vehicle	36 462	55 626	96 903
Public buses	2 168	2 842	3 203
School buses	2 076	2 268	2 736
Taxis	5 162	7 683	10 278
Others	2 362	4 063	6 107
Total	276 866	309 384	440 276

IV. THE DEVELOPMENT OF AREA LICENSING

Origins of the Licensing Scheme

In late 1973, the Singapore Government, recognising the far-reaching implications of urban transport conditions, set up an interministerial Road Transport Action Committee (RTAC) to co-cordinate transport planning measures and to formulate policies. RTAC was supported by technical staff in the Public Works Department of the Ministry of National Development and other ministries. Early in 1974, the Committee examined the problems caused by growing traffic congestion in the central area, particularly during peak hours, and, after considering alternatives, recommended the immediate introduction of restraints on car use in the central area.

In developing restraint measures, the Government decided to adopt the following guidelines:

— Accessibility and mobility within the central area were to be maintained to protect its economic viability;
— The mobility of private cars was recognised as a benefit, and restrictions on their use were to apply only when and where they were needed to combat local congestion;
— The scheme was to be easily administered and enforced;
— Efficient, reliable, and attractive alternatives should be provided for commuters who might be discouraged from driving into the city. As these alternatives would be aimed at car owners, they should provide more frequent and comfortable services than normally available to bus commuters. The diverted car-owners could and should pay higher fares for the better service provided; and
— The scheme should not require a subsidy.

Working within these guidelines, the RTAC examined such measures for reducing traffic congestion as metered road use charges, toll roads, parking fees, and area licensing.

Vehicle metering was ruled out because of anticipated technical difficulties in mass producing meters and associated problems of dealing with faulty or tampered ones. Toll roads were considered impractical since spare for high capacity toll stations could not be found along the roads into the central area. Increased parking charges were considered to be a practical way of reducing vehicle flows into the central area but unable to discourage through traffic and chauffer-driven cars both of which contribute notably to central area traffic in Singapore.

The fourth option was area licensing which in essence required a special payment to park or drive a car in the central area. The Area Licensing Scheme (ALS), adopted by the government was a package. It incuded area lincensing fees, parking charges and vehicle taxes designed to discourage the ownership and use of private cars. A park-and-ride system and expanded bus services were included as well to provide motorists with a transport alternative.

To address policy and planning issues on the traffic management, parking, public transport and vehicle ownership that have arisen in recent years, a working committee known as Committee on Land Transportation (COLT) was set up in September 1983.

The 1975 Scheme

The ALS was instituted in 1975. It hinges on a supplementary license which must be obtained and displayed by a motorist in order to enter a designated, restricted zone. This zone, which covered those parts of the central area in which it was intended to reduce congestion. The zone covered an area of 620 hectares and had 27 entry points (currently 29 points). Other entries were closed off by barriers or one-way streets. The boundaries of the zone determined primarily by land use and traffic patterns and the availability of by-pass routes.

License Fees

License fees of S$3 per day or S$60/(1) per month were arbitrarily set in an effort to create a four to one ratio between the cost of driving and parking in the city centre and the cost of going by park-and-ride. These fees were subsequently revised upwards (Table 3).

The monthly licenses are sold at the Registry of Vehicles and selected post offices. Daily licenses are sold at selected post offices and 13 roadside sales booths set up along the approach roads to the restricted zone. Daily licenses can be purchased up to 3 days in advance.

Buses, goods vehicles, motorcycles, police, military and fire fighting vehicles are exempt from buying licenses in order to favour public transport, maintain commercial activity and assist emergency and security services. To encourage high car occupancy and efficient use of road space, cars carrying at least 4 persons are also exempt from buying licenses. Taxis were exempt from June 2 to June 23, 1975 after which they were treated the same as private cars. In January 1976 licenses for taxis were increased to S$4/day or S$80/month but in April 1977 reduced to S$2/day or S$40/month.

Table 3. ALS LICENSE FEES (S$): 1975-1983

	Private Cars		Company Cars		Taxis	
	Daily	Monthly	Daily	Monthly	Daily	Monthly
1975 post ALS	3	60	3	60	3	60
1 January 1976	4	80	8	160	4	80
1 April 1977	4	80	8	160	2	40
1 March 1980-Present	5	100	10	200	2	40

From January 1976, company owned cars were required to pay S$8/day or S$160/month to enter the restricted zone. In March 1980, these were further increased to S$10/day or S$200/month.

Operating Hours

The restricted zone was originally effective from 7:30 a.m. to 9:30 a.m. daily except on Sundays and public holidays. But because congestion developed after 9:30 a.m. the restricted period was extended to 10:15 a.m. thereby removing the new peak. Extension over the whole working day was rejected for fear that it would have adverse effects on business. Furthermore, it was anticipated that the morning restrictions would promote the staggering of work hours and reduce the evening peak as well.

Parking Space and Parking Fees

When the ALS was being designed it was estimated that there were approximately 13 800 parking spaces within the restricted zone with just over 60 per cent of them privately owned and operated. The rates for these private spaces were raised to between S$50 to S$80 per space per month depending upon location. Hourly parking fees were increased as well to discourage longer-term parking (Table 4).

Table 4. URA PARKING FEES AND CHARGES IN THE ALS

	Pre-ALS 1975	Post ALS 1975	1980	1983	1984
1. Parking Fee					
A. Short Term					
i) CBD	40 cents hour	50 cents 1st hour 1.00 second hour	40 cents each 1/2 hour	40 cents each 1/2 hour	60 cents each 1/2 hour
ii) Outside CBD	40 cents hour	50 cents 1st hour 50 cents each subsequent 1/2 hour	40 cents each 1/2 hour	40 cents each 1/2 hour	60 cents each 1/2 hour
B. Monthly					
i) CBD	40 cents hour	S$70	S$70	S$70	S$70
ii) Outside CBD but inside ALS	S$30	S$60	S$70	S$70	S$70
2. Surcharge/Space/Month applied to private operators					
i) CBD	nil	S$20	S$20	S$20	S$20
ii) Outside CBD but inside ALS	nil	S$10	S$10	S$10	S$10
3. Deficiency Charges applied to private developers					
i) Per space inside CBD	S$10 000	S$10 000	S$20 000	S$32 000	S$32 000
ii) Outside CBD but inside ALS	S$ 6 000	S$ 6 000	S$12 000	S$16 000	S$16 000

The restricted zone was divided into core and non-core areas with the core designated to include the most dense and congested areas of the central business district.

The new rates were designed to reflect the geographical distribution of congestion and to favour short-term as opposed to all-day parking. The government also introduced a surcharge on private car park operators, to induce them to raise their charges and prevent them from profitting on the difference between public and private charges. This surcharge amounts to S$20 per month per space in the CBD and S$10 per month per space in the central area outside the CBD and it has not been changed since it was introduced in 1975.

Prior to the ALS in 1974, the parking policy of the URA emphasized zonal balances. If a developer did not wish to build required parking, all he had to do was to pay a deficiency charge. It was presumed that the URA would collect such charges and provide sufficient parking nearby in the same zone. This was quite appropriate for small-scale developments and ones requiring less parking.

The zonal plan was abandoned in 1979 and the URA began to oblige all developers to provide the full parking requirements on their own sites, or pay a penalty. During 1975-83, about 1 950 new parking spaces provided in the central area of which about 950 came from private developers and the remaining 1 000 from the URA. The URA proposes to build another 4 000 spaces in the central area during 1983-88. This is in addition to new spaces coming on stream in private developments.

Charges for failing to provide parking space in new buildings have been revised progressively upwards. In 1975, the charge was S$10 000 per space within the CBD and S$6 000 per space outside it. These were doubled in 1980 and increased again in 1981 to S$32 000 per space for all developments within the central area and S$16 000 per space outside it. As the construction cost of parking spaces varies from S$32 000 to 40 000 (excluding land) it is cheaper for developers to pay deficiency charges to the URA unless additional parking enhances their business opportunities.

Parking rates which were 50S¢/hour in the ALS in 1975 were raised to 40S¢/half hour in 1980 and to 60S¢/half hour in 1984. Apart from one minor change, monthly rates have remained the same since 1975 at S$70.

Park-and-Ride

The Park-and-Ride services offered to motorists as an alternative to driving comprised of parking for 7 700 cars at 10 sites on the edge of the restricted zone and for 2 400 more in existing car parks. Monthly parking fees were set at S$10. A fleet of 90 shuttle buses with a total of 2 050 seats was then provided on eleven new routes. They were licensed to carry only seated passengers and made limited stops in an attempt to provide a fast, comfortable alternative to cars. The shuttle bus fare was 50S¢ a ride or S$20 a month and the combined cost of parking and riding S$1.50 per day or S$30 per month.

Park-and-Ride proved to be unpopular from the outset. Contrary to expectation, car commuters switched to the regular buses rather than to the shuttle services. The shuttle bus routes were therefore extended beyond the fringe car parks to nearby residential areas. The Park-and-Ride service is still being provided but on a much more limited scale.

Administration and Enforcement

Administration of the ALS proved to be comparatively simple. This was largely due to the care devoted to design and planning, the gradual implementation of complementary transport policies and pre-ALS publicity.

Enforcement has also worked well. Infringements occurred at the rate of about 200 a day when licensing began but they fell to between 50 and 60 a day as drivers gained familiarity with the system.

Parking Availability

Both area licensing and the increases in parking charges made short term parking space more available within the restricted zone.

Use of Fringe Car Parks

The unpopularity of Park-and-Ride, created spare capacity and the spaces provided were turned over to use such as night garages for buses lorries, mobile cranes, and a used-car sales centre.

Road Scheme and Traffic Management

The biggest road improvement since 1975 was construction of the East Coast Parkway which was completed in 1981. It enables cross-town traffic to by-pass the central business district and the restricted zone.

Substantial improvements have also been made to the road system to assist traffic flows in and out of the restricted. Several streets have been made one-way and a number of roundabouts have been converted into signal controlled crossings. Parking has been prohibited along major arterial streets. Bus-only lanes are provided where peak hour bus volumes exceed 100.

Another major change in traffic management since 1975 was the introduction of the Area Traffic Control System in August 1981. This brings 150 traffic signals in the central area within the inner ring road under the control of a central computer.

Vehicle Taxation

Taxes of various types are levied on motor vehicle owners in Singapore and are the highest in South East Asia. They have accordingly proved effective in slowing down the rate of increase of car-ownership. Vehicles coming into the country are subject to import duties and those on cars were increased from 10 per cent of open market value in 1968 to 45 per cent in October 1972, since then they have remained the same.

Annual vehicle registration fees were only S$15/year until February 1980, when they were increased to S$1 000/year for private cars and S$5 000/year for company cars. Up to December 1973, additional registration fees were 25 per cent ad valorem. In January 1976, following the introduction of ALS, they were raised again to 100 per cent and are now (1983) at 150 per cent ad valorem. A system of preferential additional registration fees was introduced in December 1975 for buyers of new cars who were scrapping old ones. These are set to favour the purchase of new vehicles with small engines and range from 35 per cent ad valorem for cars with engines below 1 000 cc capacity to 55 per cent for those over 3 000 cc.

Annual road tax on motor cars varies with the capacity of the engine. For engines with a capcity up to 1 000 cc the taxes, which were 20¢ per cc in December 1975, increased to 40¢ in February 1980. The increases for larger engines were as follows: 1 001 to 1 601 cc from 25¢ to 50¢m 1 601 to 2 000 cc from 30¢ to 60¢, 2 001 to 3 000cc from 80¢ to 1.00S$.

Import duties, initial registration fees and additional registration fees are paid only at the time of import and first registration of a vehicle. Road tax is paid annually as a road user license fee. For a new 2 000 cc car (not replacing an old one) which has an open market price of S$10 000, the duties and taxes amount to S$21 700 bringing the total cost of registering and licensing for private use to S$31 700.

Public Transport

The scheduled services of the Singapore Bus Service (SBS) provide the backbone of public transport in Singapore. They are complemented by the Supplementary Public Transport Scheme (SPTS) which uses school buses during designated peak periods along certain routes. For this purpose school hours are staggered with commuter times, so that the school buses can be made available for commuter services.

SBS was formed November 1973 by the amalgamation of three existing bus companies. Services are provided daily from 6:00 a.m. to midnight with a fleet that has been expanded from 2 188 buses in 1974 to 3 203 buses in 1982.

SPTS operates two sets of services, linking public housing estates to the CBD and to industrial estates. Both services are run only during peak hours on working days. Not all school buses are part of the Supplementary Scheme but starting with only 200 buses in 1974, it now comprises 594 buses on 43 peak-hour routes which complement services of the SBS. The SPTS enabled the number of buses available for peak-hour service to be increased by 9 per cent in 1974 and 19 per cent in 1982. It is estimated that in 1983 the supplementary buses carried 40 000 peak-hour trips per day.

Private and school buses also provide journey-to-work service under monthly contract. No stops or pickups are allowed except at one origin and destination and, although the fares are slightly higher than on SBS buses, contract services are very popular with commuters because they provide a virtual door-to-door service. In 1983 contract service was provided by 2 190 buses.

In an attempt to induce car-owners to comute by bus, an Air-conditioned Coach Service (ACCS) was introduced in June 1975. ACCS made use of air-conditioned buses owned by tour operators to link certain housing estates with the central business district and charged a flat fare of 1S$. This service, though comfortable and convenient, attracted few passengers initially and patronage subsequently declined. It was therefore discontinued in September 1982.

In 1975 there was a well established school bus service and between 1974 and 1982 the number of vehicles in use increased from 2 076 to 2 736. All school buses are privately owned and operated but as mentioned above many are also used for STPS or contract work.

V. IMPACT OF THE AREA LICENSING SCHEME

When the ALS was introduced in 1975, it was a comprehensive package of measures to control growth in the ownership of cars and in particular to discourage their use during peak hours. The bus system was at the same time expanded to provide a travel alternative and car-pooling was encouraged by exempting pools from the ALS fee. The package also included duties, taxes, parking fees and surcharges and additional traffic control and enforcement. The impact of the ALS is the result of all these closely inter-related measures.

This impact assessment is based on published data and research studies on travel behaviour and traffic patterns, together with on the results of discussions with members of the business community such as hotel operators, shopping centre owners, real estate developers, trade associations and bankers.

Initial Impacts – 1975

When area licensing was implemented in June 1975, the immediate effect was a 45 per cent decrease in the number of motor vehicles and a 76 per cent cut in cars entering the restricted zone (RZ) during licensing hours. The decline in car travel is attributed to:

a) The diversion of cross-town traffic to routes which by-pass the restricted zone; (this diversion resulted in heavy congestion on the ring road, a problem solved by modifying the timing of traffic signals to favour ring rather than radial traffic; improvement were also initiated on other by-pass routes);

b) The exemption given to car-pools of four or more persons which represented 50 per cent of the cars entering the restricted zone during the hours of restriction.

It was expected that the morning traffic restrictions would produce a mirror image effect on the evening peak hours. In the event evening traffic was only slightly reduced.

Car-Ownership

The additional taxes and duties levied, concurrently with the ALS, on the ownership and use of cars were substantial, their effect may be judged by comparing increases in gross domestic product (GDP) in Singapore with the growth of the car fleet.

Per capita GDP increased from S$6 045 in 1975 to S$12 520 in 1982, an increase of 107 per cent. Over the same period the number of private cars increased from 143 000 (65 per 1 000 persons) to 184 000 (74 per 1 000 persons) or only 28.6 per cent. Pay increases amongst the general labour force were (19 per cent in 1979, 20 per cent in 1980, 14 to 18 per cent in 1981 and 18 per cent in 1982). The car-ownership growth rate nevertheless remained at only about 3.5 per cent per year.

Studies[3] done in connection with the 1972 transport plan led to estimates that the car population in Singapore would grow from 136 000 in 1973 to about 400 000 in 1992 assuming no restraints on ownership and use. This rate of growth would have led to a fleet of 270 000 cars in 1982. In fact, with the introduction of higher taxes and restraint measures in 1975, there were only 184 000 cars in Singapore in 1982.

Estimates of car-ownership and of per capita incomes in ASEAN capital cities in 1982 are shown in Figure 2. Although incomes in Singapore are substantially higher than elsewhere, the car-ownership rates are about the same as in Bangkok and Kuala Lumpur. This is attributed primarily to the high taxes and duties introduced with the ALS. If these taxes had been absent or had been as low as another ASEAN countries, it is likely, given growth in incomes, that car-ownership rates would by now be in the range of 150 to 200 cars per 1 000 persons. This would give result in the Singapore car fleet being 345 000 to 460 000.

Changes in Travel Patterns

A substantial shift in the modal split took place after the establishment of the ALS in 1975. Initially, nearly one-half of all work trips were made by bus and nearly one-half by car, including car-pools. Compared with pre-ALS conditions this represented an increase in travel by bus from 33 to 46 per cent and a fall in car travel from 56 to 46 per cent (Figure 3).

By 1983 a total of 69 per cent of trips to work in the RZ were by bus and only 23 per cent by car (Figure 4). Most of the recent increase in bus patronage has come from new, white-collar, low-salary employees within the RZ.

Employment in the RZ increased from 2 000 000 to 270 000, i.e. by 34 per cent between 1975 and 1983 (Table 1). Almost all of this increase was in non-manufacturing jobs, primarily in finance and tourism and most of the commuting employees do not have high enough incomes to own or have the use of cars. They therefore travel by bus. This is reflected in

Figure 2. **CAR OWNERSHIP IN ASEAN**

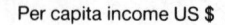

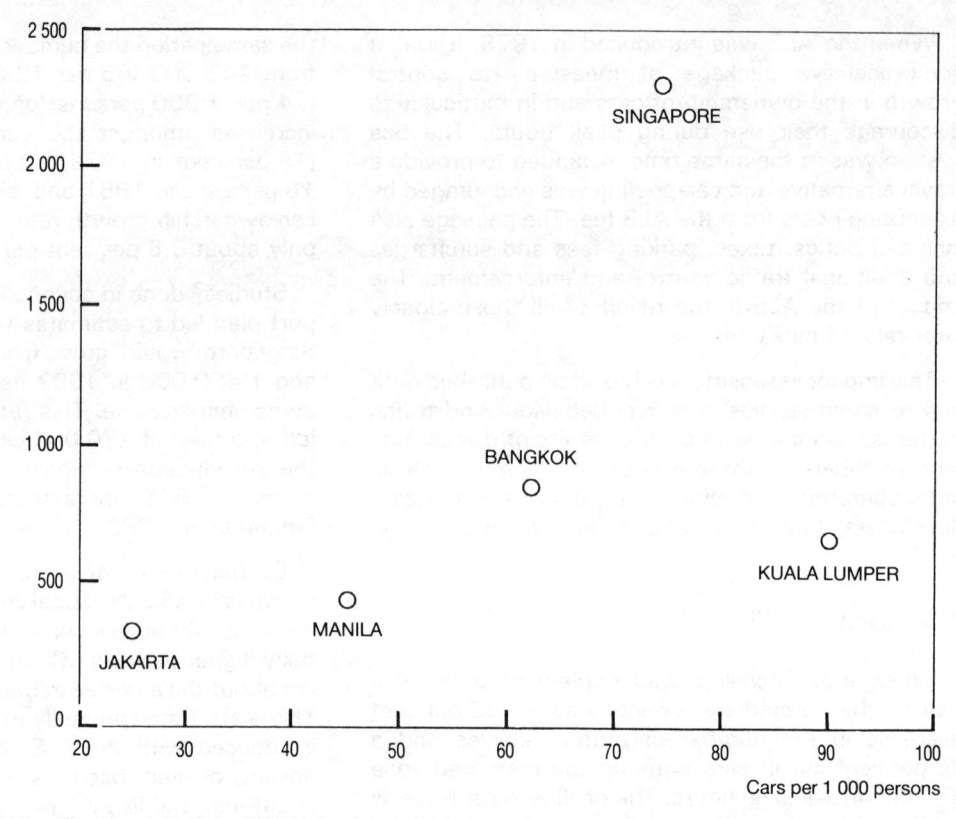

Source: V. Setty Pendaker, Urban Transportations in ASEAN, Institute of South East Asian Studies, Singapore 1984.

Figure 3. **1975 HOME TO WORK, POST ALS**
7.30 – 10.15 a.m.

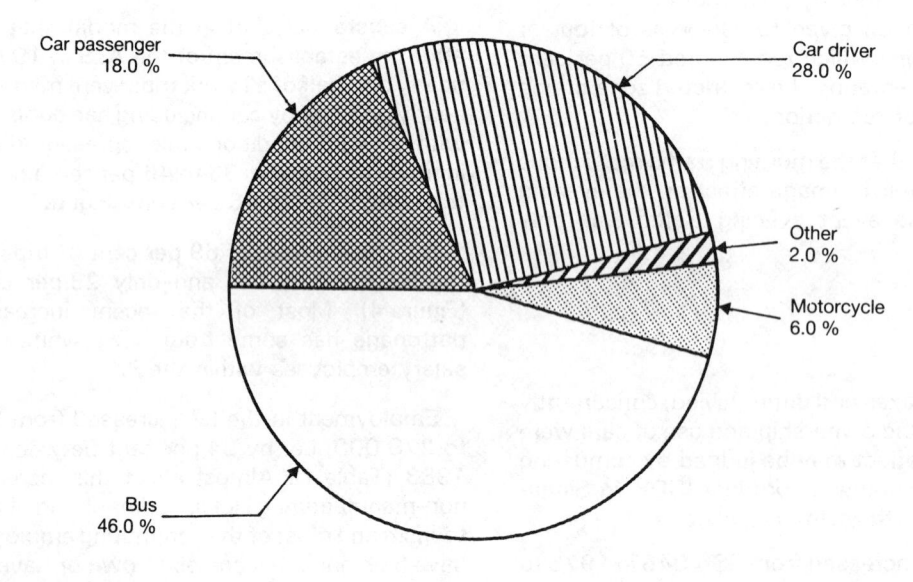

- Car passenger 18.0 %
- Car driver 28.0 %
- Other 2.0 %
- Motorcycle 6.0 %
- Bus 46.0 %

Source: World Bank.

Figure 4. 1983 HOME TO WORK (THE ALS ZONE)
7.30 – 10.15 a.m.

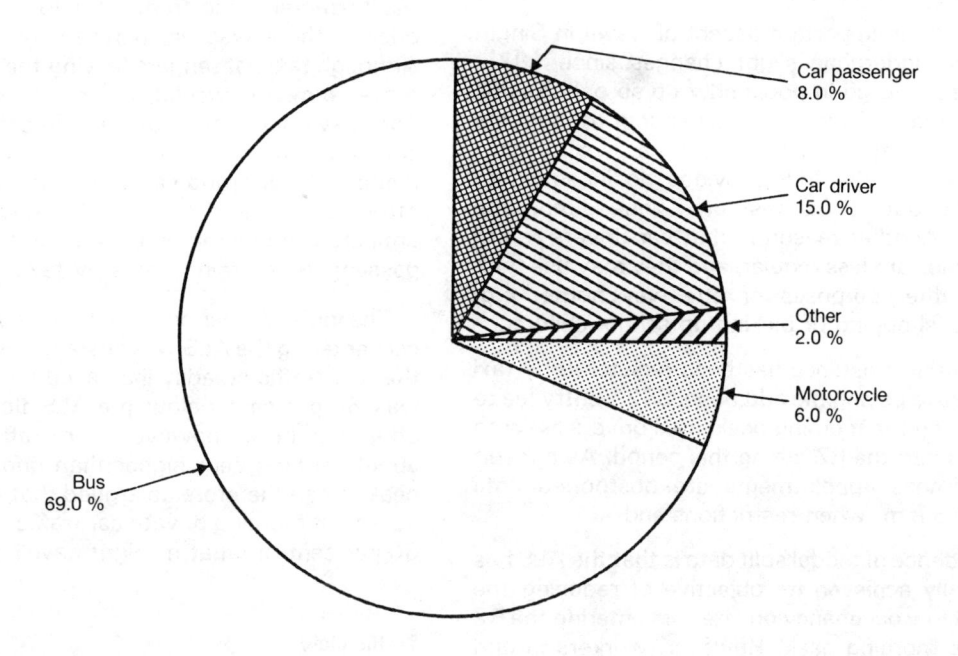

- Car passenger: 8.0 %
- Car driver: 15.0 %
- Other: 2.0 %
- Motorcycle: 6.0 %
- Bus: 69.0 %

Source: PMRTA – Household survey – Daily average.

Figure 5. 1983 ALL TRIPS TO ALS

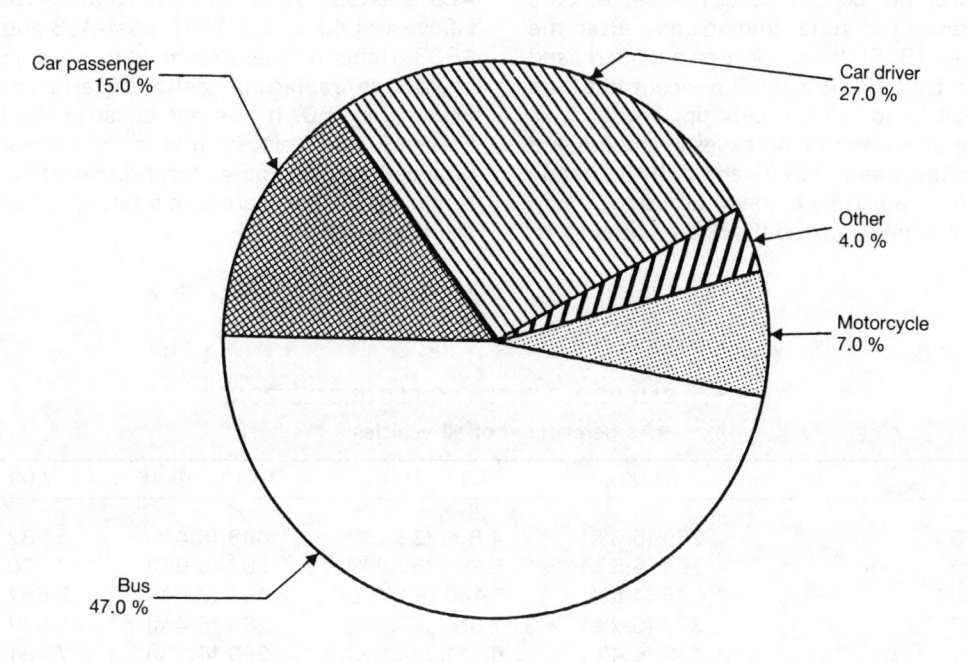

- Car passenger: 15.0 %
- Car driver: 27.0 %
- Other: 4.0 %
- Motorcycle: 7.0 %
- Bus: 47.0 %

Source: PMRTA – Household survey – Daily average.

the increase in the proportion of bus trips to the RZ throughout the day, i.e. both during and outside the restricted time; the split being 47 per cent by bus and 42 per cent by car (Figure 5).

Car-pools, an important aspect of travel in Singapore, have undergone slight changes since 1975. Most riders who go in pools now do so only for the journey to work. Fewer than 10 per cent of them are estimated to return home in a pool as well. This is largely because the ALS provides no incentive to share rides out of the restricted zone during the afternoon. Another reason is that car drivers return journey times are less regular, and they may use their cars for other purposes on the way home, e.g. recreation, shopping or picking up school children.

Substantial resistance has been registered by taxi passengers to paying the additional S$2/-entry fee to the RZ during the morning peak. Taxi drivers likewise avoid trips into the RZ during this period. As a result many business appointments are postponed until after 10:15 a.m. when restrictions end.

The evidence of modal split data is that the ALS has substantially achieved its objective of reducing the number of low occupancy private cars entering the RZ during the morning peak. However, workers in and visitors to the RZ have made unique adjustments to their travel patterns resulting in a greater concentration of movements in the afternoon.

Car Use

Since 1975 there has been a steady increase in morning peak period traffic. Prior to the ALS, car-pool vehicles were 23 per cent of total number of cars entering the restricted zone. Immediately after the ALS, (December 1975) this proportion had increased to 40 per cent. By 1980 it was 52 per cent but then decreased slightly to 47 per cent in 1982. It is suspected that as a result of increases in income and car-ownership between 1979 and 1983, many former car-poolers are likely to have shifted to driving themselves and either adjusted their hours of work to enter the ALS before 7:30 a.m. or are now able to pay the fee of S$100 per month (Table 5).

Initially, the ALS cut the number of taxis entering the restricted zone from 11 100 to 3 900. Subsequently there was an increase to 6 100 in 1982. Although taxi passengers leaving the ALS between 4 p.m.-7 p.m. on weekdays and 12 noon-3 p.m. on Saturdays pay a peak period surcharge of S$1, it is estimated that at least twice as many taxi trips are made in the afternoon as in the morning peak. This is attributed to the delaying of business appointments until after the restricted period and to some car-pool passengers returning home by taxi.

The initial 71 per cent fall in the number of private cars entering the ALS was sustained until 1977. After that car traffic steadily increased so that by 1982 it was 64 per cent below pre-ALS flows. During the afternoon peak, however, car traffic volumes are about 100 per cent higher than during the morning peak. It can therefore be argued that ALS restrictions have kept morning private car traffic to roughly about 50 per cent of what it might have been.

Traffic Flow

The impact of the ALS on traffic, given the context of a growing vehicle population, is made clear by comparing conditions in 1975 and 1983. Morning traffic conditions are today generally very good with inbound flows moving smoothly at between 40 and 50 kph. There has nevertheless been a 34 per cent increase in the total number of vehicles entering the ALS between 7:30 and 10:15 a.m (Table 6). The inflow was 43 252 in 1975 post-ALS and 57 035 in 1983. Under normal circumstances such an increase could have created congestion at certain intersections within the CBD. It has not because the East Coast Parkway and related free flow connectors have provided by-pass routes for through traffic while area traffic control has increased intersection capacity and travel speeds.

Table 5. CAR-POOLS ENTERING RESTRICTED ZONE
As percentage of all vehicles

Period	7.00	7.30 - 10.15	10.15 - 10.45	7.00 - 11.15
December 1975	259 (5.3%)	4 809 (39.7%)	448 (6.4%)	5 932 (20.3%)
November 1977	360 (6.5%)	5 902 (53.9%)	387 (5.9%)	7 020 (24.5%)
November 1980	278 (4.6%)	7 460 (52/0%)	451 (6.6%)	8 687 (26.4%)
November 1982	377 (6.7%)	7 356 (47.2%)	383 (5.4%)	8 532 (24.8%)
May 1983	413 (6.4%)	6 783 (43.8%)	290 (4.1%)	7 731 (22.2%)

Table 6. MOTOR VEHICLES ENTERING THE RESTRICTED ZONE

Month	Private cars 7:00 7:30[1]	Private cars 7:30 10:15[1]	Private cars 10:15 10:45[2]	Other vehicles (goods vehicles taxis, etc.) 7:00 7:30[1]	Other vehicles 7:30 10:45[2]	Other vehicles 10:15 10:45[2]	Total 7:00 7:30[1]	Total 7:30 10:15[1]	Total 10:15 10/45[2]
Mar 75	5 384	42 790	n.a.	4 416	31 224	n.a.	9 800	7 4014	n.a
Dec 75	4 866 (−9.6%)	12 106 (−71.1%)	6 529 (−11.5%)	4 516 (+2.3%)	31 146 (−0.2%)	7 075 (−9.1%)	9 382 (−4.3%)	43 252 (−41.6%)	13 604 (−10.3%)
May 76	5 675 (+5.4%)	10 754 (−74.9%)	6 459 (−12.4%)	4 657 (+5.5%)	26 833 (14.1%)	6 982 (−10.3%)	10 332 (+5.4%)	35 787 (−45.2%)	13 441 (−11.3%)
Nov 76	5 910 (+10.3%)	10 529 (−75.4%)	6 966 (−5.5%)	4 699 (+6.4%)	33 064 (+5.9%)	8 005 (+2.8%)	10 639 (+8.6%)	43 593 (−41.1%)	14 971 (−1.2%)
May 77	6 488 (+20.5%)	10 350 (−75.8%)	6 636 (−10.0%)	5 001 (+13.2%)	33 968 (+8.9%)	7 169 (−7.9%)	11 489 (+17.2%)	44 318 (−40.1%)	13 805 (−8.9%)
Nov 77	5 526 (+2.6%)	10 953 (−74.4%)	6 555 (−11.1%)	4 856 (+10.0%)	34 495 (+10.5%)	7 802 (+0.2%)	10 382 (+5.9%)	45 448 (−38.4%)	14 357 (−5.3%)
May 78	6 723 (+24.9%)	11 350 (−78.5%)	6 326 (−14.2%)	4 969 (+12.5%)	36 153 (+15.8%)	7 982 (+2.5%)	11 692 (+19.3%)	47 503 (−35.8%)	14 308 (−5.6%)
Nov 78	5 632 (+4.6%)	11 700 (−72.7%)	6 418 (12.9%)	5 002 (+13.3%)	36 658 (+17.4%)	8 457 (+8.6%)	10 634 (+8.6%)	48 358 (−34.7%)	14 875 (−1.9%)
May 79	5 723 (+6.3%)	13 181 (−69.2%)	5 527 (−25.1%)	4 873 (+10.3%)	36 425 (+16.7%)	9 652 (+24.0%)	10 596 (+8.1%)	49 606 (−33.0%)	15 179 (+0.1%)
Nov 79	5 643 (+4.8%)	13 497 (−66.5%)	6 558 (−11.1%)	5 505 (+24.7%)	41 246 (+32.1%)	8 931 (+14.7%)	11 148 (+13.8%)	5 4743 (−26.0%)	15 489 (+2.2%)
May 80	7 147 (+32.7%)	13 844 (−67.6%)	6 284 (−14.8%)	5 663 (+28.2%)	40 908 (+31.0%)	8 817 (+13.3%)	12 810 (+30.7%)	54 752 (−26.0%)	15 101 (−0.4%)
Nov 80	6 001 (+11.4%)	14 337 (−66.5%)	6 807 (−7.7%)	5 611 (+27.1%)	42 058 (+34.7%)	9 287 (+19.3%)	11 612 (+18.5%)	56 395 (−23.8%)	16 094 (+6.2%)
May 81	7 287 (+35.3%)	14 462 (−66.2%)	6 973 (−5.5%)	5 662 (+28.2%)	12 690 (+37.4%)	9 473 (+21.7%)	12 949 (+32.1%)	57 352 (−22.5%)	16 446 (+8.5%)
Nov 81	5 781 (+7.4%)	14 390 (−66.4%)	6 743 (−8.5%)	5 489 (+24.3%)	42 071 (+34.7%)	9 674 (+24.3%)	11 270 (+15.0%)	56 461 (−23.7%)	16 417 (+8.3%)
May 82	6 498 (+20.7%)	14 776 (−65.5%)	6 778 (−8.1%)	4 876 (+10.4%)	42 137 (+35.0%)	9 328 (+19.8%)	11 374 (+16.1%)	56 913 (−23.1%)	16 106 (+6.2%)
Nov 82	5 639 (+4.7%)	15 596 (−63.6%)	7 081 (−4.0%)	4 951 (+12.1%)	42 297 (+35.5%)	9 736 (+25.1%)	10 590 (+8.1%)	57 893 (+8.1%)	16 817 (+10.9%)
May 83	6 413 (+19.1%)	15 473 (−63.8%)	7 069 (−4.1%)	4 867 (+10.2%)	41 560 (33.1%)	9 421 (+21.0%)	11 280 (+15.1%)	57 035 (−22.9%)	16 490 (+8.8%)

1. The numbers in parentheses indicate changes in percentage compared with March 1975.
2. The numbers in parentheses indicate changes in percentage compared with August 1975.

Bumper to bumper congestion does however occur during the afternoon peak, particularly between 3-6 p.m., on certain roads within and adjacent to the ALS. Even with the by-passes and area traffic control flows on these roads are higher than their design capacity, thus reducing speeds and creating congestion.

Several factors account for these conditions:

a) Non-commuter trips and trips that do not have to be made during the restricted period, in particular for business appointments and shopping, are generally delayed until after the end of the ALS restricted period, thereby adding to the use of cars and taxis during the afternoon;

b) Through traffic is less inclined to avoid the restricted zone during the afternoon peak period;

c) As only 10 per cent of car-pool passengers share a ride home, many make use of buses for the return trip so that additional buses add to the p.m. rush traffic. Others use taxis or are collected by private car adding substantially to afternoon traffic;

d) Similarly, in the absence of restrictions many school children are picked up by private cars for the return journey home.

While the prices effect of each of these factors have not been measured, there is little doubt that together they make a substantial contribution to the congested conditions experienced during the p.m. peak-period. In this regard on certain routes in the restricted zone traffic speeds are reduced from 30 kph in the morning peak to about 18 kph during the afternoon peak.

Another contribution to increased traffic throughout the day is a substantial growth in the use of goods vehicles which are exempt from supplementary lincense fees (Table 2).

The difference between morning and afternoon conditions can be guaged from details of the traffic volumes. For example, in 1982 between 7:30 and 10:15 a.m., only 15 596 private cars entered the restricted zone whereas between 4:00 and 7:00 p.m., 40 210 cars (an increase of 158 per cent) left the zone. Furthermore, of the 15 596 private cars entering the zone during the morning peak 8 523 were car-pools. Bearing in mind that 90 per cent of car-pool drivers are estimated to leave without their pool passengers, 23 010 people ([8 523 x 3] x 0.9) must have travelled out by bus or by some other mode.

A different picture emerges from an examination of the figures for vehicles other than private cars. In 1982, 42 297 such vehicles entered the restricted zone between 7:30 and 10:15 a.m., whereas 47 240 left between 4:00 and 7:00 p.m. Since none of these vehicles (except for 3 435 taxis) are obliged to pay license fees data reflect conditions without restraint.

Total vehicle flows are the sum of the two parts just described. Those entering in the morning were 57 893 whereas those leaving were 87 450. The increase of 29 557 vehicles in the afternoon compared with the morning (an increase of 51 per cent) is attributed primarily to the factors discussed earlier, i.e. to some car-pool passengers, shopping trips, picking up of school children by car, through traffic in the central area and increased business trips by taxi and car.

Land Values

The supply of and demand for total office space in Singapore have followed a conventional economic cycle. During 1976-1977 demand was catching up with supply but during the following years, although office space increased from 1.4 million m^2 to 2.2 million m^2 an addition of 57 per cent, (Table 1), demand was higher than supply and led to high rents. However, it is estimated that in the period 1983 to 1986 supply will begin to outstrip demand and rents are indeed already coming down in anticipation of this situation. Any review of rents therefore needs to take this overall context into account.

In assessing the effects of the ALS on land values it would have been ideal to have been able to compare the changing values of sites with similar uses inside and outside the restricted zone. However, time series data for the value of similar sites are difficult to obtain and it was necessary instead to compare office rents drawn from secondary sources such as real estate compares developers combined with information from private firms engaged in office development and sales and from the Urban Development Authority. Some cross referencing with newpaper advertisements throughout the period was also undertaken.

Analysis of this data indicates no significant correlation between rents and locations inside or outside the RZ. Although locations just outside the ALS were initially (1975 and 1976) thought to be an advantage, analysis indicates that the ALS by itself is not a factor in rents. Neither does ALS appear to have been detrimental to office development within the restricted zone. Other locational factors and the timing of investment would appear to have been more important, and to have obscured what little influence ALS may have had on investment decisions.

Survey of Businessmen

The CBD is located around the mouth of the Singapore River and close to the sea. Historically, government and private firms established themselves in this are, though over the years the commercial centre has spread out to cover the wider areas of the restricted zone. An attempt was made to quantify the impact of the ALS on the wider area using available statistics. Members of the business community were asked for their assessment of the impact of the ALS on business and on location decisions within and outside the central business district. This evaluation was focussed on sales, labour availability, and the availability and cost of commercial property and commercial locations.

Sales

Representatives of the hotel industry felt that the ALS itself had not had a great impact on sales although hotel operators outside the zone, do advertise the absence of restrictions. Major reasons given for the recent decline in hotel business were over supply of rooms and the worldwide recession.

Retail shopkeepers expressed concern about declining sales due to competition, the drop in tourism and the deteriorating attractiveness of their shopping centres. The operators and owners of shopping centres, by contrast, felt that any loss of customers was due to the attraction of new shopping centres regardless of their location. The new shopping centres are attracting smaller retailer stores by bringing in anchor tenants with the offer of very attractive long leases. Overall, the economic activity in shopping centres appears to be down due to the recession, overbuilding and a decline in tourism. No particular influence from the ALS could be detected.

Employment Changes

The ALS did not appear adversely to affect labour availability. On the contrary its impact appear to be positive because its implementation had led to improved public transport. Junior business staff also enjoyed relatively easier travel to the central area following the introduction of car-pool stops near the housing estates. Further evidence for this view may be found in the fact that during a period when the work force in the central area increased from 200 000 to 270 000 (34 per cent increase), employment throughout Singapore increased by 32 per cent, from 830 000 to 1 100 000.

Availability and Cost of Commercial Property

The supply of commercial property has exceeded demand in Singapore in recent years because of overbuilding and the world recession. It is a measure of over building that the construction industry grew by 30 per cent in 1982 when the recession was at its peak. Hotels, offices and shopping centres are the types of building most overbuilt and affected by the recession. Between 1978 and 1982 the hotel industry experienced a 40 per cent growth in rooms and it is estimated that by 1987 this room capacity will have doubled. Meanwhile room occupancy rates are forecast to decline from levels as high as 86 per cent in the 1970s to 65 per cent in 1985-1986. This forecast is based on slow tourist growth rates and overbuilding in the hotel industry.

Discounts in the hotel industry are currently in the 20 to 30 per cent range. Low occupancy rates and discount rates are characteristic of hotels inside and outside the restricted zone.

Shopping centres have had a fate similar to that of hotels, bu not to the same extent. The average vacancy rate for shopping centres is about 10 per cent throughout Singapore. This ranges from 2 per cent at the Plaza Singapura (inside the RZ) to 30 per cent at the Golden Mile Shopping Centre (outside the RZ). Again, these rates of vacancy were not attributed to traffic limitation, but rather to the overbuilding and the recession and the related sharp decline in tourism. Table 7 shows shopping space by sector and locality.

Office Rents

No clear patterns could be identified amongst office rents inside and outside the restricted zone. Representatives of the business community felt they were very competitive everywhere.

In 1972 and 1973 there was a glut of office buildings. Between 1973 and 1978, new and existing companies increased demand from 800 000 to 900 000 square feet per year and rents remained steady at S$2 to S$2.5 per square foot, but by 1978 a shortage of office space occurred and rents soared. The Urban Redevelopment Authority then stepped in and encouraged new construction by organising and selling sites for office construction.

Since then the total supply of office buildings has rapidly increased and in 1983, 49 office buildings, representing 8.7 million square feet of floor space, were being built. Most of the new buildings were started during the boom years 1978 to 1981 and together they will increase Singapore's office capacity by about 50 per cent.

Meanwhile the worldwide recession has slowed down the number of new companies arriving from abroad and has discouraged expansion of existing ones. A resulting excess of supply has led prime office rents to fall by 30 per cent (from S$7-8 to S$5-6 since the peak in 1981.

Landlords are understood to have employed tactics other than rent reductions and financial incentives to encourage new or existing tenants to move in or stay. Table 8 presents the office space by sector and locality.

Table 7. OCCUPANCY OF SHOPPING SPACE[1] BY SECTOR AND LOCALITY
AS OF 31ST MARCH 1983

Sectors	Locality	Shopping space Completed (m^2)	Shopping space Occupied (m^2)	Occupancy rate %
Both Sectors	Whole Island	1 013 437	905 745	89.4
	Central Area	364 410	337 898	92.7
	Orchard Road Strip	221 942	194 803	87.8
	Rest of Island	427 085	373 044	87.3
Private sectors	Whole Island	588 028	519 539	88.4
	Central Area	209 147	188 985	90.4
	Orchard Road Strip	214 553	187 748	87.5
	Rest of Island	164 328	142 806	86.9
Public sectors	Whole Island	425 409	386 206	90.8
	Central Area	155 263	148 913	95.9
	Orchard Road Strip	7 389	7 055	95.5
	Rest of Island	262 757	230 238	87.6

1. Both fully and partially completed commercial developments are included in the survey. In the case of partially completed buildings or developments, occupancy rates are only applicable to shopping space that are completed.

Source: Ministry of National Development Research Statistics Unit.

Table 8. OCCUPANCY OF OFFICE SPACE[1] BY SECTOR
AND LOCALITY AS OF 31ST MARCH 1983

Sectors	Locality	Office space Completed (m^2)	Office space Occupied (m^2)	Occupancy
Both sectors	Whole Island	2 200 371	2 107 466	91.7
	Central Area	1 344 445	1 235 037	91.9
	Orchard Road Strip	301 833	277 401	91.9
	Rest of Island	554 093	505 028	91.1
Private sector	Whole Island	1 471 848	1 321 765	89.8
	Central Area	1 048 432	950 964	90.7
	Orchard Road Strip	232 278	211 456	91.0
	Rest of Island	191 138	159 345	83.4
Public sector	Whole Island	728 523	695 701	95.5
	Central Area	296 013	284 073	96.0
	Orchard Road Strip	69 555	65 945	94.8
	Rest of Island	362 955	345 683	95.2

1. Both fully and partially completed commercial developments are included in the survey. In the case of partially completed buildings or developments, occupancy rates are only applicable to office space that are completed.

Source: Ministry of National Development Research Statistics Unit.

None of the managers interviewed felt that the area licensing scheme was a factor influencing the level of rent inside or outside the zone. The perceived causes of the fall in rent are recession, overbuilding, and the cyclical nature of the construction industry.

Commercial Location

The views of businessmen on choice of location varied with their business. Those involved in activities such as banking, government and other financial institutions that involve direct dealings with the business community and the public, prefer the central business district. Companies that do not have direct contact with customers tend to locate outside of the central business district.

Property developers felt that they have limited choice over sites given land use patterns and the control of most land sales by the URA. Furthermore private land holdings are small compared with those of the government.

Adjustment Patterns of the Business Community

The responses of businessmen to the ALS were varied but mainly favourable. Those who were discontent with the scheme nevertheless conceded that something had had to be done to relieve morning traffic congestion. Over the 8 year period, people who need to be in the central area during the restricted hours have been observed to adjust their work habits and choice of transport. Shoppers have not been affected since most shops open at 10 a.m. or shortly afterwards.

Most businessmen were, in fact, not affected by the Scheme. Highlevel officials rarely had to pay to enter the restricted zone. Either they had company cars or a permit and free parking was provided to them. A few managers arrived earlier than 7:30 a.m. and avoided the restricted hours altogether. Managers felt that their junior staffs were likewise, not affected by the scheme since they were using bus services or car-pools. Some managers felt that their junior staffs might have benefitted from the scheme due to the promotion and use of car-pools.

The effect of the RZ on taxi operation was a major concern of both businessmen and hotel operators. Traffic restraint was believed to have affected taxi availability in the morning and in turn customers. The general feeling was that taxi availability had somewhat improved after the introduction of the surcharge on tariffs for any taxi ride going to or from the RZ between 4 and 7 p.m. on weekdays and 12 noon and 3 p.m. on Saturday. This surcharge has improved the taxi availability at shopping centres more than elsewhere.

Hotel managers were particularly concerned about taxi services. They believed that their customers were affected in one way or another since people staying at hotels inside or out of the restricted zone faced similar difficulties in obtaining taxis to go to or from the central area. Hotels have accordingly adopted various ways of dealing with the problem. Some provide shuttle mini bus service to the CBD shopping areas, others provide incentives for taxi drivers in the form of free monthly area licenses or parking space.

Air Pollution

There are six air quality monitoring stations in the central area of Singapore. The following pollutants are measured:

Carbon Monoxide	ppm
Total Acidity	ug/m^3 (Hydrogen peroxide method)
Smoke levels	ug/m^3 (Ambient smoke concentration)
Sulphur Dioxide	ug/m^3 (Parasoaniline method)
Nitric Oxide	ug/m^3
Nitrogen Dioxide	ug/m^3

Carbon monoxide concentration has increased in all locations and varied from 307 ppm (max. reading) reflecting increased vehicular traffic since 1976. There has been a substantial decrease in total acidity, smoke levels and nitric oxides and dioxides (Table 9).

The decrease in levels of pollutants cannot be attributed solely to area licensing. Much of the improvement is due to better control and monitoring of industrial pollutants; better automobile inspection and repair requirements; and the introduction of tax incentives to replace old cars with new cars. It should be noted, however, that carbon monoxide levels are still low compared with other ASEAN cities.

Noise

No noise measurements were available and in the view of officials and other people interviewed, noise was not considered to be a problem. Excessive noise emanating from motor vehicles is controlled by the police under the motor vehicle maintenance inspection and repair code although the primary concern of their work is with vehicles safety and not noise.

Traffic Safety

The level of enforcement of driving behaviour and penalities for infringements are high in Singapore and coupled with driver training have contributed to steady improvements in road safety. The number of

Table 9. AIR POLLUTION, 1978-82

Site	Year	Carbon monoxide (ppm) 7h00/15h00	Carbon monoxide (ppm) 15h00/23h00	Carbon monoxide (ppm) 23h00/7h00	Total acidity (ug/m3) Av.	Total acidity (ug/m3) Max.	Total acidity (ug/m3) Min.	Smoke (ug/m3) Av.	Smoke (ug/m3) Max.	Smoke (ug/m3) Min.	Nitric oxyde (ug/m3) Av.	Nitric oxyde (ug/m3) Max.	Nitric oxyde (ug/m3) Min.	Nitrogen dioxide (ug/m3) Av.	Nitrogen dioxide (ug/m3) Max.	Nitrogen dioxide (ug/m3) Min.
Princess House Lab (Alexandra Road)	1978	4	5	2	80	105	52	138	169	123	212	287	111	29	46	18
	1979	5	6	3	72	87	59	123	169	65	215	260	179	30	51	15
	1980	3	4	2	61	84	58	114	144	95	177	224	141	27	34	20
	1981	3	4	2	56	68	44	92	124	55	175	244	99	27	34	20
	1982	3	3	2	56	92	39	79	135	44	162	236	94	33	50	20
CID (Robinson Road)	1978	2	2	1	56	78	42	84	105	61	105	124	45	33	51	22
	1979	3	3	1	43	64	18	64	103	29	82	150	55	22	43	13
	1980	3	2	1	43	94	53	62	158	95	81	145	40	17	25	13
	1981	2	2	1	43	74	20	56	87	22	106	150	63	26	41	15
	1982	3	3	1	48	61	30	53	73	30	95	317	95	27	52	18
St. Joseph's Institution (Brass Basah Road)	1978	3	5	2	49	95	25	118	143	64	202	340	108	32	46	22
	1979	3	4	1	69	89	49	120	185	83	220	283	129	20	27	9
	1980	3	3	2	69	84	21	116	100	37	164	281	105	23	28	16
	1981	3	4	2	52	74	36	100	149	71	199	307	100	27	45	20
	1982	4	5	2	55	73	40	98	136	48	217	116	58	34	49	18
Hill Street Police Station (Hill Street)	1975	6	6	3	52	66	40	79	154	31	100	144	65	29	35	17
	1976	6	6	2	50	140	32	136	158	114	–	–	–	–	–	–
	1977	3	3	1	73	108	26	159	264	46	45	90	13	28	53	8
PSA Waterboard Office (Fullerton Road)	1975	10	8	4	140	175	125	248	292	207	156	197	113	26	41	16
	1976	8	5	2	125	224	52	153	218	80	158	372	8	30	132	4
	1977	4	4	2	21	64	Nul	118	160	95	180	301	94	42	85	37
Telecommunication Authority of Singapore (Robinson Road)	1975	13	12	7	74	96	40	160	186	80	113	171	50	29	34	23
	1976	7	7	2	91	129	9	135	185	109	109	213	35	24	46	17
	1977	8	7	2	84	129	59	196	355	33	289	419	158	32	37	21

Table 10. TRAFFIC ACCIDENTS WITHIN THE CENTRAL AREA[1]

Type of Accident	1975	1978	1980	1982
Fatal	n.a.	25	19	26
Serious	n.a.	195	105	50
Slight	n.a.	666	733	601
No Injury	n.a.	3 519	3 644	2 705
Total	n.a.	4 405	4 501	3 382

1. All of Central Area = Slightly larger than the ALS.

accidents in all categories has accordingly decreased since 1975 notwithstanding increases in motor vehicles and traffic (Table 10).

Transport Investments

The introduction of the ALS made it necessary to improve the bus service to provide commuters with an alternative to their cars. Initially 100 new buses were put into service and between 1975 and 1982 the fleet of the Singapore Bus Service was expanded from 2 312 to 3 203 vehicles. Over approximately the same period (1975 to 1983), the number of taxis grew from 5 368 to 10 318, the investment coming from the private sector.

Initial capital costs of establishing the area licensing scheme amounted approximately to S$6 600 000 of which S$6 100 000 (92 per cent) was for the fringe car parks. In contrast to the comparatively low cost of ALS the effect of the systems on reducing highway investment was very substantial. Without traffic restraints it would have been necessary to cater for greater volumes of vehicles. Additional investment in roadways and traffic management would therefore have been required. Analysis indicates that without the package of measures included in the ALS, by 1982 cars would have numbered in the region of 270 000 or 107 per 1 000 persons. Car use would have been higher too and accounted for between 50 per cent and 60 per cent of the peak period trips.

On the basis of these estimates of car-ownership and use, it seems likely that traffic volumes in 1982 during the morning peak period would have been approximately 80 per cent higher than the traffic volumes actually experienced. Under such circumstances bumper-to-bumper conditions would have been universal in the restricted zone and worse than the highest levels of congestion found in 1982 on certain routes during the afternoon peak.

The ALS has not only reduced traffic entering the zone during the morning peak from its pre-restricted volume of 74 000 to 58 000 vehicles during 1975-1983; but it has also prevented traffic from growing to an estimated 100 000 vehicles during the morning peak. Translated into deferred or cancelled investments, the savings implied by these differences are, indeed, large.

Studies[3] of transport in Singapore in 1972 had estimated that road improvements costing 1.0 billion S$ would have been required by 1982. However, these calculations did not take into account increases in personal incomes and their potential to bring about very high car-ownership rates. All factors considered road investment requirements might therefore have been on the order of 1.5 billion S$.

VI. LESSONS OF THE AREA LICENSING SCHEME

Lessons From Singapore

A number of valuable lessons may be drawn from the experience of area licensing and other measures to limit car use in Singapore. These include:

— Area licensing and its related measures have restrained car use, reduced growth rates of car-ownership, and have substantially reduced congestion;
— Area licensing has been responsible for an appreciable shift in peak hour travel from car to car-pool, and car to bus; and has led to the more economic use of road space, in the restricted zone;
— The area licensing package is perceived by the business community as beneficial to Singapore and it is clear that its removal would result in worsened congestion;
— The area licensing scheme operates only during the morning peak hours, and is inclined to concentrate trips during the afternoon as well as the morning peak.

Area licensing in Singapore has also been a significant contributor to improved traffic safety.

There have been few detrimental effects, and the improvement of traffic flow has enhanced business activity, and working conditions for employees in the central area. Finally, it has made possible savings, probably on the order of 1.5 billion S$, in road investment which would otherwise have been needed to maintain levels of service.

Application of Area Licensing to Other Cities

Area licensing, similar to the Singapore scheme, but suitably adapted is likely to provide a practical means of relieving traffic congestion in other cities. However, it needs to be stressed that the system developed in Singapore comprises a package of measures which are interdependent and mutually supportive. For example, restraint on ownership and use of private cars needs to be accompanied by expanded and improved public transport. Area licensing also needs to be accompanied by adequate alternative routes to enable through traffic to by-pass the restricted zone.

A key to the success of area licensing is the level of enforcement and public knowledge that the components of the system will be enforced without any exemptions. If effective enforcement is inadequate, area licensing will not work.

It needs to be stressed that area licensing provides a low-cost means of making the best use of available road space and avoids or defers heavy investment in highways and transport systems. As such it provides a particularly valuable solution to growing transport problems in developing countries faced with rapidly growing demand and a lack of resources.

NOTES AND REFERENCES

1. During the period of the study the exchange rate varied around S$2.10 for one US dollar.
2. World Bank, SWP ≠ 281 Relieving Traffic Congestion: The Singapore Area License Scheme, Washington D.C., June 1978;
   OECD, Managing Transport Chapter XIV, Paris, 1979.
3. Crooks, Mitchell, Peacock and Stewart, The Urban Renewal and Development, Singapore, Chapter VI. The Transportation Plan, UNDP/Republic of Singapore, Singapore 1972. Wilbur Smith and Associates. Singapore Mass Transit Study—Phase I, World Bank/UNDP/Republic of Singapore, Singapore 1984.

LIST OF THE MEMBERS OF THE AD HOC GROUP ON TRANSPORT AND THE ENVIRONMENT

CHAIRMAN: **Mr. P. FORTON**

AUSTRALIA	Mr. G. McALPINE	JAPAN	Mr. Y. HIRAYAMA
			Mr. H. KAJIWARA
BELGIUM	Mr. J. BEYLOOS		Mr. H. KOYANAGI
	Mr. P. FORTON		Mr. T. MATSUMURA
			Mr. N. OZAWA
CANADA	Mr. J. HOLLINS		Mr. H. TAKAGI
	Mr. A. STONE		Mr. T. TAKEUCHI
DENMARK	Mr. L. LARSEN	NETHERLANDS	Mr. R. BOEREK
	Mr. O. MUNSTER		Mr. BRAAKENBURG
	Mr. S. SCHMIDT		Mr. W. HOOGENDOORN
			Mr. M. JANSSENS
FINLAND	Mr. R. HAKANEN		Mr. S. JITTA
	Mr. J. NYRHILA		Mr. R. LANSMAN
		NORWAY	Mr. P. DOVLE
FRANCE	Mr. P. BAR		Mr. L. LANDRO
	Mr. B. DEROUBAIX		
	Mr. B. DURAND	PORTUGAL	Mr. G. CANCIO MARTINS
	Ms. A. EYSSARTIER	SWEDEN	Ms. B. BERTILSSON
	Ms. A. LE FLOCH		Mr. G. FRIBERG
	Mr. Y. LE GOFF		Mr. J. KARLSSON
	Mr. B. LEQUIME		Mr. B. KOHLMARK
	Mr. L. ROPARS		
		SWITZERLAND	Mr. J. AQUARONE
GERMANY	Ms. U. DALDRUP		Mr. A. CLERC
	Mr. W. HULSMANN		Mr. E. RATHE
	Mr. H. KEITER		Mr. F. RIEHL
	Mr. K. RIBBECK		Mr. R. ROTHEN
	Mr. H. SCHULZ		Mr. G. VERDAN
		TURKEY	Mr. A. KIZIL
GREECE	Ms. V. SOTIRIADOU		Mr. K. TEPEDELEN
	Mr. E. VASILAKOS		
		UNITED STATES	Mr. G. HAWTHORN
ITALY	Mr. S. CASADIO	CEC	Mr. E. MACKINLAY
	Mr. G. DEGIORGIS		Mr. MARK
	Mr. G. L'OCCASO KRIEGERN		Mr. H. SCHMIDT-OHLENDORF
	Mr. C. LOMONACO		
	Mr. G. PUGLISI	YUGOSLAVIA	Ms. M. ZLATIC

OECD SECRETARIAT

Mr. A. ALEXANDRE
Mr. C. AVEROUS

Consultants and experts: Mr. D. BAYLISS, Mr. R. BEHBEHANI, Mr. T. BENDIXSON, Mr. P. BOVY, Mr. M. DOWNEY, Mr. N. EMERSON, Mr. S. FALK, Mr. J. FRADIN, Mr. L. KEEFER, Ms. M. LINSTER, Mr. A. LOMBARD, Mr. T. MAY, Mr. G. MEIGHORNER, Mr. K. MORITA, Mr. B. PEARCE, Mr. R. PRUD'HOMME, Mr. H. SIMKOWITZ, Mr. E. VASILAKOS.

WHERE TO OBTAIN OECD PUBLICATIONS
OÙ OBTENIR LES PUBLICATIONS DE L'OCDE

ARGENTINA - ARGENTINE
Carlos Hirsch S.R.L.,
Florida 165, 4º Piso,
(Galeria Guemes) 1333 Buenos Aires
Tel. 33.1787.2391 y 30.7122

AUSTRALIA - AUSTRALIE
D.A. Book (Aust.) Pty. Ltd.
11-13 Station Street (P.O. Box 163)
Mitcham, Vic. 3132   Tel. (03) 873 4411

AUSTRIA - AUTRICHE
OECD Publications and Information Centre,
4 Simrockstrasse,
5300 Bonn (Germany)   Tel. (0228) 21.60.45
Gerold & Co., Graben 31, Wien 1   Tel. 52.22.35

BELGIUM - BELGIQUE
Jean de Lannoy,
Avenue du Roi 202
B-1060 Bruxelles   Tel. (02) 538.51.69

CANADA
Renouf Publishing Company Ltd/
Éditions Renouf Ltée,
1294 Algoma Road, Ottawa, Ont. K1B 3W8
Tel: (613) 741-4333
Toll Free/Sans Frais:
Ontario, Quebec, Maritimes:
1-800-267-1805
Western Canada, Newfoundland:
1-800-267-1826
Stores/Magasins:
61 rue Sparks St., Ottawa, Ont. K1P 5A6
Tel: (613) 238-8985
211 rue Yonge St., Toronto, Ont. M5B 1M4
Tel: (416) 363-3171
Federal Publications Inc.,
301-303 King St. W.,
Toronto, Ont. M5V 1J5
Tel. (416)581-1552
Les Éditions la Liberté inc.,
3020 Chemin Sainte-Foy,
Sainte-Foy, P.Q. G1X 3V6,
Tel. (418)658-3763

DENMARK - DANEMARK
Munksgaard Export and Subscription Service
35, Nørre Søgade, DK-1370 København K
Tel. +45.1.12.85.70

FINLAND - FINLANDE
Akateeminen Kirjakauppa,
Keskuskatu 1, 00100 Helsinki 10   Tel. 0.12141

FRANCE
OCDE/OECD
Mail Orders/Commandes par correspondance :
2, rue André-Pascal,
75775 Paris Cedex 16
Tel. (1) 45.24.82.00
Bookshop/Librairie : 33, rue Octave-Feuillet
75016 Paris
Tel. (1) 45.24.81.67 or/ou (1) 45.24.81.81
Librairie de l'Université,
12a, rue Nazareth,
13602 Aix-en-Provence   Tel. 42.26.18.08

GERMANY - ALLEMAGNE
OECD Publications and Information Centre,
4 Simrockstrasse,
5300 Bonn   Tel. (0228) 21.60.45

GREECE - GRÈCE
Librairie Kauffmann,
28, rue du Stade, 105 64 Athens   Tel. 322.21.60

HONG KONG
Government Information Services,
Publications (Sales) Office,
Information Services Department
No. 1, Battery Path, Central

ICELAND - ISLANDE
Snæbjörn Jónsson & Co., h.f.,
Hafnarstræti 4 & 9,
P.O.B. 1131 - Reykjavik
Tel. 13133/14281/11936

INDIA - INDE
Oxford Book and Stationery Co.,
Scindia House, New Delhi 110001
Tel. 331.5896/5308
17 Park St., Calcutta 700016   Tel. 240832

INDONESIA - INDONÉSIE
Pdii-Lipi, P.O. Box 3065/JKT.Jakarta
Tel. 583467

IRELAND - IRLANDE
TDC Publishers - Library Suppliers,
12 North Frederick Street, Dublin 1
Tel. 744835-749677

ITALY - ITALIE
Libreria Commissionaria Sansoni,
Via Lamarmora 45, 50121 Firenze
Tel. 579751/584468
Via Bartolini 29, 20155 Milano   Tel. 365083
La diffusione delle pubblicazioni OCSE viene
assicurata dalle principali librerie ed anche da :
Editrice e Libreria Herder,
Piazza Montecitorio 120, 00186 Roma
Tel. 6794628
Libreria Hœpli,
Via Hœpli 5, 20121 Milano   Tel. 865446
Libreria Scientifica
Dott. Lucio de Biasio "Aeiou"
Via Meravigli 16, 20123 Milano   Tel. 807679

JAPAN - JAPON
OECD Publications and Information Centre,
Landic Akasaka Bldg., 2-3-4 Akasaka,
Minato-ku, Tokyo 107   Tel. 586.2016

KOREA - CORÉE
Kyobo Book Centre Co. Ltd.
P.O.Box: Kwang Hwa Moon 1658,
Seoul   Tel. (REP) 730.78.91

LEBANON - LIBAN
Documenta Scientifica/Redico,
Edison Building, Bliss St.,
P.O.B. 5641, Beirut   Tel. 354429-344425

**MALAYSIA/SINGAPORE -
MALAISIE/SINGAPOUR**
University of Malaya Co-operative Bookshop
Ltd.,
7 Lrg 51A/227A, Petaling Jaya
Malaysia   Tel. 7565000/7565425
Information Publications Pte Ltd
Pei-Fu Industrial Building,
24 New Industrial Road No. 02-06
Singapore 1953   Tel. 2831786, 2831798

NETHERLANDS - PAYS-BAS
SDU Uitgeverij
Christoffel Plantijnstraat 2
Postbus 20014
2500 EA's-Gravenhage   Tel. 070-789911
Voor bestellingen: Tel. 070-789880

NEW ZEALAND - NOUVELLE-ZÉLANDE
Government Printing Office Bookshops:
Auckland: Retail Bookshop, 25 Rutland Stseet,
Mail Orders, 85 Beach Road
Private Bag C.P.O.
Hamilton: Retail: Ward Street,
Mail Orders, P.O. Box 857
Wellington: Retail, Mulgrave Street, (Head
Office)
Cubacade World Trade Centre,
Mail Orders, Private Bag
Christchurch: Retail, 159 Hereford Street,
Mail Orders, Private Bag
Dunedin: Retail, Princes Street,
Mail Orders, P.O. Box 1104

NORWAY - NORVÈGE
Narvesen Info Center - NIC,
Bertrand Narvesens vei 2,
P.O.B. 6125 Etterstad, 0602 Oslo 6
Tel. (02) 67.83.10, (02) 68.40.20

PAKISTAN
Mirza Book Agency
65 Shahrah Quaid-E-Azam, Lahore 3 Tel. 66839

PHILIPPINES
I.J. Sagun Enterprises, Inc.
P.O. Box 4322 CPO Manila
Tel. 695-1946, 922-9495

PORTUGAL
Livraria Portugal,
Rua do Carmo 70-74,
1117 Lisboa Codex   Tel. 360582/3

**SINGAPORE/MALAYSIA -
SINGAPOUR/MALAISIE**
See "Malaysia/Singapor". Voir
«Malaisie/Singapour»

SPAIN - ESPAGNE
Mundi-Prensa Libros, S.A.,
Castelló 37, Apartado 1223, Madrid-28001
Tel. 431.33.99
Libreria Bosch, Ronda Universidad 11,
Barcelona 7   Tel. 317.53.08/317.53.58

SWEDEN - SUÈDE
AB CE Fritzes Kungl. Hovbokhandel,
Box 16356, S 103 27 STH,
Regeringsgatan 12,
DS Stockholm   Tel. (08) 23.89.00
Subscription Agency/Abonnements:
Wennergren-Williams AB,
Box 30004, S104 25 Stockholm Tel. (08)54.12.00

SWITZERLAND - SUISSE
OECD Publications and Information Centre,
4 Simrockstrasse,
5300 Bonn (Germany)   Tel. (0228) 21.60.45
Librairie Payot,
6 rue Grenus, 1211 Genève 11
Tel. (022) 31.89.50
United Nations Bookshop/Librairie des Nations-
Unies
Palais des Nations,
1211 - Geneva 10
Tel. 022-34-60-11 (ext. 48 72)

TAIWAN - FORMOSE
Good Faith Worldwide Int'l Co., Ltd.
9th floor, No. 118, Sec.2
Chung Hsiao E. Road
Taipei   Tel. 391.7396/391.7397

THAILAND - THAILANDE
Suksit Siam Co., Ltd., 1715 Rama IV Rd.,
Samyam Bangkok 5   Tel. 2511630
INDEX Book Promotion & Service Ltd.
59/6 Soi Lang Suan, Ploenchit Road
Patjumawman, Bangkok 10500
Tel. 250-1919, 252-1066

TURKEY - TURQUIE
Kültur Yayinlari Is-Türk Ltd. Sti.
Atatürk Bulvari No: 191/Kat. 21
Kavaklidere/Ankara   Tel. 25.07.60
Dolmabahce Cad. No: 29
Besiktas/Istanbul   Tel. 160.71.88

UNITED KINGDOM - ROYAUME-UNI
H.M. Stationery Office,
Postal orders only: (01)211-5656
P.O.B. 276, London SW8 5DT
Telephone orders: (01) 622.3316, or
Personal callers:
49 High Holborn, London WC1V 6HB
Branches at: Belfast, Birmingham,
Bristol, Edinburgh, Manchester

UNITED STATES - ÉTATS-UNIS
OECD Publications and Information Centre,
2001 L Street, N.W., Suite 700,
Washington, D.C. 20036 - 4095
Tel. (202) 785.6323

VENEZUELA
Libreria del Este,
Avda F. Miranda 52, Aptdo. 60337,
Edificio Galipan, Caracas 106
Tel. 951.17.05/951.23.07/951.12.97

YUGOSLAVIA - YOUGOSLAVIE
Jugoslovenska Knjiga, Knez Mihajlova 2,
P.O.B. 36, Beograd   Tel. 621.992

Orders and inquiries from countries where
Distributors have not yet been appointed should be
sent to:
OECD, Publications Service, 2, rue André-Pascal,
75775 PARIS CEDEX 16.

Les commandes provenant de pays où l'OCDE n'a
pas encore désigné de distributeur doivent être
adressées à :
OCDE, Service des Publications. 2, rue André-
Pascal, 75775 PARIS CEDEX 16.

71784-07-1988

OECD PUBLICATIONS, 2, rue André-Pascal, 75775 PARIS CEDEX 16 - No. 44275 1988
PRINTED IN FRANCE
(97 88 09 1) ISBN 92-64-13183-3